Mathematics for Industry

Volume 6

Aims & Scope

The meaning of "Mathematics for Industry" (sometimes abbreviated as MI or MfI) is different from that of "Mathematics in Industry" (or of "Industrial Mathematics"). The latter is restrictive: it tends to be identified with the actual mathematics that specifically arises in the daily management and operation of manufacturing. The former, however, denotes a new research field in mathematics that may serve as a foundation for creating future technologies. This concept was born from the integration and reorganization of pure and applied mathematics in the present day into a fluid and versatile form capable of stimulating awareness of the importance of mathematics in industry, as well as responding to the needs of industrial technologies. The history of this integration and reorganization indicates that this basic idea will someday find increasing utility. Mathematics can be a key technology in modern society.

The series aims to promote this trend by (1) providing comprehensive content on applications of mathematics, especially to industry technologies via various types of scientific research, (2) introducing basic, useful, necessary and crucial knowledge for several applications through concrete subjects, and (3) introducing new research results and developments for applications of mathematics in the real world. These points may provide the basis for opening a new mathematics-oriented technological world and even new research fields of mathematics.

More information about this series at http://www.springer.com/series/13254

Jing Yao Zhang · Makoto Ohsaki

Tensegrity Structures

Form, Stability, and Symmetry

Springer

Jing Yao Zhang
Nagoya City University
Nagoya
Japan

Makoto Ohsaki
Hiroshima University
Higashi-Hiroshima
Japan

ISSN 2198-350X ISSN 2198-3518 (electronic)
Mathematics for Industry
ISBN 978-4-431-54812-6 ISBN 978-4-431-54813-3 (eBook)
DOI 10.1007/978-4-431-54813-3

Library of Congress Control Number: 2015932232

Springer Tokyo Heidelberg New York Dordrecht London

Printed on acid-free paper

Springer Japan KK is part of Springer Science+Business Media (www.springer.com)

Preface

Aims and Scope

Tensegrity structures are now more than 60-years old, since their birth as artworks. However, they are not "old" nor out of fashion! On the contrary, they are becoming more and more present in many different fields, including but not limited to engineering, biomedicine, and mathematics. These applications make use of the unique mechanical as well as mathematical properties of tensegrity structures in contrast to conventional structural forms such as trusses and frames.

Our primary objective in writing this book is to provide a textbook for self-study which is easily accessible not only to engineers and scientists, but also to upper-level undergraduate and graduate students. Both students and professionals will find material of interest to them in the book. With this objective in mind, the presentation of this book is detailed with many examples, and moreover, it is self-contained.

There are already several existing books on tensegrity structures; most of them present approaches to realization and practical applications of those structures. By contrast, this book is devoted to helping the readers achieve a deeper understanding of *fundamental* mechanical and mathematical properties of tensegrity structures. In particular, emphasis is placed on the two key problems in preliminary design of tensegrity structures—*self-equilibrium* and *(super-)stability*, by extensively utilizing the concept of force density and high level of symmetry of the structures.

Subjects and Contents

Tensegrity structures are similar in appearance to conventional bar-joint structures (trusses), however, their members carry forces (prestresses) even when no external load is applied. This means that their nodes and members have to be balanced by the prestresses so as to maintain their equilibrium. Furthermore, most tensegrity structures are intrinsically unstable in the absence of prestresses, and it is the introduction of prestresses that makes them stable. For these reasons, finding the *self-equilibrated configuration* and investigation of *stability* are the two key problems in the preliminary design of tensegrity structures.

Finding the configuration associated with prestresses, in the state of self-equilibrium, is called *form-finding* or shape-finding. It is a common design problem for tension structures, including tensile membrane structures and cable-nets. The problem is difficult because the configuration and prestresses cannot be determined separately as a result of the high interdependency between them. Further difficulties arise from the fact that tensegrity structures maintain their stability without any support.

A structure is stable if and only if it has the locally minimum total potential energy, or strain energy in the absence of external loads. Stability investigation of tensegrity structures is necessary because their stability cannot be guaranteed as can that of cable-nets or membrane structures carrying tension only in their structural elements. This comes from the fact that tensegrity structures are composed of (continuous) tensile members and (discontinuous) compressive members. Moreover, it is possible for tensegrity structures to be *super-stable*, which is a more robust stability criterion, if proper prestresses are associated with the proper connectivity pattern.

In this book, basic concepts and applications of tensegrity structures are introduced in Chap. 1. Chapter 2 formulates the matrices and vectors necessary for the study of self-equilibrium and stability. The analytical conditions for self-equilibrium of several highly symmetric tensegrity structures with simple geometries are given in Chap. 3. Chapter 4 defines the three stability criteria—stability, prestress-stability, and super-stability—and derives the necessary conditions and sufficient conditions for super-stability. The force density method, which guarantees super-stability, is presented in Chap. 5 for numerical form-finding of relatively complex tensegrity structures. Utilizing the analytical formulations for highly symmetric structures given in Appendix D, the self-equilibrium and super-stability conditions are derived for the prismatic tensegrity structures in Chap. 6 and those for the star-shaped structures in Chap. 7; both these classes of structures are of dihedral symmetry. Additionally, Chap. 8 presents the self-equilibrium and super-stability conditions for structures with tetrahedral symmetry.

At the end of the preface, we have to give our deepest thanks to our families, friends, and former and current students for their supports. Part of the work on symmetry has been conducted in close collaboration with Dr. Simon D. Guest of the University of Cambridge and Professor Robert Connelly of Cornell University; they showed us a new way to study tensegrity structures. Mr. Masaki

Okano of Nagoya City University read the first half of the book carefully and found many mistakes, which we then were able to correct. We also appreciate the proposal of writing this book by Dr. Yuko Sumino of Springer Japan; she has always been helpful during the preparation and publication of the book.

Nagoya, December 2014 Jing Yao Zhang
Higashi-Hiroshima Makoto Ohsaki

Contents

Chapter 1
Introduction

Abstract In this introductory chapter, we first introduce the basic concepts and some applications of tensegrity structures, and then present their design problems which motivates our study in this book. Finally, a brief review of the existing researches on the design problems is given.

Keywords Applications · Form-finding methods · Stability criteria

1.1 General Introduction

The term *tensegrity* was created by Richard Buckminster Fuller as a contraction of '*tens*ional' and 'int*egrity*' [15]. It refers to the integrity of a stable structure balanced by *continuous* structural members (cables) in tension and *discontinuous* structural members (struts) in compression. Moreover, the cables are flexible and global components, while the struts are stiff and local components.

The first tensegrity structure, called X-column, is considered to be built by Kenneth Snelson in 1948 [35].[1] Snelson came up with the idea of building this structure as an answer to the question posted by Fuller, who was his teacher at Black Mountain College at that time: "Is it possible to build a structure to illustrate the structural principle of nature, which was observed to rely on that continuous tension embraces isolated compression elements?"

There is no strict definition of tensegrity structures up to now that is accepted by all people. Instead of giving a strict definition of our own, we generally accept that a tensegrity structure should have the following characteristics:

[1] More details on the birth of tensegrity structures can be found in the papers [25–27] by Motro.

J.Y. Zhang and M. Ohsaki, *Tensegrity Structures*, Mathematics for Industry 6,
DOI 10.1007/978-4-431-54813-3_1

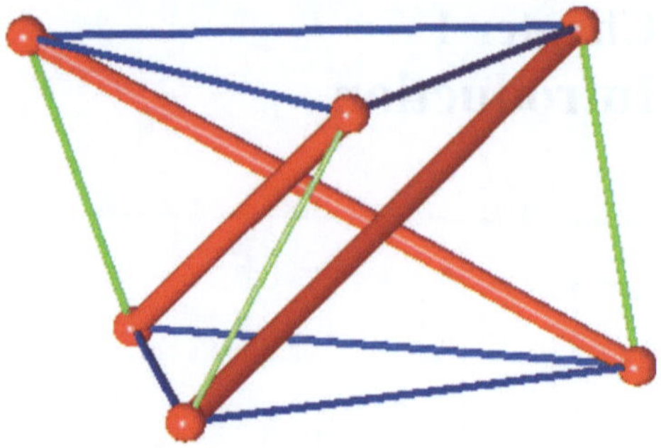

Fig. 1.1 The simplest tensegrity structure in three-dimensional space. The struts in compression are denoted by *thick lines*, and the cables in tension are denoted by *thin lines*

Characteristics of a tensegrity structure:

- The structure is *free-standing*, without any support.
- The structural members are straight.
- There are only two different types of structural members: *struts* carrying compression and *cables* carrying tension.
- The struts do not contact with each other at their ends.[2]

Moreover, we will persist in the entire book that the members in thick lines indicate struts in compression, and the members in thin lines indicate cables in tension, because tensile members are generally flexible and slender.

Figure 1.1 shows the simplest three-dimensional tensegrity structure. The structures having similar appearances are called *prismatic structures*, which will be studied in detail in Chaps. 3 and 6. The struts of the structure do not contact with each other. Moreover, supports or fixed nodes are unnecessary to maintain its (super-)stability with the exclusion of rigid-body motions.

1.2 Applications

Tensegrity structures were originally born in arts; however, they 'exist' universally, from the micro scale to the macro scale. In the micro scale, for example, response of living cells subjected to environmental changes can be interpreted and predicted by tensegrity models; in the intermediate scale, the human body can be modeled as a tensegrity structure; and in the macro scale, structure of the cosmos can also be regarded as a tensegrity structure, where the planets are the nodes and their interactions are the invisible members.

[2] With a very limited exceptions, the struts of some tensegrity structures are allowed to share common nodes, especially in the two-dimensional cases.

Owing to universalness of tensegrity structures, applications of their principles have been consecutively increasing in a great variety of fields, since their birth as art works. This section introduces some of the interesting applications. However, this introduction is no way exhaustive, since any such attempt would become out-of-date very soon due to the rapidly increasing number. The up-to-date information is provided on our homepage.[3]

1.2.1 Applications in Architecture

The members of tensegrity structures are the simplest possible ones, because they are straight and carry only axial forces. Moreover, a tensegrity structure maintains its stability with the minimum possible number of structural members, which is much less than the necessary number for a conventional bar-joint structure (truss) consisting of the same number of nodes. Therefore, tensegrity structures are considered as one of the optimal structural systems, in particular in the engineering view.

Furthermore, tensegrity structures have many other advantages when they are used as long-span structures to cover a large space without columns inside. Some of these advantages by comparison to some other structural systems are listed below.

- Introduction of prestresses into the structures could significantly enhances their structural stiffness, although this is not always the case. Therefore, they can be built with much smaller amounts of materials while having the same capacity of resisting external loads; as a reward, this can also significantly reduce the gravitational loads, which are usually dominant in the design of long-span structures.
- The structural members are of very high mechanical efficiency, because they carry only axial forces such that the stresses (normal stresses only) in a member are uniform.
- The struts in compression that are prone to member buckling might be more slender, because they are local components and much shorter in length than cables. This can also effectively reduce the gravitational loads.
- The cables in tension can make full use of high-strength materials, because large cross sections due to member buckling are not necessary.
- Complex joints connecting different members are not necessary, since the flexible cables are much easier to be attached to the struts, while the struts do not contact with each other.

Using the concept of tensegrity structures, David Geiger designed a permanent long-span structure, called the Georgia Dome [5, 21]. The structure was constructed in 1992 as the main hall for the 1996 Atlanta Summer Olympic Games in U.S. It has a height of 82.5 m, a length of 227 m, a width of 185 m, and a total floor area of 9,490 m^2.

[3] Online sources on tensegrity structures collected by the authors are consecutively updated at http://zhang.AIStructure.net/links/tensegritylinks/.

The great success of Georgia Dome aroused the interests and enthusiasms of many structural engineers and researchers, and a number of tensegrity-domes have been built around the world [46, 49]. However, it should be noted that tensegrity-domes are not 'real' tensegrity structures in strict definition, because they are not free-standing and maintain their stability by being attached to supports at the boundary.

The experimental facility built in Chiba, Japan in 2001 as shown in Fig. 1.2 is one of the earliest attempts to use 'real' tensegrity structure in architectural engineering [20]. Two tensegrity units are used as structural components, and one isolated strut at the top of each unit is used to support the membrane roof. One of the units is 10 m high and the other is 7 m high. The units have the similar shape to the simplest (prismatic) structure as shown in Fig. 1.1, with three additional 'vertical' cables to attain proper rigidity for practical applications.

1.2.2 Applications in Mechanical Engineering

In the filed of mechanical engineering, tensegrity structures are utilized as 'smart' structures [2, 6, 16, 34] and deployable structures [38], the shapes of which are

Fig. 1.2 Example of a pair of tensegrity structures used as structural components to support a membrane roof. The structure was constructed in Chiba, Japan in 2001. The *left* photo is the interior view of the building, the *upper-right* photo is its exterior night view, and the *lower-right* photo is one of the tensegrity structures under construction. (Courtesy: Dr. K. Kawaguchi at the University of Tokyo)

actively adjusted to satisfy different requirements in different circumstances. The reasons that tensegrity structures are suitable for smart structures rely on the facts that

- The self-equilibrated configuration and prestresses of a tensegrity structure are highly interdependent—its configuration can be actively controlled by adjusting the prestresses or member lengths.
- Tensegrity structures have predictable responses over a wide range of different shapes [38].
- The control systems (sensors and actuators) can be easily embedded and implemented in the members, because they are straight (one-dimensional).

1.2.3 Applications in Biomedical Engineering

Principles of tensegrity structures are also found in biomedical engineering. Researchers in biomedical engineering were initially interested in using tensegrity structure as a model for the structure of viruses [4] to interpret their structural behaviors subject to change of external environment.

Increasing interests were extended to the macroscopic level as well as microscopic level, including those in the human body [32]:

- At the macroscopic level, the 206 bones that constitute our skeleton are pulled up against the force of gravity and stabilized in a vertical form by the pull of tensile muscles, tendons, and ligaments [17].
- At the other end of the scale, proteins and other key molecules in the body stabilize themselves through the principles of tensegrity [3, 13, 23, 42, 44, 45].

1.2.4 Applications in Mathematics

Other than arts and engineering, tensegrity structures are also studied in mathematics, mainly on their stability in the filed of structural rigidity [7–12, 18, 43]. Moreover, their principles have also been applied to solve some challenging mathematical problems, such as packing problems.

The particle packing problem studies how the particles can be packed together as tightly as possible to occupy the minimum space [22], which has been a persistent scientific problem for many years [40]. The problem also leads to better understanding of the behavior of disordered materials ranging from powders to glassy solids.

In the packing problem, the centers of hard-particles must keep a minimum distance but can be as far apart as desired. Hence, it can be regarded as a tensegrity with invisible compressive members [14]. Furthermore, it can be formulated as a problem of detecting stability (rigidity in mathematics) of the tensegrity structures associated with the contact graph of the packing.

1.3 Form-Finding and Stability

In aid of the applications in Sect. 1.2, the following two preliminary design problems are the keys to understanding tensegrity structures.

Two key problems in the preliminary design of tensegrity structures:

- How to achieve their self-equilibrated configurations satisfying (geometrical and mechanical) requirements by the designers; and
- How to guarantee their (super-)stability.

This book mainly presents some of our studies on these two problems.

1.3.1 General Background

The members of a tensegrity structure carry (only) axial forces, even when no external load is applied. The forces in a member with external load absent is called *prestresses* or *self-stresses*. Struts carrying compressive prestresses push the nodes at their ends away, while cables carrying tensile prestresses intend to pull them back.

The interaction of compressions and tensions in the structure makes all the nodes stay in the state of force balance. This state is called the *self-equilibrium state*, and its configuration in the self-equilibrium state is called *self-equilibrated configuration*.

The difficulty in designing a tensegrity structure comes from that its (self-equilibrated) configuration cannot be arbitrarily assigned, because the configuration and the prestresses in the members are interdependent with each other. This is different from the design of conventional bar-joint structures (trusses) carrying no prestress.

The problem of determining the configuration as well as prestresses of a tensegrity structure is called *form-finding* or *shape-finding*. Other tension structures, such as cable-nets and membrane structures, have similar form-finding problems. In fact, some of the existing methods for form-finding problems of tensegrity structures were originally developed for cable-nets.

What causes more difficulties in designing a tensegrity structure is that it is usually unstable in the absence of prestresses. This comes from the fact that a tensegrity structure maintains its stability with the minimum possible number of members, and it is usually kinematically indeterminate. Hence, it is the introduction of prestresses into the members that stabilizes the structure. However, prestresses does not guarantee a stable tensegrity structure as in the case of cable-nets and membrane structures.

Therefore, in the process of form-finding for a tensegrity structure, it is important to find a *stable* configuration in the state of *self-equilibrium*.

1.3.2 Existing Form-finding Methods

For the form-finding problem of tensegrity structures, there have been a great number of methods developed so far, and the number is still rapidly increasing. Many of these methods were reviewed by several review papers, see, for example, the Refs. [19, 39].

In the following, we classify the existing methods into several general categories: intuition methods, analytical methods, and numerical methods. Note that this classification is not unique, and we do not attempt to exhaust them.

1.3.2.1 Intuition Methods

In the early stage of development, tensegrity structures were studied by purely intuitive approaches, mainly conducted by artists. The intuitive methods made the first and one of the most important contributions to the development of tensegrity structures by exploring and spreading the special structural philosophy behind them.

In the intuition method, the well-known regular and convex polyhedra were usually used as reference [31]. For example, the simplest three-dimensional tensegrity structure as shown in Fig. 1.1 is generated from the twisted 3-gonal dihedron as shown in Fig. 1.3—the cables of the structure correspond to the edges of the twisted dihedron, and the struts correspond to the diagonals.

Moreover, the regular truncated tetrahedral tensegrity structure as shown in Fig. 1.4a is generated from the truncated tetrahedron in Fig. 1.4b. More details on generation of these structures can be found in Chap. 3.

New self-equilibrated configurations of a tensegrity structure were usually found by making physical models by trial and error; however, this is strongly restricted within our knowledge and intuition on geometry of existing objects. More systematic ways for design of tensegrity structures had not been developed until they attracted attentions of researchers and engineers.

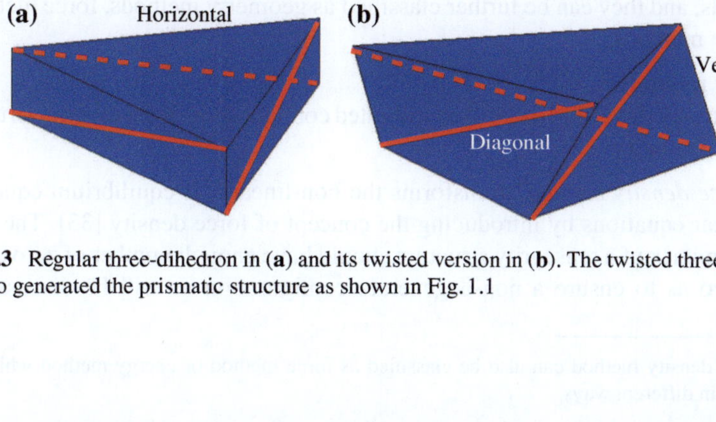

Fig. 1.3 Regular three-dihedron in (**a**) and its twisted version in (**b**). The twisted three-dihedron is used to generated the prismatic structure as shown in Fig. 1.1

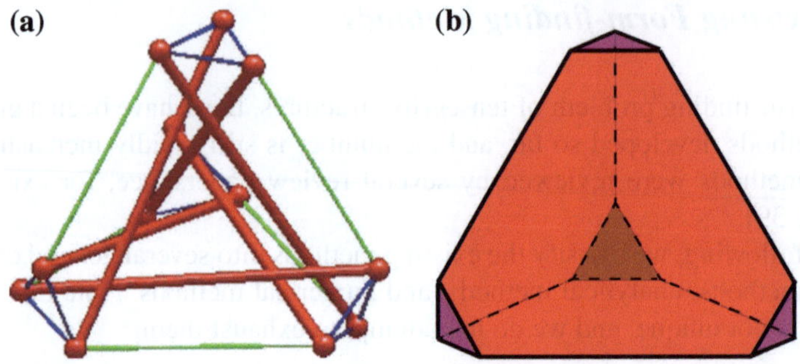

Fig. 1.4 An example of tensegrity structure in (**a**) intuitively made from the truncated tetrahedron in (**b**)

1.3.2.2 Analytical Methods

Analytical solutions for form-finding of tensegrity structures are available only for simple cases, or when the structures have high level of symmetry [8, 29].

For the highly symmetric structures, self-equilibrium analysis of the whole structure can be simplified to that of a limited number of nodes. It is, therefore, possible to derive analytical solutions even for complex structures with a large number of nodes and members. In Chap. 3, we will study self-equilibrium of several classes of highly symmetric tensegrity structures in detail.

Furthermore, we will use a powerful mathematical tool, group representation theory, to study self-equilibrium as well as stability of these structures in Chaps. 6–8.

1.3.2.3 Numerical Methods

Because analytical methods are limited to some specific structures, most of the existing form-finding methods for tensegrity structures are numerical methods. Numerical methods are more flexible for free-form design of tensegrity structures than analytical methods, and they can be further classified as geometry methods, force methods, and energy methods.

Geometry methods:

Geometry methods find the self-equilibrated configuration of a tensegrity structure in view of geometry.

- The *force density method*[4] transforms the non-linear self-equilibrium equations into linear equations by introducing the concept of force density [33]. The force density matrix of a tensegrity structure should have enough number of zero eigenvalues so as to ensure a non-degenerate configuration as will be discussed in

[4] The force density method can also be classified as force method or energy method while it is approached in different ways.

Chap. 2. There can be many solutions for such purpose, for example, a semi-symbolic approach was presented in [41]; in Chap. 5, we will show that eigenvalue analysis enables us to find the feasible force densities.

- With the fixed lengths for cables, self-equilibrated configuration of a tensegrity structure can be determined by solving an optimization problem where total length of struts is to be maximized [30].

Force methods:

Force methods find the self-equilibrated configuration of a tensegrity structure by considering force balances of the nodes or members.

- The *dynamic relaxation method* initially developed for tensile structures [1] has been extended to form-finding of tensegrity structures [48]. The structure deforms from an initial configuration with zero velocities subjected to the unbalanced loads due to assigned prestresses. Deformation of the structure obeys the fictitious dynamic equations, until it settles down with low enough velocities or kinetic energy.
- The idea of *static nonlinear structural analysis* has also been applied to form-finding problem of tensegrity structures [47]. Moore-Penrose generalized inverse of the tangent stiffness matrix is utilized to avoid its non-invertibility due to being free-standing.
- In the *internal coordinate method*, self-equilibrium equations of a structure is formulated as a product of the equilibrium matrix in internal coordinate system and prestresses. The equilibrium matrix is enforced to be singular so as to let the structure carry non-trivial prestresses [36].

Energy methods:

As will discussed in Chap. 4, a stable structure with local minimum of total potential energy (or strain energy when no external load is considered) is in the state of (self-)equilibrium. Therefore, self-equilibrium of the structure can be guaranteed by a stable configuration. Among the energy methods, many make use of *optimization techniques* to search for locally minimum (strain) energy of the structure [28].

1.3.3 Stability

A structure is said to be *stable* if it returns to its initial configuration after being released from any small disturbance. There is a simple rule, called Maxwell's rule, for preliminary study on stability (kinematic determinacy) of bar-joint structures (trusses).

In the paper [24] by Maxwell, he showed that a three-dimensional truss having n joints (nodes) requires in general at least $(m{=})3n - 6$ bars (members without prestresses) to render it stable; i.e., a truss is stable if the number of bars m satisfies

$$m \geq 3n - 6, \tag{1.1}$$

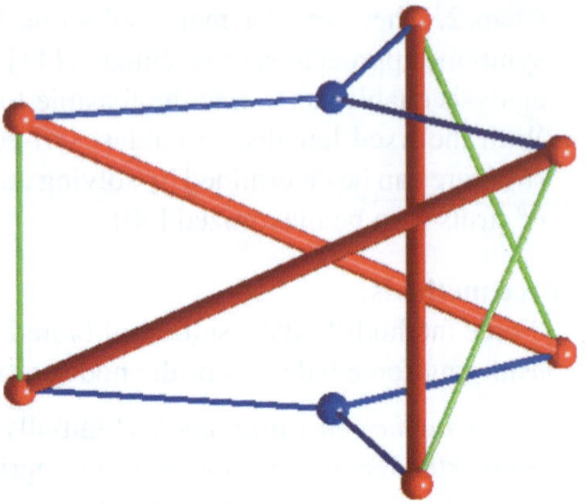

Fig. 1.5 The simplest star-shaped tensegrity structure. This structure is stable although the number of members is smaller than the necessary number required by Maxwell's rule

where 6 is the number of rigid-body motions of the structure in three-dimensional space.

However, Maxwell's rule is usually not applicable to tensegrity structures, although they have similar appearance and properties to trusses except for the existence of prestresses. Tensegrity structures can be stable with less members required by Maxwell's rule for trusses. Consider, for instance, the structure called star-shaped tensegrity structure as shown in Fig. 1.5.

Example 1.1 Comparison of the number of members of a stable star-shaped tensegrity structure as shown in Fig. 1.5 to the number needed by Maxwell's rule.

The star-shaped tensegrity structure as shown in Fig. 1.5 consists of eight nodes and twelve members; i.e., $n = 8$ and $m = 12$. According to Maxwell's rule in Eq. (1.1), the structure cannot be stable if the number of members is less than 18:

$$3n - 6 = 3 \times 8 - 6 = 18. \tag{1.2}$$

However, this structure is stable,[5] although the number of its members is much smaller than that is needed by Maxwell's rule. Stability of a class of structures having the similar symmetry properties to this structure will be discussed in Chap. 7 in detail.

In structural engineering, stability of a structure is usually investigated by verification of positive definiteness of its tangent stiffness matrix, which is the second-order increment of the total potential energy with respect to displacements [37]. Stability

[5] The simplest star-shaped tensegrity structure is indeed super-stable; i.e., it is stable irrespective of selection of materials as well as magnitude of prestresses.

investigation of tensegrity structures is complicated due to the fact that prestresses are also involved, through the geometrical stiffness.

To clarify this influence, there are two other stability criteria adopted in the community of tensegrity structures: *prestress-stability* and *super-stability* [12]. In Chap. 4, we will present the formulations of the stiffness matrices for a tensegrity structure, and then present the equivalent conditions for its super-stability to those in mathematics [7, 9].

1.4 Remarks

Tensegrity structures were born in the field of art, and then find their applications in many different fields, due to their distinct properties compared to other structural systems.

In the (preliminary) design of a tensegrity structure, form-finding and stability are the two key problems. There is no perfect method for finding the self-equilibrated configuration as designed, mainly due to the highly interdependent of geometry and prestresses. On the other hand, super-stability has not been well recognized in the studies of tensegrity structures.

This book is devoted to these two design problems of tensegrity structures. We hope the contents following this introductory chapter could let the readers have deeper understanding of the properties of tensegrity structures, and therefore, push forward their applications.

References

1. Barnes, M. R. (1999). Form finding and analysis of tension structures by dynamic relaxation. *International Journal of Space Structures*, *14*(2), 89–104.
2. Bohm, V., & Zimmermann, K. (2013). Vibration-driven mobile robots based on single actuated tensegrity structures. In *IEEE International Conference on Robotics and Automation (ICRA)* (pp. 5475–5480) IEEE.
3. Broers, J. L. V., Peeters, E. A. G., Kuijpers, H. J. H., Endert, J., Bouten, C. V. C., Oomens, C. W. J., et al. (2004). Decreased mechanical stiffness in LMNA-/- cells is caused by defective nucleo-cytoskeletal integrity: implications for the development of laminopathies. *Human Molecular Genetics*, *13*(21), 2567–2580.
4. Caspar, D. L. D., & Klug, A. (1962). Physical principles in the construction of regular viruses. *Cold Spring Harbor symposia on quantitative biology* (pp. 1–24). New York: Cold Spring Harbor Laboratory Press.
5. Castro, G., & Levy, M. P. (1992). Analysis of the Georgia Dome cable roof. In *Computing in Civil Engineering and Geographic Information Systems Symposium*, (pp. 566–573) ASCE.
6. Chan, W. L., Arbelaez, D., Bossens, F., & Skelton, R. E. (2004). Active vibration control of a three-stage tensegrity structure. In K-W. Wang (Ed.), *11th Annual International Symposium on Smart Structures and Materials*, (pp. 340–346).
7. Connelly, R. (1982). Rigidity and energy. *Inventiones Mathematicae*, *66*(1), 11–33.
8. Connelly, R. (1995). Globally rigid symmetric tensegrities. *Structural Topology*, *21*, 59–78.

9. Connelly, R. (1999). In M. F. Thorpe & P. M. Duxbury (Eds.), Tensegrity structures: why are they stable? Rigidity theory and applications. (pp. 47–54). New York: Kluwer Academic/Plenum Publishers
10. Connelly, R. (2004). Generic global rigidity. *Discrete & Computational Geometry*, *33*(4), 549–563.
11. Connelly, R., & Schlenker, J. M. (2010). On the infinitesimal rigidity of weakly convex polyhedra. *European Journal of Combinatorics*, *31*(4), 1080–1090.
12. Connelly, R., & Whiteley, W. (1996). Second-order rigidity and prestress stability for tensegrity frameworks. *SIAM Journal on Discrete Mathematics*, *9*(3), 453–491.
13. Denton, M. J., Dearden, P. K., & Sowerby, S. J. (2003). Physical law not natural selection as the major determinant of biological complexity in the subcellular realm: new support for the pre-Darwinian conception of evolution by natural law. *Biosystems*, *71*(3), 297–303.
14. Donev, A., Torquato, S., Stillinger, F. H., & Connelly, R. (2004). A linear programming algorithm to test for jamming in hard-sphere packings. *Journal of Computational Physics*, *197*(1), 139–166.
15. Fuller, R. B. (1975). *Synergetics, explorations in the geometry of thinking*. London: Collier Macmillan.
16. Hirai, S., Koizumi, Y., Shibata, M., Wang, M. H., & Bin, L. (2013). Active shaping of a tensegrity robot via pre-pressure. In *IEEE/ASME International Conference on Advanced Intelligent Mechatronics (AIM)* (pp. 19–25) IEEE.
17. Ingber, D. E. (1998). The architecture of life. *Scientific American*, *278*, 48–57.
18. Jordán, T., & Szabadka, Z. (2009). Operations preserving the global rigidity of graphs and frameworks in the plane. *Computational Geometry*, *42*(6–7), 511–521.
19. Juan, S. H., & Mirats Tur, J. M. (2008). Tensegrity frameworks: static analysis review. *Mechanism and Machine Theory*, *43*(7), 859–881.
20. Kawaguchi, K., & Ohya, S. (2004). Preliminary report of observation of real scale tensegrity skeletons under temperature change. In *Proceedings of Annual Symposium of International Association for Shell and Spatial Structures (IASS)*, Montpellier, France.
21. Levy, M.P. (1994). The Georgia Dome and beyond: achieving lightweight-longspan structures. In *Proceedings of the IASS-ASCE International Symposium*, April 1994 (pp. 560–562)
22. López, C. O., & Beasley, J. E. (2011). A heuristic for the circle packing problem with a variety of containers. *European Journal of Operational Research*, *214*(3), 512–525.
23. Luo, Y. Z., Xu, X., Lele, T., Kumar, S., & Ingber, D. E. (2008). A multi-modular tensegrity model of an actin stress fiber. *Journal of Biomechanics*, *41*(11), 2379–2387.
24. Maxwell, J. C. (1864). On the calculation of the equilibrium and stiffness of frames. *Philosophical Magazine*, *27*(182), 294–299.
25. Motro, R. (1992). Tensegrity systems: the state of the art. *International Journal of Space Structures*, *7*(2), 75–83.
26. Motro, R. (1996). Structural morphology of tensegrity systems. *International Journal of Space Structures*, *11*(1, 2), 233–240.
27. Motro, R., & Raducanu, V. (2003). Tensegrity systems. *International Journal of Space Structures*, *18*(2), 77–84.
28. Ohsaki, M., Zhang, J. Y., & Taguchi, T. (2014). Form-finding and stability analysis of tensegrity structures using nonlinear programming and fictitious material properties. In *Proceedings of the International Conference on Computational Methods*, July 2014. Cambridge.
29. Oppenheim, I. J., & Williams, W. O. (2000). *Geometric effects in an elastic tensegrity structure*. In Advances in Continuum Mechanics and Thermodynamics of Material Behavior (pp. 51–65). The Netherlands: Springer.
30. Pellegrino, S., & Calladine, C. R. (1986). Matrix analysis of statically and kinematically indeterminate frameworks. *International Journal of Solids and Structures*, *22*(4), 409–428.
31. Pugh, A. (1976). *An introduction to tensegrity*. Berkeley: University of California Press.
32. Scarr, G. (2012). A consideration of the elbow as a tensegrity structure. *International Journal of Osteopathic Medicine*, *15*(2), 53–65.

33. Schek, H.-J. (1974). The force density method for form finding and computation of general networks. *Computer Methods in Applied Mechanics and Engineering*, *3*(1), 115–134.
34. Shea, K., Fest, E., & Smith, I. F. C. (2002). Developing intelligent tensegrity structures with stochastic search. *Advanced Engineering Informatics*, *16*(1), 21–40.
35. Snelson, K. Kenneth Snelson: art and ideas. available online: http://kennethsnelson.net/.
36. Sultan, C., Corless, M., & Skelton, R. E. (2001). The prestressability problem of tensegrity structures: some analytical solutions. *International Journal of Solids and Structures*, *38*(30–31), 5223–5252.
37. Thompson, J. M. T., & Hunt, G. W. (1984). *Elastic instability phenomena*. Chichester: Wiley.
38. Tibert, G. (2002). *Deployable tensegrity structures for space applications*. Ph.D. thesis, Department of Mechanics, Royal Institute of Technology, Sweden.
39. Tibert, A. G., & Pellegrino, S. (2003). Review of form-finding methods for tensegrity structures. *International Journal of Space Structures*, *18*(4), 209–223.
40. Uhler, C., & Wright, S. J. (2013). Packing ellipsoids with overlap. *SIAM Review*, *55*(4), 671–706.
41. Vassart, N., & Motro, R. (1999). Multiparametered formfinding method: application to tensegrity systems. *International Journal of Space Structures*, *14*(2), 147–154.
42. Wang, N., Naruse, K., Stamenović, D., Fredberg, J. J., Mijailovich, S. M., Tolić-Nørrelykke, I. M., et al. (2001). Mechanical behavior in living cells consistent with the tensegrity model. *Proceedings of the National Academy of Sciences of the United States of America*, *98*(14), 7765–7770.
43. Whiteley, W. (1997). *Rigidity and scene analysis* (pp. 893–916). Boca Raton: CRC Press.
44. Zanotti, G., & Guerra, C. (2003). Is tensegrity a unifying concept of protein folds? *FEBS Letters*, *534*(1–3), 7–10.
45. Zavodszky, M. I., Lei, M., Thorpe, M. F., Day, A. R., & Kuhn, L. A. (2004). Modeling correlated main-chain motions in proteins for flexible molecular recognition. *Proteins: Structure, Function, and Bioinformatics*, *57*(2), 243–261.
46. Zhang, W.-D., & Dong, S.-L. (2004). Advances in cable domes. *Journal of Zhejiang University (Engineering Science)*, *10*, 14.
47. Zhang, J. Y., & Ohsaki, M. (2013). Free-form design of tensegrity structures by non-rigid-body motion analysis. In *Proceedings of Annual Symposium of International Association for Shell and Spatial Structures (IASS)*, September 2013. Wroclaw, Poland.
48. Zhang, L., Maurin, B., & Motro, R. (2006). Form-finding of nonregular tensegrity systems. *Journal of Structural Engineering*, *132*(9), 1435–1440.
49. Zhang, G. J., Ge, J. Q., Wang, S., Zhang, A. L., Wang, W. S., Wang, M. Z., et al. (2012). Design and research on cable dome structural system of the national fitness center in Ejin Horo Banner, Inner Mongolia. *Jianzhu Jiegou Xuebao (Journal of Building Structures)*, *33*(4), 12–22.

Chapter 2
Equilibrium

Abstract Tensegrity structures are classified as prestressed pin-jointed structures, and they have distinct properties compared to other pin-jointed structures: (1) they are free-standing, without any support; and (2) they have both tensile and compressive members. Prior to further studies on tensegrity structures in the following chapters, this chapter presents the formulations of (self-)equilibrium for general prestressed pin-jointed structures. The equilibrium equations are formulated in two ways: (1) using the *equilibrium matrix* associated with prestresses or axial forces, and (2) using the *force density matrix* associated with nodal coordinates. Conditions for static as well as kinematic determinacy of a prestressed pin-jointed structure are then given in terms of rank of the equilibrium matrix. Furthermore, the *non-degeneracy condition* for a prestressed free-standing pin-jointed structure is presented in terms of rank deficiency of the force density matrix.

Keywords Equilibrium equations · Static determinacy · Kinematic determinacy · Force density matrix · Non-degeneracy condition

2.1 Definition of Configuration

This section first introduces the basic mechanical assumptions for a general prestressed[1] pin-jointed structure. Configuration of a prestressed pin-jointed structure is described by its connectivity and geometry: *connectivity* defines how the nodes are connected by the members; and *geometry* (realization) of the structure is described by coordinates of the nodes.

In the category of general prestressed pin-jointed structures, the following structures are included:

- *Truss*, which carries no prestress;
- *Cable-net*, which is attached to supports, and consists of only tensile members;
- *Tensegrity-dome*, which is also attached to supports, and consists of both tensile and compressive members;
- *Tensegrity*, which has *NO* support, and consists of both tensile and compressive members.

[1] 'Prestressed' means that prestresses are introduced to the structure a priori. Prestresses are the internal forces in the members when no external load is applied.

J.Y. Zhang and M. Ohsaki, *Tensegrity Structures*, Mathematics for Industry 6,
DOI 10.1007/978-4-431-54813-3_2

2.1.1 Basic Mechanical Assumptions

There are two types of structural elements in a general prestressed pin-jointed structure:

- *Members*, which are straight; and
- *Nodes*, or joints, that connect the members.

Moreover, there are two types of nodes:

- *Fixed nodes*, or supports, which cannot have any displacement even subjected to external loads; and
- *Free nodes*, displacements of which are not constrained.

Note that there exist only free nodes in a tensegrity structure, such that it is free-standing. This makes its (self-)equilibrium distinct from other prestressed pin-jointed structures, and leads to the necessary non-degeneracy condition presented in Sect. 2.5.

Furthermore, in order to study the mechanical properties of a prestressed pin-jointed structure making use of the existing powerful mathematical tools, we adopt the following mechanical assumptions for its members and nodes (joints).

Mechanical assumptions for a prestressed pin-jointed structure:

1. Members are connected to the nodes (joints) at their two ends. The joints are *pin-joints*, also called hinge joints, allowing the members to rotate freely about them.
2. Self-weight of the structure is neglected. External loads, if exist, are applied at the nodes.
3. Member failure, such as yielding or buckling, is not considered.

From the first two assumptions, it is obvious that the members carry only axial forces, either compression or tension. The axial forces in the members are called *prestresses* or self-stresses, when no external load is applied. Some pin-jointed structures do not carry prestress, e.g., trusses. Based on the type of prestress, there are also two types of members:

Two types of members in a prestressed pin-jointed structure:

- *Struts*, which carry compressive prestresses; and
- *Cables*, which carry tensile prestresses.

A tensegrity structure consists of both struts and cables. It will be shown in Chap. 4 that this fact makes its stability properties much different from, and more complicated than, the cable-nets which consist of only cables.

In the following, we consider a prestressed pin-jointed structure, which consists of m members, n free nodes, and n^{f} fixed nodes in d-dimensional space. For simplicity, displacements of the fixed nodes are constrained in every direction, and we do not consider any other types of supports, such as roller supports, displacements of which are constrained in specified directions. Such constraints can indeed be incorporated into the formulations presented hereafter without any difficulty.

2.1.2 Connectivity

Connectivity, or *topology*, of a structure defines how its members connect the nodes. Since the members are assumed to be straight, the structure can be modeled as a directed graph in graph theory [5, 6]. The vertices and edges of the directed graph respectively represent the nodes and members of the structure.

The connectivity and directions of the members can be described by the so-called *connectivity matrix*, denoted by $\mathbf{C}^{\mathrm{s}}$. There are only two non-zero entries, 1 and -1, in each row of the connectivity matrix, corresponding to the two nodes connected by the specific member; and all other entries in the same row are zero.

Suppose that a member numbered as k $(k = 1, 2, \dots, m)$ connects node i and node j $(i, j = 1, 2, \dots, n + n^{\mathrm{f}})$. The components of the kth row $C^{\mathrm{s}}_{(k,p)}$ of the connectivity matrix $\mathbf{C}^{\mathrm{s}} \in \mathbb{R}^{m\times(n+n^{\mathrm{f}})}$ is defined as

$$C^{\mathrm{s}}_{(k,p)} = \begin{cases} \mathrm{sign}(j-p), & \text{if } p = i; \\ \mathrm{sign}(i-p), & \text{if } p = j; \\ 0, & \text{otherwise}; \end{cases} \qquad (p = 1, 2, \dots, n + n^{\mathrm{f}}), \tag{2.1}$$

where

$$\mathrm{sign}(j-i) = \begin{cases} 1, & \text{if } j > i; \\ -1, & \text{if } j < i. \end{cases} \tag{2.2}$$

Hence, for the two nodes i and j, where $j < i$, the direction[2] of member k connecting these two nodes is defined as pointing from node j toward node i.

For convenience, the fixed nodes are preceded by the free nodes in the numbering sequence. Thus, the connectivity matrix $\mathbf{C}^{\mathrm{s}}$ can be partitioned into two parts as follows:

$$\mathbf{C}^{\mathrm{s}} = \left(\mathbf{C},\ \mathbf{C}^{\mathrm{f}}\right), \tag{2.3}$$

where $\mathbf{C} \in \mathbb{R}^{m\times n}$ and $\mathbf{C}^{\mathrm{f}} \in \mathbb{R}^{m\times n^{\mathrm{f}}}$ respectively correspond to connectivity of the free nodes and that of the fixed nodes.

---

2 Definition of member directions is not unique. Using opposite definition necessarily leads to the same equilibrium equation.

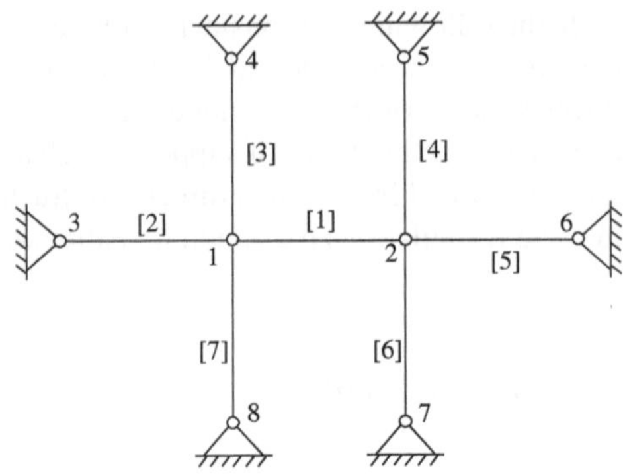

Fig. 2.1 A prestressed pin-jointed structure (cable-net) with fixed nodes considered in Example 2.1. The structure consists of two free nodes 1 and 2, six fixed nodes (supports) 3–8, and seven members [1]–[7]

Example 2.1 Connectivity matrix of a two-dimensional prestressed pin-jointed structure (cable-net) as shown in Fig. 2.1.

The prestressed pin-jointed structure as shown in Fig. 2.1 consists of eight nodes and seven members; i.e., $n + n^{\mathrm{f}} = 8$ and $m = 7$. The structure consists of both free nodes and fixed nodes (supports), and it is usually used as a simple example of cable-nets. Nodes 1 and 2 are free nodes, and nodes 3–8 are fixed nodes, where the free nodes are numbered preceding the fixed nodes. Thus, we have

$$n = 2, \ n^{\mathrm{f}} = 6, \ \text{and}\ m = 7. \tag{2.4}$$

The connectivity matrices corresponding to the entire structure $\mathbf{C}^{\mathrm{s}} \in \mathbb{R}^{7\times 8}$, the free nodes $\mathbf{C} \in \mathbb{R}^{7\times 2}$, and the fixed nodes $\mathbf{C}^{\mathrm{f}} \in \mathbb{R}^{7\times 6}$ are written as follows according to Eqs. (2.1) and (2.3):

$$\mathbf{C}^{\mathrm{s}} = \begin{array}{c} \begin{array}{cc|cccccc} 1 & 2 & 3 & 4 & 5 & 6 & 7 & 8 \end{array} \\ \left(\begin{array}{rr|rrrrrr} 1 & -1 & 0 & 0 & 0 & 0 & 0 & 0 \\ 1 & 0 & -1 & 0 & 0 & 0 & 0 & 0 \\ 1 & 0 & 0 & -1 & 0 & 0 & 0 & 0 \\ 0 & 1 & 0 & 0 & -1 & 0 & 0 & 0 \\ 0 & 1 & 0 & 0 & 0 & -1 & 0 & 0 \\ 0 & 1 & 0 & 0 & 0 & 0 & -1 & 0 \\ 1 & 0 & 0 & 0 & 0 & 0 & 0 & -1 \end{array}\right) \\ \mathbf{C} \qquad\qquad \mathbf{C}^{\mathrm{f}} \end{array} \begin{array}{c} [1] \\ [2] \\ [3] \\ [4] \\ [5] \\ [6] \\ [7] \end{array}. \tag{2.5}$$

A structure is said to be *free-standing*, if it has no fixed node. A free-standing structure can freely transform while preserving distance between any pair of nodes. Moreover, its connectivity matrix becomes

$$\mathbf{C} = \mathbf{C}^{\mathrm{s}}, \tag{2.6}$$

with the vanishing sub-matrix $\mathbf{C}^{\mathrm{f}}$ corresponding to fixed nodes. From the definition of the connectivity matrix $\mathbf{C}$ for free-standing structures in Eq. (2.1), we have an important property:

$$\mathbf{C}\mathbf{i}_n = \mathbf{C}\begin{pmatrix}1\\1\\\vdots\\1\end{pmatrix} = \begin{pmatrix}0\\0\\\vdots\\0\end{pmatrix} = \mathbf{0}, \tag{2.7}$$

where all entries in the vector $\mathbf{i}_n \in \mathbb{R}^n$ are one. This comes from the fact that each row of $\mathbf{C}$ has only two non-zero entries, 1 and -1, such that sum of the entries in each row by means of multiplying the vector $\mathbf{i}_n$ turns out to be zero.

Example 2.2 Connectivity matrix of the two-dimensional free-standing structure as shown in Fig. 2.2.

The two-dimensional free-standing structure as shown in Fig. 2.2 has no fixed node. There are in total five free nodes and eight members in the structure; i.e., $n = 5$, $n^{\mathrm{f}} = 0$, and $m = 8$.

According to Eq. (2.1), the connectivity matrix $\mathbf{C}$ ($=\mathbf{C}^{\mathrm{s}} \in \mathbb{R}^{8\times 5}$) of the structure is

$$\mathbf{C} = \mathbf{C}^{\mathrm{s}} = \begin{array}{c} \begin{matrix} 1 & 2 & 3 & 4 & 5 \end{matrix} \\ \begin{pmatrix} 1 & -1 & 0 & 0 & 0 \\ 1 & 0 & -1 & 0 & 0 \\ 1 & 0 & 0 & -1 & 0 \\ 1 & 0 & 0 & 0 & -1 \\ 0 & 1 & 0 & -1 & 0 \\ 0 & 0 & 1 & -1 & 0 \\ 0 & 1 & 0 & 0 & -1 \\ 0 & 0 & 1 & 0 & -1 \end{pmatrix} \end{array} \begin{matrix} [1] \\ [2] \\ [3] \\ [4] \\ [5] \\ [6] \\ [7] \\ [8] \end{matrix}. \tag{2.8}$$

It is obvious that sum of the entries in each row of $\mathbf{C}$ is zero.

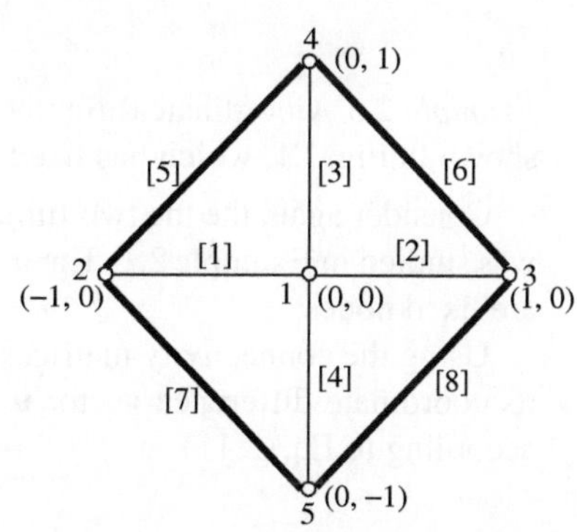

Fig. 2.2 A two-dimensional free-standing structure considered in Example 2.2. The structure consists of five free nodes 1–5 and eight members [1]–[8]. It is not attached to any fixed node or support

2.1.3 Geometry Realization

Geometry realization of a pin-jointed structure is described by coordinates of the nodes, or nodal coordinates. Let $\mathbf{x}$, $\mathbf{y}$, $\mathbf{z}$ ($\in \mathbb{R}^n$), and $\mathbf{x}^{\mathrm{f}}$, $\mathbf{y}^{\mathrm{f}}$, $\mathbf{z}^{\mathrm{f}}$ ($\in \mathbb{R}^{n^{\mathrm{f}}}$) denote the vectors of coordinates of the free nodes and the fixed nodes in x-, y-, and z-directions, respectively.

The *coordinate differences* u_k, v_k, and w_k in x-, y-, and z-directions of member k connecting node i and node j can be respectively calculated as follows:

$$\begin{aligned} u_k &= \mathrm{sign}(j-i)\cdot x_i + \mathrm{sign}(i-j)\cdot x_j = -\mathrm{sign}(j-i)\cdot(x_j - x_i),\\ v_k &= \mathrm{sign}(j-i)\cdot y_i + \mathrm{sign}(i-j)\cdot y_j = -\mathrm{sign}(j-i)\cdot(y_j - y_i),\\ w_k &= \mathrm{sign}(j-i)\cdot z_i + \mathrm{sign}(i-j)\cdot z_j = -\mathrm{sign}(j-i)\cdot(z_j - z_i). \end{aligned} \tag{2.9}$$

From the definition of connectivity matrix in Eq. (2.1), we know that $\mathrm{sign}(j-i)$ and $\mathrm{sign}(i-j)$ with $i \neq j$ in the kth row $\mathbf{C}^{\mathrm{s}}_k$ are the only two non-zero entries, which are respectively equal to 1 and -1, or -1 and 1. Thus, Eq. (2.9) can be rewritten in a matrix-vector form as follows:

$$\begin{aligned} u_k &= \mathbf{C}^{\mathrm{s}}_k \begin{pmatrix} \mathbf{x} \\ \mathbf{x}^{\mathrm{f}} \end{pmatrix} = \mathbf{C}_k\mathbf{x} + \mathbf{C}^{\mathrm{f}}_k\mathbf{x}^{\mathrm{f}},\\ v_k &= \mathbf{C}^{\mathrm{s}}_k \begin{pmatrix} \mathbf{y} \\ \mathbf{y}^{\mathrm{f}} \end{pmatrix} = \mathbf{C}_k\mathbf{y} + \mathbf{C}^{\mathrm{f}}_k\mathbf{y}^{\mathrm{f}},\\ w_k &= \mathbf{C}^{\mathrm{s}}_k \begin{pmatrix} \mathbf{z} \\ \mathbf{z}^{\mathrm{f}} \end{pmatrix} = \mathbf{C}_k\mathbf{z} + \mathbf{C}^{\mathrm{f}}_k\mathbf{z}^{\mathrm{f}}, \end{aligned} \tag{2.10}$$

where $\mathbf{C}_k$ and $\mathbf{C}^{\mathrm{f}}_k$ are the kth rows of $\mathbf{C}$ and $\mathbf{C}^{\mathrm{f}}$, respectively.

Furthermore, *coordinate difference vectors* $\mathbf{u}$, $\mathbf{v}$, and $\mathbf{w}$ ($\in \mathbb{R}^m$) are expressed in a matrix-vector form as follows:

$$\begin{aligned} \mathbf{u} &= \mathbf{C}\mathbf{x} + \mathbf{C}^{\mathrm{f}}\mathbf{x}^{\mathrm{f}},\\ \mathbf{v} &= \mathbf{C}\mathbf{y} + \mathbf{C}^{\mathrm{f}}\mathbf{y}^{\mathrm{f}},\\ \mathbf{w} &= \mathbf{C}\mathbf{z} + \mathbf{C}^{\mathrm{f}}\mathbf{z}^{\mathrm{f}}. \end{aligned} \tag{2.11}$$

Example 2.3 Coordinate difference vectors of the two-dimensional structure as shown in Fig. 2.1, which has fixed nodes.

Consider again the the two-dimensional structure as shown in Fig. 2.1, which was studied in Example 2.1. The nodes 1 and 2 are free nodes, and the nodes 3–8 are fixed nodes.

Using the connectivity matrices $\mathbf{C} \in \mathbb{R}^{7\times 2}$ and $\mathbf{C}^{\mathrm{f}} \in \mathbb{R}^{7\times 6}$ given in Eq. (2.5), its coordinate difference vector $\mathbf{u} \in \mathbb{R}^7$ in x-direction is calculated as follows according to Eq. (2.11):

$$
\begin{aligned}
\mathbf{u} &= \mathbf{C}\mathbf{x} + \mathbf{C}^{\mathrm{f}}\mathbf{x}^{\mathrm{f}} \\
&= \begin{pmatrix} 1 & -1 \\ 1 & 0 \\ 1 & 0 \\ 0 & 1 \\ 0 & 1 \\ 0 & 1 \\ 1 & 0 \end{pmatrix} \begin{pmatrix} x_1 \\ x_2 \end{pmatrix} + \begin{pmatrix} 0 & 0 & 0 & 0 & 0 & 0 \\ -1 & 0 & 0 & 0 & 0 & 0 \\ 0 & -1 & 0 & 0 & 0 & 0 \\ 0 & 0 & -1 & 0 & 0 & 0 \\ 0 & 0 & 0 & -1 & 0 & 0 \\ 0 & 0 & 0 & 0 & -1 & 0 \\ 0 & 0 & 0 & 0 & 0 & -1 \end{pmatrix} \begin{pmatrix} x_3 \\ x_4 \\ x_5 \\ x_6 \\ x_7 \\ x_8 \end{pmatrix} \\
&= \begin{pmatrix} x_1 - x_2 \\ x_1 - x_3 \\ x_1 - x_4 \\ x_2 - x_5 \\ x_2 - x_6 \\ x_2 - x_7 \\ x_1 - x_8 \end{pmatrix}.
\end{aligned} \tag{2.12}
$$

In a similar manner, the coordinate difference vector $\mathbf{v} \in \mathbb{R}^7$ in y-direction is calculated as

$$
\mathbf{v} = \mathbf{C}\mathbf{y} + \mathbf{C}^{\mathrm{f}}\mathbf{y}^{\mathrm{f}} = \begin{pmatrix} y_1 - y_2 \\ y_1 - y_3 \\ y_1 - y_4 \\ y_2 - y_5 \\ y_2 - y_6 \\ y_2 - y_7 \\ y_1 - y_8 \end{pmatrix}. \tag{2.13}
$$

For free-standing structures, the terms $\mathbf{C}^{\mathrm{f}}\mathbf{x}^{\mathrm{f}}$ and $\mathbf{C}^{\mathrm{f}}\mathbf{y}^{\mathrm{f}}$ corresponding to fixed nodes in Eq. (2.11) vanish. Therefore, the coordinate difference vectors are

$$
\begin{aligned}
\mathbf{u} &= \mathbf{C}\mathbf{x}, \\
\mathbf{v} &= \mathbf{C}\mathbf{y}, \\
\mathbf{w} &= \mathbf{C}\mathbf{z}.
\end{aligned} \tag{2.14}
$$

Example 2.4 Coordinate difference vectors of the two-dimensional free-standing structure as shown in Fig. 2.2.

The two-dimensional free-standing structure as shown in Fig. 2.2 was studied in Example 2.2. All of its five nodes are free nodes.

The coordinate difference vector $\mathbf{u} \in \mathbb{R}^8$ in x-direction of the two-dimensional free-standing structure as shown in Fig. 2.2 is calculated as follows, by using the connectivity matrix $\mathbf{C} \in \mathbb{R}^{8\times5}$ in Eq. (2.8):

$$\mathbf{u} = \mathbf{C}\mathbf{x} = \begin{pmatrix} 1 & -1 & 0 & 0 & 0 \\ 1 & 0 & -1 & 0 & 0 \\ 1 & 0 & 0 & -1 & 0 \\ 1 & 0 & 0 & 0 & -1 \\ 0 & 1 & 0 & -1 & 0 \\ 0 & 0 & 1 & -1 & 0 \\ 0 & 1 & 0 & 0 & -1 \\ 0 & 0 & 1 & 0 & -1 \end{pmatrix} \begin{pmatrix} x_1 \\ x_2 \\ x_3 \\ x_4 \\ x_5 \end{pmatrix} = \begin{pmatrix} x_1 - x_2 \\ x_1 - x_3 \\ x_1 - x_4 \\ x_1 - x_5 \\ x_2 - x_4 \\ x_3 - x_4 \\ x_2 - x_5 \\ x_3 - x_5 \end{pmatrix}. \tag{2.15}$$

Similarly, the coordinate difference vector $\mathbf{v} \in \mathbb{R}^8$ in y-direction is calculated as

$$\mathbf{v} = \mathbf{C}\mathbf{y} = \begin{pmatrix} y_1 - y_2 \\ y_1 - y_3 \\ y_1 - y_4 \\ y_1 - y_5 \\ y_2 - y_4 \\ y_3 - y_4 \\ y_2 - y_5 \\ y_3 - y_5 \end{pmatrix}. \tag{2.16}$$

By using the coordinate differences u_k, v_k, and w_k, the following relation holds for the square of length l_k of member k:

$$l_k^2 = u_k^2 + v_k^2 + w_k^2. \tag{2.17}$$

Let $\mathbf{l} \in \mathbb{R}^m$ denote the vector of member lengths, the kth entry of which is the length of member k. Let $\mathbf{U}$, $\mathbf{V}$, $\mathbf{W}$, and $\mathbf{L}$ ($\in\mathbb{R}^{m\times m}$) denote the diagonal versions of the coordinate difference vectors $\mathbf{u}$, $\mathbf{v}$, $\mathbf{w}$, and the member length vector $\mathbf{l}$; i.e.,

$$\begin{aligned} \mathbf{U} &= \mathrm{diag}(\mathbf{u}), & \mathbf{V} &= \mathrm{diag}(\mathbf{v}), \\ \mathbf{W} &= \mathrm{diag}(\mathbf{w}), & \mathbf{L} &= \mathrm{diag}(\mathbf{l}). \end{aligned} \tag{2.18}$$

Hence, square of member length matrix $\mathbf{L}$ is expressed as

$$\mathbf{L}^2 = \mathbf{U}^2 + \mathbf{V}^2 + \mathbf{W}^2, \tag{2.19}$$

where the diagonal entries of the diagonal matrix $\mathbf{L}^2$ are l_k^2. Alternatively, we can write l_k^2 as entries of the vector $\mathbf{L}\mathbf{l}$ expressed by the following equation

$$\mathbf{L}\mathbf{l} = \mathbf{U}\mathbf{u} + \mathbf{V}\mathbf{v} + \mathbf{W}\mathbf{w}. \tag{2.20}$$

2.2 Equilibrium Matrix

In this section, we obtain the (self-)equilibrium equations, in the form of equilibrium matrix associated with axial forces (prestresses) of the members, in two different ways: (1) those by assembling force balance of each node; and (2) those directly derived by applying the principle of virtual work. These equations are essentially consistent with each other. It will be shown later in Sect. 2.3 that the equilibrium matrix and its transpose, called the compatibility matrix, are the key matrices to understand static and kinematic determinacy of a pin-jointed structure. Furthermore, they are directly related to the linear stiffness matrix as will be presented in Sect. 4.2 in Chap. 4. An equivalent formulation of the equilibrium equations, using the force density matrix associated with the nodal coordinates, will be given in Sect. 2.4.

2.2.1 Equilibrium Equations by Balance of Forces

Consider a single node, for instance numbered as i, as a reference node. For simplicity, we consider only equilibrium of free nodes in the following, while equilibrium of fixed nodes can be derived in a similar manner.

Suppose the reference node i is connected to node i_j by member k_j, and there are in total m_i such members. The axial force (or prestress when no external load) carried in member k_j is denoted by s_{k_j} $(j = 1, 2, \dots, m_i)$. External loads applied at the free nodes (or reaction forces at the fixed nodes) in x-, y-, and z-directions are denoted by load vectors $\mathbf{p}^x$, $\mathbf{p}^y$, and $\mathbf{p}^z$ $(\in \mathbb{R}^{n+n^{\mathrm{f}}})$, respectively. The ith entries of $\mathbf{p}^x$, $\mathbf{p}^y$, $\mathbf{p}^z$ are the loads p_i^x, p_i^y, p_i^z applied at node i in each direction. If node i is a fixed node, then p_i^x, p_i^y, p_i^z refer to the reaction forces.

Equilibrium equation of the reference node i in x direction is

$$
\begin{aligned}
p_i^x + \sum_{j=1}^{m_i} s_{k_j} \frac{x_{i_j} - x_i}{l_{k_j}} &= p_i^x - \sum_{j=1}^{m_i} \operatorname{sign}(i_j - i) \frac{u_{k_j} s_{k_j}}{l_{k_j}} \\
&= p_i^x - \sum_{j=1}^{m_i} C_{(k_j,i)} \frac{u_{k_j} s_{k_j}}{l_{k_j}} \\
&= 0,
\end{aligned}
\tag{2.21}
$$

where $x_{i_j} - x_i = -\operatorname{sign}(i_j - i) \cdot u_{k_j}$ and the (k_j, i)th entry $C_{(k_j,i)}$ of the connectivity matrix $\mathbf{C}$ of free nodes have been used.

Example 2.5 Equilibrium equation of a single node of a two-dimensional structure as shown in Fig. 2.3.

As shown in Fig. 2.3, the free node i of a two-dimensional structure is connected to three other nodes i_1, i_2, and i_3 by members k_1, k_2, and k_3, respectively. Thus, we have $m_i = 3$ for this case. Moreover, p_i^x and p_i^y are the external loads applied at node i in x- and y-directions, respectively.

Using Eq. (2.21), equilibrium equation of free node i in x-direction is written as

$$\begin{aligned} & p_i^x + s_{k_1}\frac{x_{i_1}-x_i}{l_{k_1}} + s_{k_2}\frac{x_{i_2}-x_i}{l_{k_2}} + s_{k_3}\frac{x_{i_3}-x_i}{l_{k_3}} \\ &= p_i^x - \text{sign}(i_1 - i)\frac{s_{k_1}u_{k_1}}{l_{k_1}} - \text{sign}(i_2 - i)\frac{s_{k_2}u_{k_2}}{l_{k_2}} - \text{sign}(i_3 - i)\frac{s_{k_3}u_{k_3}}{l_{k_3}} \\ &= p_i^x - \sum_{j=1}^{3}\text{sign}(i_j - i)\frac{u_{k_j}s_{k_j}}{l_{k_j}} \\ &= 0. \end{aligned} \tag{2.22}$$

Similar equilibrium equation can be written for node i in y-direction.

Let $\mathbf{s} \in \mathbb{R}^m$ denote *member force vector*, the kth element s_k of which is the axial force in member k. Because the non-zero entries in the ith column of $\mathbf{C}$ correspond to the nodes that are connected to node i by the corresponding members, the equilibrium equation of the free node i in x-direction in Eq. (2.21) can be written in a matrix form as

$$(\mathbf{C}^\top)_i\mathbf{U}\mathbf{L}^{-1}\mathbf{s} = p_i^x, \tag{2.23}$$

where $(\mathbf{C}^\top)_i$ denotes the ith row of $\mathbf{C}^\top$; i.e., the transpose of the ith column of $\mathbf{C}$; moreover, $\mathbf{L}^{-1}$ denotes inverse matrix of the member length matrix $\mathbf{L}$, and the kth entry of $\mathbf{L}^{-1}$ is $1/l_k$.

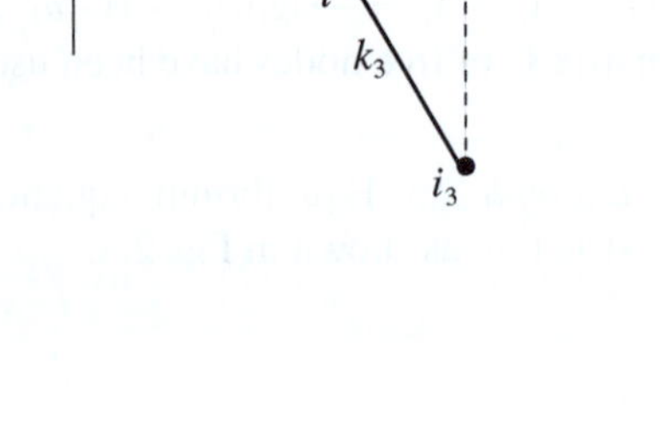

Fig. 2.3 Equilibrium of a reference node i of a two-dimensional structure subjected to external loads p_i^x and p_i^y. Nodes i_1, i_2, and i_3 are connected to node i by members k_1, k_2, and k_3, respectively

Example 2.6 Equilibrium equation of node 3 of the two-dimensional free-standing structure as shown in Fig. 2.2 in a matrix form.

Consider the equilibrium equation of the free node 3 of the two-dimensional free-standing structure as shown in Fig. 2.2. Using the third row $(\mathbf{C}^\top)_3$ of the transpose of the connectivity matrix $\mathbf{C}$ given in Eq. (2.8), we have

$$\begin{aligned}
&(\mathbf{C}^\top)_3\mathbf{U}\mathbf{L}^{-1}\mathbf{s} \\
&= \left(0,\ -\frac{u_2}{l_2},\ 0,\ 0,\ 0,\ \frac{u_6}{l_6},\ 0,\ \frac{u_8}{l_8}\right)(s_1,\ s_2,\ s_3,\ s_4,\ s_5,\ s_6,\ s_7,\ s_8)^\top \\
&= -\frac{u_2}{l_2}s_2 + \frac{u_6}{l_6}s_6 + \frac{u_8}{l_8}s_8 \\
&= \frac{x_3 - x_1}{l_2}s_2 + \frac{x_3 - x_4}{l_6}s_6 + \frac{x_3 - x_5}{l_8}s_8,
\end{aligned} \tag{2.24}$$

where

$$\begin{aligned}
(\mathbf{C}^\top)_3 &= \left(0,\ -1,\ 0,\ 0,\ 0,\ 1,\ 0,\ 1\right), \\
\mathbf{U} &= \text{diag}\left(u_1,\ u_2,\ u_3,\ u_4,\ u_5,\ u_6,\ u_7,\ u_8\right), \\
\mathbf{L}^{-1} &= \text{diag}\left(\frac{1}{l_1},\ \frac{1}{l_2},\ \frac{1}{l_3},\ \frac{1}{l_4},\ \frac{1}{l_5},\ \frac{1}{l_6},\ \frac{1}{l_7},\ \frac{1}{l_8}\right), \\
\mathbf{s}^\top &= \left(s_1,\ s_2,\ s_3,\ s_4,\ s_5,\ s_6,\ s_7,\ s_8\right),
\end{aligned} \tag{2.25}$$

and moreover, the following coordinate differences of the members connected to node 3 have been used:

$$\begin{aligned}
u_2 &= x_1 - x_3, \\
u_6 &= x_3 - x_4, \\
u_8 &= x_3 - x_5.
\end{aligned} \tag{2.26}$$

From the equilibrium equation of a single node in x-direction as defined in Eq. (2.21), we have

$$p_3^x - \left(\frac{x_3 - x_1}{l_2}s_2 + \frac{x_3 - x_4}{l_6}s_6 + \frac{x_3 - x_5}{l_8}s_8\right) = 0, \tag{2.27}$$

where p_3^x denotes the external load applied at node 3 in x-direction. From Eqs. (2.24) and (2.27), we have

$$(\mathbf{C}^\top)_3\mathbf{U}\mathbf{L}^{-1}\mathbf{s} = p_3^x, \tag{2.28}$$

which validates the equilibrium equation (2.23) in the matrix form.

The equilibrium equations of all free nodes of the structure in x-direction are summarized in a matrix form as

$$\mathbf{D}^x\mathbf{s} = \mathbf{p}^x, \tag{2.29}$$

where

$$\mathbf{D}^x = \mathbf{C}^\top\mathbf{U}\mathbf{L}^{-1}. \tag{2.30}$$

In a similar manner, the equilibrium equations in y- and z-directions are

$$\mathbf{D}^y\mathbf{s} = \mathbf{p}^y \quad \text{and} \quad \mathbf{D}^z\mathbf{s} = \mathbf{p}^z, \tag{2.31}$$

with

$$\mathbf{D}^y = \mathbf{C}^\top\mathbf{V}\mathbf{L}^{-1} \quad \text{and} \quad \mathbf{D}^z = \mathbf{C}^\top\mathbf{W}\mathbf{L}^{-1}. \tag{2.32}$$

The equilibrium equations of free nodes of a pin-jointed structure with respect to member force vector $\mathbf{s}$ can then be combined as

$$\mathbf{D}\mathbf{s} = \mathbf{p}, \tag{2.33}$$

where

$$\mathbf{D} = \begin{pmatrix} \mathbf{D}^x \\ \mathbf{D}^y \\ \mathbf{D}^z \end{pmatrix} \quad \text{and} \quad \mathbf{p} = \begin{pmatrix} \mathbf{p}^x \\ \mathbf{p}^y \\ \mathbf{p}^z \end{pmatrix}. \tag{2.34}$$

In the above equations, $\mathbf{D} \in \mathbb{R}^{3n\times m}$ is called the *equilibrium matrix*. For a two-dimensional structure, the size of $\mathbf{D}$ is $2n \times m$, since it becomes

$$\mathbf{D} = \begin{pmatrix} \mathbf{D}^x \\ \mathbf{D}^y \end{pmatrix}. \tag{2.35}$$

Define the reaction force vectors in x-, y-, and z-directions as $\mathbf{f}^x, \mathbf{f}^y, \mathbf{f}^z$ $(\in\mathbb{R}^{n^\mathrm{f}})$, respectively. Equilibrium equations of the fixed nodes can be written in a similar way as the free nodes as

$$\mathbf{D}^\mathrm{f}\mathbf{s} = \mathbf{f}, \tag{2.36}$$

where

$$\mathbf{D}^\mathrm{f} = \begin{pmatrix} (\mathbf{C}^\mathrm{f})^\top\mathbf{U}\mathbf{L}^{-1} \\ (\mathbf{C}^\mathrm{f})^\top\mathbf{V}\mathbf{L}^{-1} \\ (\mathbf{C}^\mathrm{f})^\top\mathbf{W}\mathbf{L}^{-1} \end{pmatrix} \quad \text{and} \quad \mathbf{f} = \begin{pmatrix} \mathbf{f}^x \\ \mathbf{f}^y \\ \mathbf{f}^z \end{pmatrix}. \tag{2.37}$$

2.2.2 Equilibrium Equations by the Principle of Virtual Work

In this subsection, the equilibrium equations are derived in a more systematic way than those obtained by considering force balance in the previous subsection.

Suppose that the structure is subjected to virtual displacements δx_i, δy_i, and δz_i $(i = 1, 2, \dots, 3n)$, which cause virtual member length extensions δl_k $(k = 1, 2, \dots, m)$. Thus, the total virtual work $\delta \Pi$, due to the virtual displacements and member length extensions, can be written as follows:

$$\delta \Pi = \sum_{k=1}^{m} s_k \delta l_k - \sum_{i=1}^{n} p_i^x \delta x_i - \sum_{i=1}^{n} p_i^y \delta y_i - \sum_{i=1}^{n} p_i^z \delta z_i, \tag{2.38}$$

which should be zero, when the structure is in equilibrium, according to the principle of virtual work [7]; i.e.,

$$\delta \Pi = 0, \tag{2.39}$$

for arbitrary δx_i, δy_i, δz_i, and δl_k satisfying compatibility conditions.

Using the coordinate differences u_k, v_k, and w_k in each direction of member k, the following relation holds:

$$\delta(l_k^2) = \delta(u_k^2) + \delta(v_k^2) + \delta(w_k^2), \tag{2.40}$$

which comes from the definition of member length. Equation (2.40) results in

$$l_k \delta l_k = u_k \delta u_k + v_k \delta v_k + w_k \delta w_k. \tag{2.41}$$

Substituting Eq. (2.41) into Eqs. (2.38) and (2.39) gives

$$\begin{aligned} \delta \Pi = & \left(\sum_{k=1}^{m} \frac{s_k u_k}{l_k} \delta u_k - \sum_{i=1}^{n} p_i^x \delta x_i \right) \\ & + \left(\sum_{k=1}^{m} \frac{s_k v_k}{l_k} \delta v_k - \sum_{i=1}^{n} p_i^y \delta y_i \right) \\ & + \left(\sum_{k=1}^{m} \frac{s_k w_k}{l_k} \delta w_k - \sum_{i=1}^{n} p_i^z \delta z_i \right) \\ = & \ 0. \end{aligned} \tag{2.42}$$

Because u_k, v_k, and w_k are functions of coordinates in x-, y-, and z-directions, respectively, and moreover, δx_i, δy_i, and δz_i are arbitrary (virtual) values, the following three independent equations should be true at the same time so that Eq. (2.42) holds:

$$\begin{aligned}
x\text{-direction: } & \sum_{k=1}^{m} \frac{s_k u_k}{l_k} \delta u_k - \sum_{i=1}^{n} p_i^x \delta x_i = 0, \\
y\text{-direction: } & \sum_{k=1}^{m} \frac{s_k v_k}{l_k} \delta v_k - \sum_{i=1}^{n} p_i^y \delta y_i = 0, \\
z\text{-direction: } & \sum_{k=1}^{m} \frac{s_k w_k}{l_k} \delta w_k - \sum_{i=1}^{n} p_i^z \delta z_i = 0.
\end{aligned} \tag{2.43}$$

In the following, we consider the equation only in x-direction, for clarity. Those in y- and z-directions can be obtained in a similar way.

If member k is connected by nodes i and j, then its coordinate difference u_k can be calculated by using the components $C_{(k,i)}$ and $C_{(k,j)}$ in the kth row of the connectivity matrix $\mathbf{C}$ defined in Eq. (2.1):

$$u_k = C_{(k,i)} x_i + C_{(k,j)} x_j, \tag{2.44}$$

and therefore, its variation is

$$\delta u_k = C_{(k,i)} \delta x_i + C_{(k,j)} \delta x_j. \tag{2.45}$$

Because all entries in the kth row of $\mathbf{C}$, except for $C_{(k,i)}$ and $C_{(k,j)}$, are zero, Eq. (2.45) can be rewritten as

$$\delta u_k = \sum_{i=1}^{n} C_{(k,i)} \delta x_i. \tag{2.46}$$

Substituting Eq. (2.46) into the first equation in Eq. (2.43) for node i in x-direction, we have

$$\begin{aligned}
& \sum_{k=1}^{m} \sum_{i=1}^{n} \frac{s_k u_k}{l_k} C_{(k,i)} \delta x_i - \sum_{i=1}^{n} p_i^x \delta x_i \\
&= \sum_{i=1}^{n} \left(\sum_{k=1}^{m} \frac{s_k u_k}{l_k} C_{(k,i)} - p_i^x \right) \delta x_i \\
&= 0.
\end{aligned} \tag{2.47}$$

Because δx_i are arbitrary values, we then have n equilibrium equations

$$\sum_{k=1}^{m} \frac{s_k u_k}{l_k} C_{(k,i)} - p_i^x = 0, \quad (i = 1, 2, \ldots, n), \tag{2.48}$$

which can be further summarized in a matrix form as

$$\begin{aligned}\sum_{k=1}^{m} \frac{s_k u_k}{l_k} C_{(k,i)} &= (\mathbf{C}^\top)_i \mathbf{U}\mathbf{L}^{-1}\mathbf{s} \\ &= p_i^x, \qquad (i = 1, 2, \ldots, n).\end{aligned} \tag{2.49}$$

Assembling the equilibrium equations in x-direction for all nodes, we have the equilibrium equation in a matrix form as

$$\mathbf{D}^x \mathbf{s} = \mathbf{p}^x, \tag{2.50}$$

which coincides with the equilibrium equation derived in Eq. (2.29). In a similar way, we can derive the equilibrium equations in y- and z-directions, which will not be repeated here.

Since the discussions on equilibrium and stability do not depend on the units, we will omit the units in the following examples.

Example 2.7 Equilibrium matrix of the two-dimensional free-standing structure as shown in Fig. 2.2.

For simplicity, we consider a symmetric geometry realization of the two-dimensional free-standing structure as shown in Fig. 2.2. Its nodal coordinates are given in vector forms as

$$\begin{aligned}\mathbf{x} &= (0,\ -1,\ 1,\ 0,\ 0)^\top, \\ \mathbf{y} &= (0,\ 0,\ 0,\ 1,\ -1)^\top.\end{aligned} \tag{2.51}$$

From Eq. (2.15), the coordinate difference vectors $\mathbf{u}$ and $\mathbf{v}$ of the structure are

$$\mathbf{u} = \mathbf{C}\mathbf{x} = \begin{pmatrix} x_1 - x_2 \\ x_1 - x_3 \\ x_1 - x_4 \\ x_1 - x_5 \\ x_2 - x_4 \\ x_3 - x_4 \\ x_2 - x_5 \\ x_3 - x_5 \end{pmatrix} = \begin{pmatrix} 1 \\ -1 \\ 0 \\ 0 \\ -1 \\ 1 \\ -1 \\ 1 \end{pmatrix},$$

$$\mathbf{v} = \mathbf{C}\mathbf{y} = \begin{pmatrix} y_1 - y_2 \\ y_1 - y_3 \\ y_1 - y_4 \\ y_1 - y_5 \\ y_2 - y_4 \\ y_3 - y_4 \\ y_2 - y_5 \\ y_3 - y_5 \end{pmatrix} = \begin{pmatrix} 0 \\ 0 \\ -1 \\ 1 \\ -1 \\ -1 \\ 1 \\ 1 \end{pmatrix}. \tag{2.52}$$

Note that we have $\mathbf{C} = \mathbf{C}^{\mathrm{s}}$ for free-standing structures. Moreover, the member length matrix $\mathbf{L} \in \mathbb{R}^{8\times 8}$ is

$$\mathbf{L} = \mathrm{diag}(1,\ 1,\ 1,\ 1,\ \sqrt{2},\ \sqrt{2},\ \sqrt{2},\ \sqrt{2}). \tag{2.53}$$

Therefore, the equilibrium matrix $\mathbf{D} \in \mathbb{R}^{10\times 8}$ is calculated as follows by using Eq. (2.34):

$$\mathbf{D} = \begin{pmatrix} \mathbf{C}^{\top}\mathbf{U}\mathbf{L}^{-1} \\ \mathbf{C}^{\top}\mathbf{V}\mathbf{L}^{-1} \end{pmatrix}$$
$$= \frac{1}{2}\begin{pmatrix} 2 & -2 & 0 & 0 & 0 & 0 & 0 & 0 \\ -2 & 0 & 0 & 0 & -\sqrt{2} & 0 & -\sqrt{2} & 0 \\ 0 & 2 & 0 & 0 & 0 & \sqrt{2} & 0 & \sqrt{2} \\ 0 & 0 & 0 & 0 & \sqrt{2} & -\sqrt{2} & 0 & 0 \\ 0 & 0 & 0 & 0 & 0 & 0 & \sqrt{2} & -\sqrt{2} \\ 0 & 0 & -2 & 2 & 0 & 0 & 0 & 0 \\ 0 & 0 & 0 & 0 & -\sqrt{2} & 0 & \sqrt{2} & 0 \\ 0 & 0 & 0 & 0 & 0 & -\sqrt{2} & 0 & \sqrt{2} \\ 0 & 0 & 2 & 0 & \sqrt{2} & \sqrt{2} & 0 & 0 \\ 0 & 0 & 0 & -2 & 0 & 0 & -\sqrt{2} & -\sqrt{2} \end{pmatrix}. \tag{2.54}$$

2.3 Static and Kinematic Determinacy

Tensegrity structures are always statically indeterminate, so that their members can carry prestresses, when no external load is applied. Moreover, tensegrity structures are usually kinematically indeterminate, such that they are unstable in the absence of prestresses. Kinematically indeterminate structures can be stabilized by proper prestresses in the self-equilibrium state. The criteria and conditions for stability of tensegrity structures will be given in the next chapter. In this section, we will focus on conditions for static and kinematic determinacy of general (prestressed) pin-jointed structures.

A structure is said to be *statically indeterminate*, if the member forces and reaction forces of the structure cannot be uniquely determined by using only (static) equilibrium equations, while subjected to external loads. By contrast, member forces and reaction forces of the structure can be uniquely determined by considering only equilibrium equations, if it is *statically determinate*. Static indeterminacy of a structure enables it possess member forces (prestresses) in the members, even though no external load is involved.

A structures is said to be *kinematically indeterminate*, if there exists a nodal motion, except for the rigid-body motions, keeping all member lengths unchanged. Such motion is called *mechanism*, including infinitesimal mechanism and finite mechanism. *Infinitesimal mechanism* implies that the motion (nodal displacement) is sufficiently small, while *finite mechanism* can have large displacements. Two simple example structures illustrating the difference between infinitesimal and finite mechanisms are shown in Fig. 2.4. Finite mechanisms are widely used for machinery or deployable structures that can largely change their shapes. In this book, we will focus only on infinitesimal mechanisms.

By contrast to mechanisms, *rigid-body motions* refer to the motions that do no change the distance between any pair of nodes. Note that a pair of nodes does not limit to the two nodes connected by a member. Rigid-body motions include translation in each direction, rotation about an arbitrary axis, and the combination of these motions. Figure 2.5 shows the three rigid-body motions of a structure in the two-dimensional space, while there exist six rigid-body motions for a structure in the three-dimensional space.

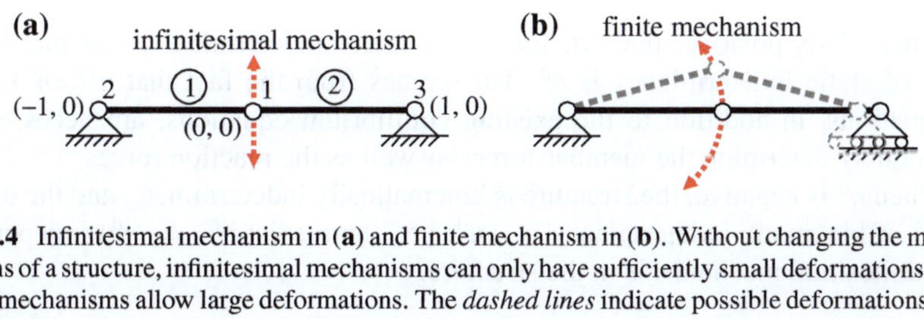

Fig. 2.4 Infinitesimal mechanism in (**a**) and finite mechanism in (**b**). Without changing the member lengths of a structure, infinitesimal mechanisms can only have sufficiently small deformations, while finite mechanisms allow large deformations. The *dashed lines* indicate possible deformations of the structures

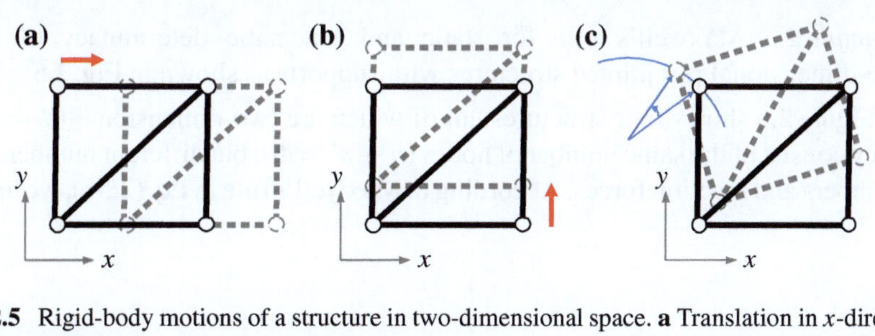

Fig. 2.5 Rigid-body motions of a structure in two-dimensional space. **a** Translation in x-direction, **b** translation in y-direction, **c** rotation about any point

2.3.1 Maxwell's Rule

There is a simple rule, called *Maxwell's rule*, to identify degrees of static and kinematic determinacy of a pin-jointed structure. The rule was proposed by Maxwell [10] in the 19th century. Using only the number m of member, the numbers n and n^{f} of nodes, the number n^{r} of reaction forces, static as well as kinematic determinacy of a d-dimensional structure consisting of fixed nodes can be identified as follows by observing the sign of the number n^{d}:

Maxwell's Rule:

$$n^{\mathrm{d}} = m + n^{\mathrm{r}} - d(n + n^{\mathrm{f}}) \begin{cases} < 0, & \text{kinematically indeterminate,} \\ = 0, & \text{statically and kinematically determinate,} \\ > 0, & \text{statically indeterminate.} \end{cases} \tag{2.55}$$

The principle of Maxwell's rule lies on whether the linear equilibrium equations can be uniquely solved: if the number of unknown parameters, including m member forces and n^{r} reaction forces, is equal to the total number $d(n + n^{\mathrm{f}})$ of (equilibrium) equations, with d equations at each of the $n + n^{\mathrm{f}}$ nodes, then the unknown parameters can be 'uniquely'[3] determined; and therefore, the structure is statically determinate. This in fact corresponds to $n^{\mathrm{d}} = 0$.

Moreover, there are two other cases concerning the sign of n^{d}:

1. When n^{d} is positive, the structure is statically indeterminate, and the degree n^{s} of static indeterminacy is n^{d}. This comes from the fact that $n^{\mathrm{s}}(=n^{\mathrm{d}})$ more equations, in addition to the existing equilibrium equations, are necessary to uniquely determine the member forces as well as the reaction forces.
2. When n^{d} is negative, the structure is kinematically indeterminate, and the degree n^{m} of kinematic indeterminacy is $-n^{\mathrm{d}}$; i.e., there exist $n^{\mathrm{m}}(=-n^{\mathrm{d}})$ independent infinitesimal mechanisms in the structure.

Example 2.8 Maxwell's rule for static and kinematic determinacy of the two-dimensional pin-jointed structures with supports as shown in Fig. 2.6.

Figure 2.6 shows four structures, all of which are two-dimensional ($d = 2$). They consist of the same number of nodes ($n + n^{\mathrm{f}} = 5$), but different numbers of members and reaction forces. According to Maxwell's rule in Eq. (2.55), we have

[3] The existence of unique solution is subjected to independence of the equilibrium equations.

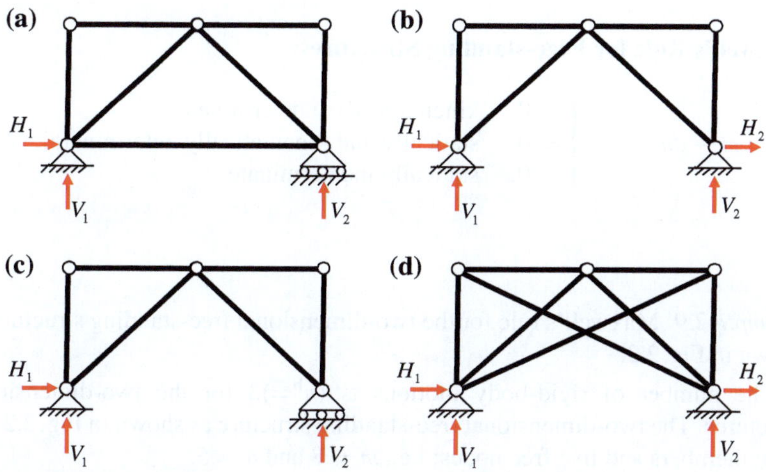

Fig. 2.6 Static and kinematic determinacy of the two-dimensional pin-jointed structures in Example 2.8. H_1, H_2 and V_1, V_2 are the reaction forces in horizontal and vertical directions, respectively. **a, b** Statically and kinematically determinate structures, **c** kinematically indeterminate structure, **d** statically indeterminate structure

$$
\begin{aligned}
&\text{(a) } m = 7,\ n^{\mathrm{r}} = 3: \quad n^{\mathrm{d}} = 7 + 3 - 2 \times 5 = 0 \\
&\qquad \Longrightarrow \text{statically and kinematically determinate;} \\
&\text{(b) } m = 6,\ n^{\mathrm{r}} = 4: \quad n^{\mathrm{d}} = 6 + 4 - 2 \times 5 = 0 \\
&\qquad \Longrightarrow \text{statically and kinematically determinate;} \\
&\text{(c) } m = 6,\ n^{\mathrm{r}} = 3: \quad n^{\mathrm{d}} = 6 + 3 - 2 \times 5 = -1 \\
&\qquad \Longrightarrow \text{kinematically indeterminate with degree of one; i.e., } n^{\mathrm{m}} = 1; \\
&\text{(d) } m = 8,\ n^{\mathrm{r}} = 4: \quad n^{\mathrm{d}} = 8 + 4 - 2 \times 5 = 2 \\
&\qquad \Longrightarrow \text{statically indeterminate with degree of two; i.e., } n^{\mathrm{s}} = 2.
\end{aligned}
\tag{2.56}
$$

In the case of free-standing structures without any fixed node; i.e., $n^{\mathrm{f}} = 0$, or reaction force; i.e., $n^{r} = 0$, rigid-body motions should be considered in the Maxwell's rule to verify internal static indeterminacy. The number of rigid-body motions n^{b} is given by

$$
n^{\mathrm{b}} = \frac{d^2 + d}{2}, \tag{2.57}
$$

from which n^{b} is equal to 3 for a two-dimensional case, and it is equal to 6 for a three-dimensional case.

Maxwell's Rule for Free-standing Structures:

$$n^{\mathrm{d}} = m - dn + n^{\mathrm{b}} \begin{cases} < 0, & \text{kinematically indeterminate,} \\ = 0, & \text{statically and kinematically determinate,} \\ > 0, & \text{statically indeterminate.} \end{cases} \tag{2.58}$$

Example 2.9 Maxwell's rule for the two-dimensional free-standing structure as shown in Fig. 2.2.

The number of rigid-body motions is (n^{b}=)3 for the two-dimensional structures. The two-dimensional free-standing structure as shown in Fig. 2.2 has eight members and five free nodes; i.e., $m = 8$ and $n = 5$.

According to Maxwell's rule for free-standing structures in Eq. (2.58), we have

$$\begin{aligned} n^{\mathrm{d}} &= m - dn + n^{\mathrm{b}} \\ &= 8 - 2 \times 5 + 3 \\ &= 1; \end{aligned} \tag{2.59}$$

i.e., the degree n^{d} of static indeterminacy of the structure is one. This means that there exists only one prestress mode in the structure.

It should be noted that Maxwell's rule is *NOT* a sufficient condition for identification of static or kinematic determinacy of a pin-jointed structure, because neither connectivity nor geometry realization of the structure has been taken into consideration. It has been well recognized that there are many exceptions while applying Maxwell's rule. Hence, Maxwell's rule can only be used for preliminary study; exact investigation can be conducted by checking rank of the equilibrium matrix, details on which will be given in the next two subsections.

Example 2.10 The two-dimensional pin-jointed structure in Fig. 2.7 as an exceptional example of Maxwell's rule.

The structure as shown in Fig. 2.7 consists of six members, four reaction forces, and five nodes; i.e., $m = 6$, $n^{\mathrm{r}} = 4$, and $n + n^{\mathrm{f}} = 5$. It is statically and kinematically determinate, according to Maxwell's rule:

$$\begin{aligned} & n^{\mathrm{d}} = 6 + 4 - 2 \times 5 = 0 \\ \Longrightarrow\ & \text{statically and kinematically determinate?} \end{aligned} \tag{2.60}$$

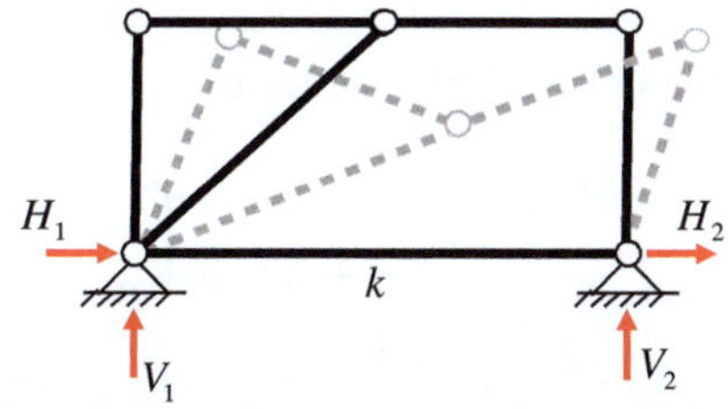

Fig. 2.7 Exceptional example of Maxwell's rule in Example 2.10. According to Maxwell's rule, it is statically and kinematically determinate. However, the structure is in fact statically and kinematically indeterminate, because there exists one (finite) mechanism as indicated by *dashed lines* in the figure, and it has one prestress mode

However, the structure is actually statically and kinematically indeterminate. It can deform as indicated by grey dashed lines in the figure, keeping lengths of all members unchanged. Moreover, member k directly connecting the two supports can have arbitrary prestress.

2.3.2 *Modified Maxwell's Rule*

To incorporate the exceptions in Maxwell's rule, a modified version of the rule was presented by Calladine [1] as follows:

Modified Maxwell's Rule for Free-standing Structures:

The following relation holds for the number n^{s} of independent prestress modes and the number n^{m} of independent infinitesimal mechanisms:

$$n^{\mathrm{s}} - n^{\mathrm{m}} = m - dn + n^{\mathrm{b}}. \tag{2.61}$$

Remember that n^{s} is the number of independent prestress modes or degree of static indeterminacy, and n^{m} is the number of independent infinitesimal mechanisms or degree of kinematic indeterminacy.

Example 2.11 Modified Maxwell's rule for the two-dimensional free-standing structure as shown in Fig. 2.2.

The two-dimensional free-standing structure as shown in Fig. 2.2 consists of five free nodes and eight members; i.e., $n = 5$ and $m = 8$. According to the modified Maxwell's rule in Eq. (2.61), we have

$$\begin{aligned} n^{\mathrm{s}} - n^{\mathrm{m}} &= m - dn + n^{\mathrm{b}} \\ &= 8 - 2 \times 5 + 3 \\ &= 1, \end{aligned} \tag{2.62}$$

which indicates that there exists at least one prestress mode, because the number n^{m} of (infinitesimal) mechanisms has to be non-negative; i.e., $n^{\mathrm{m}} \geq 0$.

Using the above relation, we can derive the number of prestress modes or infinitesimal mechanisms, if either of them is known. From Example 2.12 studied in the next subsection, we know that the structure has only one prestress mode; i.e., $n^{\mathrm{s}} = 1$. Therefore, from Eq. (2.62), the number n^{m} of independent (infinitesimal) mechanisms is

$$n^{\mathrm{m}} = n^{\mathrm{s}} - 1 = 0; \tag{2.63}$$

i.e., the structure in Fig. 2.2 consists of only one prestress mode and no mechanism. Calculation of number of mechanisms of the structure will be revisited in a formal way in Example 2.13.

Readers are encouraged to use the modified Maxwell's rule to revisit the exceptional example of Maxwell's rule presented in Example 2.10.

2.3.3 Static Determinacy

To correctly evaluate static and kinematic determinacy of a pin-jointed structure, it is necessary to investigate rank of the equilibrium matrix, or equivalently, that of the compatibility matrix. We will first prove that the compatibility matrix, which relates member extensions to nodal displacements, is the transpose of the equilibrium matrix. It will be further shown that rank of the equilibrium matrix reveals whether there are enough number of linear self-equilibrium and compatibility equations, so as to uniquely derive the non-trivial solutions for member forces and nodal displacements, respectively. In the following, we will concentrate on free-standing structures, because tensegrity structures belong to this category.

When there is no external load applied to the structure; i.e., $\mathbf{p} = \mathbf{0}$, the *self-equilibrium equation* with respect to the vector of self-equilibrium prestresses (or axial member forces) $\mathbf{s}$ of a free-standing structure is written as follows from Eq. (2.33):

$$\mathbf{D}\mathbf{s} = \mathbf{0}. \tag{2.64}$$

Denote rank of the equilibrium matrix $\mathbf{D} \in \mathbb{R}^{dn \times m}$ by r^{D}; i.e.,

$$r^{\mathrm{D}} = \mathrm{rank}(\mathbf{D}). \tag{2.65}$$

It is obvious that the rank of a matrix cannot be larger than its dimensions; i.e.,

$$r^{\mathrm{D}} \leq \min(dn, m), \tag{2.66}$$

where $\min(dn, m)$ refers to the smaller value of dn and m.

Rank r^{D} of the equilibrium matrix is indeed the number of independent equations. Thus, the number n^{s} of independent prestress modes, or degree of static indeterminacy, is calculated as

$$n^{\mathrm{s}} = m - r^{\mathrm{D}}. \tag{2.67}$$

Moreover, there are two cases concerning number of possible solutions for prestresses, by considering the value of r^{D} in comparison to the number m of prestresses:

- $r^{\mathrm{D}} < m$ or $n^{\mathrm{s}} > 0$ (Statically indeterminate):
  In this case, there exist $m - r^{\mathrm{D}}$ independent non-trivial solutions (prestress modes in members); i.e., $\mathbf{s} \neq \mathbf{0}$, satisfying the self-equilibrium equation (2.64). The possible solutions lie in the null-space of $\mathbf{D}$ spanned by the independent prestress modes $\bar{\mathbf{s}}_i$ $(i = 1, 2, \ldots, m - r^{\mathrm{D}})$ as follows:

$$\mathbf{s} = \sum_{i=1}^{n^{\mathrm{s}}} \alpha_i \bar{\mathbf{s}}_i, \tag{2.68}$$

  where α_i are arbitrary coefficients, and the independent prestress modes $\bar{\mathbf{s}}_i$ satisfy

$$\mathbf{D}\bar{\mathbf{s}}_i = \mathbf{0} \quad \text{and} \quad \bar{\mathbf{s}}_i^{\top}\bar{\mathbf{s}}_j = \delta_{ij}, \tag{2.69}$$

  with δ_{ij} referring to Kronecker's delta:

$$\delta_{ij} = \begin{cases} 1, & \text{if } i = j, \\ 0, & \text{if } i \neq j. \end{cases} \tag{2.70}$$

  Hence, the structure is *statically indeterminate* in this case, because we cannot uniquely determine the prestresses in the structure without additional information (equations).
- $r^{\mathrm{D}} = m$ or $n^{\mathrm{s}} = 0$ (Statically determinate):
  On the other hand, for the case with $r^{\mathrm{D}} = m$, the structure cannot contain any non-zero prestresses while no external load is applied. Therefore, the structure is *statically determinate* in this case.

Example 2.12 Static determinacy of the two-dimensional free-standing structure as shown in Fig. 2.2 and its single independent prestress mode.

For the two-dimensional free-standing structure in Fig. 2.2, the geometry realization of the structure in Example 2.7 is adopted. The structure consists of eight members; i.e., $m = 8$. Investigation of rank[4] r^{D} of its equilibrium matrix $\mathbf{D} \in \mathbb{R}^{10\times 8}$ in Eq. (2.54) gives

$$r^{\mathrm{D}} = 7. \tag{2.71}$$

Hence, the number n^{s} of independent prestress modes can be calculated as follows by using Eq. (2.67):

$$\begin{aligned} n^{\mathrm{s}} &= m - r^{\mathrm{D}} = 8 - 7 \\ &= 1, \end{aligned} \tag{2.72}$$

which means that there exists only one prestress mode. The normalized prestress mode $\bar{\mathbf{s}}$ of the structure is obtained from the null-space of $\mathbf{D}$ as

$$\bar{\mathbf{s}} = \frac{1}{\sqrt{12}} \begin{pmatrix} \sqrt{2} \\ \sqrt{2} \\ \sqrt{2} \\ \sqrt{2} \\ -1 \\ -1 \\ -1 \\ -1 \end{pmatrix}. \tag{2.73}$$

It is easy to verify that the following equation holds

$$\mathbf{D}\bar{\mathbf{s}} = \mathbf{0}. \tag{2.74}$$

Hence, prestresses of the structure in proportion to $\bar{\mathbf{s}}$ would necessarily satisfy the self-equilibrium equations.

2.3.4 Kinematic Determinacy

Let $\mathbf{d} \in \mathbb{R}^{dn}$ denote the vector of (infinitesimal) nodal displacements due to the external loads $\mathbf{p}$ applied to the structure, and let $\mathbf{e} \in \mathbb{R}^{m}$ denote the vector of member extensions, which are related to the small displacements $\mathbf{d}$ by the kinematic relations

[4] Rank and null-space of a matrix $\mathbf{D}$ can be found by using, for example, the commands *rank*($\mathbf{D}$) and *null*($\mathbf{D}$) in Octave or Matlab, respectively.

in terms of the *compatibility matrix* $\mathbf{H} \in \mathbb{R}^{m \times dn}$ [2, 3]:

$$\mathbf{Hd} = \mathbf{e}. \tag{2.75}$$

From the principle of virtual work, the virtual work done by the external loads is equal to virtual internal work done by member forces. Hence, for the virtual displacements $\delta\mathbf{d}$, which can have arbitrary values, and their corresponding virtual member extensions $\delta\mathbf{e}$, we have

$$\mathbf{p}^\top(\delta\mathbf{d}) = \mathbf{s}^\top(\delta\mathbf{e}). \tag{2.76}$$

Substituting Eq. (2.75) into Eq. (2.76), we have

$$\mathbf{p}^\top(\delta\mathbf{d}) = \mathbf{s}^\top\mathbf{H}(\delta\mathbf{d}), \tag{2.77}$$

since $\mathbf{H}$ is a constant matrix subjected to virtual displacements. From the relationship between the external loads $\mathbf{p}$ and the prestresses $\mathbf{s}$ in Eq. (2.33), we obtain that

$$\mathbf{p}^\top = \mathbf{s}^\top\mathbf{D}^\top. \tag{2.78}$$

Because the virtual displacements $\delta\mathbf{d}$ are arbitrary, Eqs. (2.77) and (2.78) lead to

$$\mathbf{s}^\top\mathbf{D}^\top = \mathbf{s}^\top\mathbf{H}. \tag{2.79}$$

Equation (2.79) is always true only if the compatibility matrix $\mathbf{H}$ is equal to the transpose $\mathbf{D}^\top$ of the equilibrium matrix $\mathbf{D}$ [1, 8]:

$$\mathbf{H} = \mathbf{D}^\top. \tag{2.80}$$

Hence, the transpose $\mathbf{D}^\top$ of the equilibrium matrix $\mathbf{D}$ is exactly the compatibility matrix, and the kinematic relation of the structure in Eq. (2.75) can be rewritten as

$$\mathbf{D}^\top\mathbf{d} = \mathbf{e}. \tag{2.81}$$

The rigid-body motions of a free-standing structure obviously do not lead to member extensions, therefore, they should be excluded in the discussions of kinematic indeterminacy. Accordingly, the total degree of freedom of a free-standing structure is $dn - n^{\mathrm{b}}$, when the number n^{b} of rigid-body motions have been excluded.

If a non-rigid-body motion $\mathbf{d}$ leads to no member extension; i.e.,

$$\mathbf{D}^\top\mathbf{d} = \mathbf{0}, \tag{2.82}$$

then $\mathbf{d}$ is a mechanism, which is usually denoted by $\mathbf{d}_{\mathrm{m}}$.

Moreover, it is easy to see that the rank of the compatibility matrix $\mathbf{D}^\top$ is equal to that of the equilibrium matrix $\mathbf{D}$:

$$\text{rank}(\mathbf{D}^\top) = \text{rank}(\mathbf{D}) = r^\text{D}. \tag{2.83}$$

Therefore, the number of independent infinitesimal mechanisms, or degree of kinematic indeterminacy, n^m of a free-standing structure is computed by

$$n^\text{m} = dn - n^\text{b} - r^\text{D}. \tag{2.84}$$

Note that for a structure, of which the rigid-body motions are constrained by the fixed nodes, n^b in the equation vanishes. Furthermore, we have the following two cases for the value of n^m:

- $r^\text{D} < dn - n^\text{b}$ or $n^\text{m} > 0$ (Kinematically indeterminate):
  There exists n^m independent non-trivial displacements $\mathbf{m}_i$ $(i = 1, 2, \ldots, n^\text{m})$, which are not rigid-body motions, preserving the member lengths; i.e.,

  $$\mathbf{D}^\top \mathbf{m}_i = \mathbf{0}. \tag{2.85}$$

  Therefore, the structure is *kinematically indeterminate*. Moreover, a mechanism $\mathbf{d}_\text{m}$ can be written as a linear combination of the independent mechanisms $\mathbf{m}_i$ as follows:

  $$\mathbf{d}_\text{m} = \sum_{i=1}^{n^\text{m}} \beta_i \mathbf{m}_i, \tag{2.86}$$

  where β_i are arbitrary coefficients, and the independent mechanisms $\mathbf{m}_i$ are normalized as

  $$\mathbf{m}_i^\top \mathbf{m}_j = \delta_{ij}. \tag{2.87}$$

- $r^\text{D} = dn - n^\text{b}$ or $n^\text{m} = 0$ (Kinematically determinate):
  The structure is *kinematically determinate*, because there exists no non-trivial displacement vector preserving the member lengths, except for the rigid-body motions.

Example 2.13 Kinematic determinacy of the two-dimensional free-standing structure as shown in Fig. 2.2.

The two-dimensional free-standing structure as shown in Fig. 2.2 consists of five free nodes; i.e., $n = 5$. The same geometry realization as in Example 2.7 is adopted for the structure. Rank r^D of the compatibility matrix $\mathbf{D}^\top$, or equivalently that of the equilibrium matrix as in Example 2.12, is $(r^\text{D}=)$ 7.

Therefore, degree n^{m} of kinematic indeterminacy, or number of infinitesimal mechanisms, is calculated as follows according to Eq. (2.84) for free-standing structures:

$$\begin{aligned} n^{\mathrm{m}} &= dn - n^{\mathrm{b}} - r^{\mathrm{D}} \\ &= 2 \times 5 - 3 - 7 \\ &= 0, \end{aligned} \tag{2.88}$$

which means that there exists no mechanism in the structure.

Example 2.14 Kinematic indeterminacy of the two-dimensional structure with fixed nodes as shown in Fig. 2.4a and its infinitesimal mechanism.

The two-dimensional structure in Fig. 2.4a consists of two members, one free node, and two fixed nodes; i.e., $m = 2$, $n = 1$, and $n^{\mathrm{f}} = 2$. The connectivity matrices $\mathbf{C}^{\mathrm{s}} \in \mathbb{R}^{2\times 3}$, $\mathbf{C} \in \mathbb{R}^{2\times 1}$, and $\mathbf{C}^{\mathrm{f}} \in \mathbb{R}^{2\times 2}$ of the structure respectively are

$$\mathbf{C}^{\mathrm{s}} = \overset{\begin{matrix}1 & 2 & 3\end{matrix}}{\left(\begin{array}{c|cc} 1 & -1 & 0 \\ 1 & 0 & -1 \end{array}\right)} \begin{matrix}[1] \\ [2]\end{matrix}, \quad \mathbf{C} = \begin{pmatrix} 1 \\ 1 \end{pmatrix}, \quad \mathbf{C}^{\mathrm{f}} = \begin{pmatrix} -1 & 0 \\ 0 & -1 \end{pmatrix}. \tag{2.89}$$

Geometry realization of the structure is adopted as follows:

$$\mathbf{x} = (0)\,,\quad \mathbf{x}^{\mathrm{f}} = \begin{pmatrix} -1 \\ 1 \end{pmatrix}, \quad \mathbf{y} = (0)\,, \quad \mathbf{y}^{\mathrm{f}} = \begin{pmatrix} 0 \\ 0 \end{pmatrix}. \tag{2.90}$$

Thus, the coordinate difference vectors $\mathbf{u}$ and $\mathbf{v}$ $(\in \mathbb{R}^2)$ are

$$\begin{aligned} \mathbf{u} &= \begin{pmatrix} 1 \\ 1 \end{pmatrix} (0) + \begin{pmatrix} -1 & 0 \\ 0 & -1 \end{pmatrix} \begin{pmatrix} -1 \\ 1 \end{pmatrix} \\ &= \begin{pmatrix} 1 \\ -1 \end{pmatrix}, \\ \mathbf{v} &= \begin{pmatrix} 0 \\ 0 \end{pmatrix}; \end{aligned} \tag{2.91}$$

and the member length vector $\mathbf{l} \in \mathbb{R}^2$ is

$$\mathbf{l} = \begin{pmatrix} 1 \\ 1 \end{pmatrix}. \tag{2.92}$$

From Eq. (2.34), the equilibrium matrix $\mathbf{D} \in \mathbb{R}^{2\times 2}$ corresponding to the only one free node is

$$\begin{aligned}\mathbf{D} &= \begin{pmatrix} \mathbf{D}^x \\ \mathbf{D}^y \end{pmatrix} = \begin{pmatrix} \mathbf{C}^\top \mathbf{U}\mathbf{L}^{-1} \\ \mathbf{C}^\top \mathbf{V}\mathbf{L}^{-1} \end{pmatrix} \\ &= \begin{pmatrix} 1 & -1 \\ 0 & 0 \end{pmatrix}. \end{aligned} \tag{2.93}$$

It is obvious that the rank r^{D} of $\mathbf{D}$ is equal to one:

$$r^{\mathrm{D}} = \mathrm{rank}(\mathbf{D}) = 1. \tag{2.94}$$

According to Eqs. (2.67) and (2.84), there exist one prestress mode and one infinitesimal mechanism in the structure; i.e., $n^{\mathrm{s}} = n^{\mathrm{m}} = 1$, since we have

$$\begin{aligned} n^{\mathrm{s}} &= m - r^{\mathrm{D}} = 2 - 1 \\ &= 1, \\ n^{\mathrm{m}} &= dn - r^{\mathrm{D}} = 1 \times 2 - 1 \\ &= 1, \end{aligned} \tag{2.95}$$

where n^{b} is not included because rigid-body motions of the structure have been constrained by the two fixed nodes.

The normalized prestress mode $\bar{\mathbf{s}} \in \mathbb{R}^2$ and infinitesimal mechanism $\mathbf{d}_{\mathrm{m}} \in \mathbb{R}^2$, lying in the null-space of $\mathbf{D}$ and $\mathbf{D}^\top$, respectively, are calculated as

$$\bar{\mathbf{s}} = \frac{1}{\sqrt{2}} \begin{pmatrix} 1 \\ 1 \end{pmatrix}, \quad \mathbf{m} = \begin{pmatrix} 0 \\ 1 \end{pmatrix}. \tag{2.96}$$

Therefore, the structure is in equilibrium, if its two members carry the same prestresses, either tension or compression; and its only infinitesimal mechanism is illustrated in Fig. 2.4a.

2.3.5 Remarks

Section 2.3 presented existing methodologies for identification of static as well as kinematic determinacy of a (prestressed) pin-jointed structure. The original Maxwell's rule is neither a sufficient nor a necessary condition, but it is in a very simple form such that it can be used for preliminary studies. The modified Maxwell's rule by Calladine provides deeper understanding of static and kinematic determinacy of a structure, through the number n^{s} of independent prestress modes as well as the number n^{m} of independent mechanisms.

Degrees of static and kinematic determinacy of a d-dimensional free-standing pin-jointed structure can be identified by using the numbers of nodes n and members m, along with the rank r^{D} of the equilibrium matrix.

Static and kinematic determinacy of a free-standing pin-jointed structure:

	Statically		Kinematically	
	Determinate	Indeterminate	Determinate	Indeterminate
$n^{\mathrm{s}} = m - r^{\mathrm{D}}$	=0	>0		
$n^{\mathrm{m}} = dn - n^{\mathrm{b}} - r^{\mathrm{D}}$			=0	>0

2.4 Force Density Matrix

In this section, equilibrium equations of a prestressed pin-jointed structure are formulated with respect to nodal coordinates associated with the force density matrix. The characteristics of force density matrix are critical for understanding self-equilibrium and stability of tensegrity structures, as will be extensively used throughout the remaining of this book.

2.4.1 Definition of Force Density Matrix

Force density q_k of member k is defined as the ratio of its member force s_k to its member length l_k; i.e.,

$$q_k = \frac{s_k}{l_k}. \tag{2.97}$$

Moreover, the *force density vector* $\mathbf{q} \in \mathbb{R}^m$, consisting of force densities of all members, is calculated by

$$\mathbf{q} = \mathbf{L}^{-1}\mathbf{s}, \tag{2.98}$$

the kth entry of which is q_k.

To rewrite the equilibrium equation in Eq. (2.29) with respect to member forces $\mathbf{s}$ in x-direction into those with respect to nodal coordinates $\mathbf{x}$ and $\mathbf{x}^{\mathrm{f}}$, we have

$$\begin{aligned}\mathbf{D}^x\mathbf{s} &= \mathbf{C}^\top\mathbf{U}\mathbf{L}^{-1}\mathbf{s} = \mathbf{C}^\top\mathbf{U}\mathbf{q} = \mathbf{C}^\top\mathbf{Q}\mathbf{u}\\ &= \mathbf{C}^\top\mathbf{Q}\mathbf{C}\mathbf{x} + \mathbf{C}^\top\mathbf{Q}\mathbf{C}^{\mathrm{f}}\mathbf{x}^{\mathrm{f}},\end{aligned} \tag{2.99}$$

where $\mathbf{Q} = \text{diag}(\mathbf{q})$ is the diagonal version of the force density vector $\mathbf{q}$. Moreover, the above equations hold because we have the following relations for any vectors $\mathbf{a}$, $\mathbf{b}$ of the same size and their diagonal versions $\mathbf{A}$, $\mathbf{B}$

$$\mathbf{Ab} = \mathbf{Ba}, \text{ with } \mathbf{A} = \text{diag}(\mathbf{a}) \text{ and } \mathbf{B} = \text{diag}(\mathbf{b}). \tag{2.100}$$

Let K_i denote the set of members connected to the free node i. Equation (2.99) can also be derived from Eq. (2.21) as follows:

$$\begin{aligned} \sum_{k \in K_i} C_{(k,i)} \frac{u_k s_k}{l_k} &= \sum_{k \in K_i} C_{(k,i)} q_k u_k = (\mathbf{C}^\top)_i \mathbf{Qu} \\ &= (\mathbf{C}^\top)_i \mathbf{QCx}, \end{aligned} \tag{2.101}$$

because the (k, i)th entry of $C_{(k,i)}$ is non-zero only if member k is connected to node i.

In a similar way to Eq. (2.99), Eq. (2.31) can be rewritten as

$$\begin{aligned} \mathbf{D}^y \mathbf{s} &= \mathbf{C}^\top \mathbf{V} \mathbf{L}^{-1} \mathbf{s} = \mathbf{C}^\top \mathbf{Vq} = \mathbf{C}^\top \mathbf{Qv} \\ &= \mathbf{C}^\top \mathbf{QCy} + \mathbf{C}^\top \mathbf{QC}^\mathrm{f} \mathbf{y}^\mathrm{f}, \\ \mathbf{D}^z \mathbf{s} &= \mathbf{C}^\top \mathbf{W} \mathbf{L}^{-1} \mathbf{s} = \mathbf{C}^\top \mathbf{Wq} = \mathbf{C}^\top \mathbf{Qw} \\ &= \mathbf{C}^\top \mathbf{QCz} + \mathbf{C}^\top \mathbf{QC}^\mathrm{f} \mathbf{z}^\mathrm{f}. \end{aligned} \tag{2.102}$$

Example 2.15 Equilibrium equation of free node 3 of the two-dimensional free-standing structure as shown in Fig. 2.2.

Consider equilibrium of free node 3 of the two-dimensional free-standing structure as shown in Fig. 2.2. Using the third row $(\mathbf{C}^\top)_3$ of transpose of the connectivity matrix $\mathbf{C}$ defined in Eq. (2.89), and according to Eq. (2.101), its equilibrium equation in x-direction is

$$\begin{aligned} (\mathbf{C}_3)^\top \mathbf{Qu} &= \begin{pmatrix} 0 \\ -1 \\ 0 \\ 0 \\ 0 \\ 1 \\ 0 \\ 1 \end{pmatrix}^\top \begin{pmatrix} q_1 &&&&&&& \\ & q_2 &&&&&& \\ && q_3 &&&&& \\ &&& q_4 &&&& \\ &&&& q_5 &&& \\ &&&&& q_6 && \\ &&&&&& q_7 & \\ &&&&&&& q_8 \end{pmatrix} \begin{pmatrix} u_1 \\ u_2 \\ u_3 \\ u_4 \\ u_5 \\ u_6 \\ u_7 \\ u_8 \end{pmatrix} \\ &= -q_2 u_2 + q_6 u_6 + q_8 u_8 \end{aligned}$$

$$= (x_3 - x_1)\frac{s_2}{l_2} + (x_3 - x_4)\frac{s_6}{l_6} + (x_3 - x_5)\frac{s_6}{l_8}$$
$$= \frac{x_3 - x_1}{l_2}s_2 + \frac{x_3 - x_4}{l_6}s_6 + \frac{x_3 - x_5}{l_8}s_8, \quad (2.103)$$

which coincides with Eq. (2.24) derived in a different way.

Define $\mathbf{E} \in \mathbb{R}^{n\times n}$ and $\mathbf{E}^{\mathrm{f}} \in \mathbb{R}^{n\times n^{\mathrm{f}}}$ as

$$\begin{aligned}\mathbf{E} &= \mathbf{C}^{\top}\mathbf{Q}\mathbf{C},\\ \mathbf{E}^{\mathrm{f}} &= \mathbf{C}^{\top}\mathbf{Q}\mathbf{C}^{\mathrm{f}},\end{aligned} \quad (2.104)$$

where the matrix $\mathbf{E}$ is called *force density matrix* of the free nodes, and $\mathbf{E}^{\mathrm{f}}$ is that of the fixed nodes. $\mathbf{E}$ is also called small stress matrix, for example in paper [4]. Using $\mathbf{E}$ and $\mathbf{E}^{\mathrm{f}}$, the equations in Eqs. (2.99) and (2.102) are simplified as

$$\begin{aligned}\mathbf{D}^{x}\mathbf{s} &= \mathbf{E}\mathbf{x} + \mathbf{E}^{\mathrm{f}}\mathbf{x}^{\mathrm{f}},\\ \mathbf{D}^{y}\mathbf{s} &= \mathbf{E}\mathbf{y} + \mathbf{E}^{\mathrm{f}}\mathbf{y}^{\mathrm{f}},\\ \mathbf{D}^{z}\mathbf{s} &= \mathbf{E}\mathbf{z} + \mathbf{E}^{\mathrm{f}}\mathbf{z}^{\mathrm{f}}.\end{aligned} \quad (2.105)$$

Therefore, the equilibrium equations in Eqs. (2.29) and (2.31) are rewritten as follows by using the force density matrices $\mathbf{E}$ and $\mathbf{E}^{\mathrm{f}}$ associated with the nodal coordinate vectors $\mathbf{x}$, $\mathbf{y}$, $\mathbf{z}$ and $\mathbf{x}^{\mathrm{f}}$, $\mathbf{y}^{\mathrm{f}}$, $\mathbf{z}^{\mathrm{f}}$:

$$\begin{aligned}\mathbf{E}\mathbf{x} + \mathbf{E}^{\mathrm{f}}\mathbf{x}^{\mathrm{f}} &= \mathbf{p}^{x},\\ \mathbf{E}\mathbf{y} + \mathbf{E}^{\mathrm{f}}\mathbf{y}^{\mathrm{f}} &= \mathbf{p}^{y},\\ \mathbf{E}\mathbf{z} + \mathbf{E}^{\mathrm{f}}\mathbf{z}^{\mathrm{f}} &= \mathbf{p}^{z}.\end{aligned} \quad (2.106)$$

2.4.2 Direct Definition of Force Density Matrix

Instead of formulating the force density matrix $\mathbf{E}$ through Eq. (2.104), it can also be assembled directly using force densities of the members. The (i, j)th entry $E_{(i,j)}$ of the force density matrix $\mathbf{E}$ is defined as

$$E_{(i,j)} = \begin{cases} \sum\limits_{k\in K_i} q_k & \text{for } i = j,\\ -q_k & \text{if nodes } i \text{ and } j \text{ are connected by member } k,\\ 0 & \text{for other cases.}\end{cases} \quad (2.107)$$

Example 2.16 Force density matrices of the two-dimensional cable-net as shown in Fig. 2.1 and the free-standing structure as shown in Fig. 2.2.

First, consider the two-dimensional structure (cable-net) with fixed nodes as shown in Fig. 2.1. There are in total two free nodes 1 and 2 in the structure. Node 1 is connected by members [1], [2], [3], and [7], thus, the (1, 1)th entry $E_{(1,1)}$ of the force density matrix **E** is $q_1 + q_2 + q_3 + q_7$; moreover, node 1 is connected to the other free node 2 by member [1], hence, the (1, 2)th entry $E_{(1,2)}$ of **E** is $-q_1$.

Similarly, the (2, 2)th entry $E_{(2,2)}$ of **E** is $q_1 + q_4 + q_5 + q_6$, because node 2 is connected by members [1], [4], [5], and [6]; and the (2, 1)th entry $E_{(2,1)}$ is $-q_1$ because free node 2 is connected to the other free node 1 by member [1]. Therefore, the force density matrix $\mathbf{E} \in \mathbb{R}^{2\times 2}$ corresponding to the free nodes of the structure is

$$\mathbf{E} = \begin{array}{c} \begin{array}{cc} \text{Node 1} & \text{Node 2} \end{array} \\ \begin{pmatrix} q_1 + q_2 + q_3 + q_7 & -q_1 \\ -q_1 & q_1 + q_4 + q_5 + q_6 \end{pmatrix} \end{array} \begin{array}{l} \\ \text{Node 1} \\ \text{Node 2} \end{array} \tag{2.108}$$

In comparison, the force density matrix $\mathbf{E} \in \mathbb{R}^{5\times 5}$ of the two-dimensional free-standing structure as shown in Fig. 2.2 is

$$\mathbf{E} = \begin{array}{c} \begin{array}{ccccc} \text{Node 1} & \text{Node 2} & \text{Node 3} & \text{Node 4} & \text{Node 5} \end{array} \\ \begin{pmatrix} q_1+q_2+q_3+q_4 & -q_1 & -q_2 & -q_3 & -q_4 \\ -q_1 & q_1+q_5+q_7 & 0 & -q_5 & -q_7 \\ -q_2 & 0 & q_2+q_6+q_8 & -q_6 & -q_8 \\ -q_3 & -q_5 & -q_6 & q_3+q_5+q_6 & 0 \\ -q_4 & -q_7 & -q_8 & 0 & q_4+q_7+q_8 \end{pmatrix} \end{array} \begin{array}{l} \\ \text{Node 1} \\ \text{Node 2} \\ \text{Node 3} \\ \text{Node 4} \\ \text{Node 5} \end{array} \tag{2.109}$$

It is obvious from the examples that the force density matrix **E** is square and symmetric. Furthermore, it should be noted that sum of the entries in each row or column of the force density matrix in Eq. (2.109) is zero, which is always true for free-standing structures.

2.4.3 Self-equilibrium of the Structures with Supports

When the external loads are absent, the structure is said to be in a state of self-equilibrium—the nodes are equilibrated only by the prestresses in the members. Hence, self-equilibrium equations of the structure with respect to nodal coordinates can be written as follows, by setting the external loads $\mathbf{p}^x, \mathbf{p}^y, \mathbf{p}^z$ in Eq. (2.106) to zeros:

$$\begin{aligned}\mathbf{E}\mathbf{x} + \mathbf{E}^{\mathrm{f}}\mathbf{x}^{\mathrm{f}} &= \mathbf{0},\\ \mathbf{E}\mathbf{y} + \mathbf{E}^{\mathrm{f}}\mathbf{y}^{\mathrm{f}} &= \mathbf{0},\\ \mathbf{E}\mathbf{z} + \mathbf{E}^{\mathrm{f}}\mathbf{z}^{\mathrm{f}} &= \mathbf{0}.\end{aligned} \tag{2.110}$$

The above self-equilibrium equations are actually *non-linear* with respect to the nodal coordinates $\mathbf{x}, \mathbf{y}, \mathbf{z}$, because $\mathbf{E}$ and $\mathbf{E}^{\mathrm{f}}$ depend on the member lengths, which are non-linear functions of nodal coordinates. This dependency is apparent if we revisit the definitions of the coordinate differences in Eq. (2.9), the member lengths in Eq. (2.17), the force densities in Eq. (2.97), and the force density matrix in Eq. (2.104).

However, if the force densities are assigned or determined a priori, the force density matrices $\mathbf{E}$ and $\mathbf{E}^{\mathrm{f}}$ become constant. In this way, the self-equilibrium equations in Eq. (2.110) turn out to be *linear* with respect to the nodal coordinates. The only unknown parameters in Eq. (2.110) are the nodal coordinates $\mathbf{x}, \mathbf{y}, \mathbf{z}$ of the free nodes, while those $\mathbf{x}^{\mathrm{f}}, \mathbf{y}^{\mathrm{f}}, \mathbf{z}^{\mathrm{f}}$ of the supports are given.

The process of finding appropriate nodal coordinates as well as distribution of prestresses of a prestressed structure, satisfying the self-equilibrium equations, is generally called *form-finding* or *shape-finding*. Form-finding is a common design problem for prestressed structures, because their (self-equilibrated) configurations cannot be arbitrarily assigned in contrast to trusses carrying no prestress.

For the structures with fixed nodes, such as cable-nets and tensegrity-domes, the unknown coordinates $\mathbf{x}, \mathbf{y}, \mathbf{z}$ of the free nodes can be simply obtained as

$$\begin{aligned}\mathbf{x} &= -\mathbf{E}^{-1}\mathbf{E}^{\mathrm{f}}\mathbf{x}^{\mathrm{f}},\\ \mathbf{y} &= -\mathbf{E}^{-1}\mathbf{E}^{\mathrm{f}}\mathbf{y}^{\mathrm{f}},\\ \mathbf{z} &= -\mathbf{E}^{-1}\mathbf{E}^{\mathrm{f}}\mathbf{z}^{\mathrm{f}},\end{aligned} \tag{2.111}$$

if the force density matrix $\mathbf{E}$ is full-rank, or equivalently, if $\mathbf{E}$ is invertible. Hence, geometry realization of the structure, which is described in terms of nodal coordinates, can be uniquely determined.

This is the original idea of the *force density method* for the form-finding problem of cable-nets, where the non-linear self-equilibrium equations with respect to nodal coordinates are transformed into linear equations by introducing the concept of force density. Moreover, the force density matrix $\mathbf{E}$ of a cable-net is always positive definite, because it consists of only tensile members; i.e., cables, which carry positive (tensile) prestresses. Therefore, the self-equilibrium equations in Eq. (2.110) always have unique solutions as given in Eq. (2.111) [11].

This idea of force density method for cable-nets has been successfully extended to the form-finding problems of tensile membrane structures [9], by modelling them as cable-nets. This can be done because these two types of (tensile) structures share the following two common mechanical properties:

1. Both of cable-nets and membrane structures are attached to fixed nodes;
2. There exist only (positive) tensile forces in cable-nets or tensile stresses in membranes.

However, force density method cannot be directly applied to the form-finding problem of tensegrity structures. This comes from the fact that tensegrity structures are free-standing, without any fixed node, such that the corresponding force density matrix can never be full-rank. In fact, the force density matrix of a tensegrity structure should have certain rank deficiency; i.e., it has certain number of zero eigenvalues, in order to construct a structure in the space with desired dimensions (usually dimension of three). This is called the *non-degeneracy condition*, which is a necessary condition for geometry realization of tensegrity structures, and it will be presented in the next section.

Furthermore, we will discuss in Chap. 5 on how to use the idea of force density method for the form-finding of tensegrity structures, in which appropriate force densities are searched in order to satisfy this non-degeneracy condition.

2.5 Non-degeneracy Condition for Free-standing Structures

This section presents the non-degeneracy condition for a free-standing prestressed pin-jointed structure, in order to guarantee its geometry realization in the space with desired dimensions. The condition is described as an inequality with respect to the rank deficiency of the force density matrix.

Rank deficiency of a square symmetric matrix is the number of its zero eigenvalues. From the definition of force density matrix $\mathbf{E}$ in Eqs. (2.104) or (2.107), $\mathbf{E}$ of a free-standing structure has rank deficiency of at least one, because the sum of its entries in each row or column is always equal to zero. Hence, the vector $\mathbf{i}_n \in \mathbb{R}^n$ with all entries equal to one is obviously the eigenvector corresponding to this zero eigenvalue:

$$\mathbf{E}\mathbf{i}_n = \mathbf{C}^\top \mathbf{Q}\mathbf{C}\mathbf{i}_n = \mathbf{0} = 0\mathbf{i}_n, \tag{2.112}$$

because we have $\mathbf{C}\mathbf{i}_n = \mathbf{0}$ from Eq. (2.7). This property comes from the fact that free-standing structures do not have any fixed node.

Moreover, for a free-standing prestressed pin-jointed structure, the self-equilibrium equations in each direction in Eq. (2.110) become

$$\begin{aligned} \mathbf{E}\mathbf{x} &= \mathbf{0}, \\ \mathbf{E}\mathbf{y} &= \mathbf{0}, \\ \mathbf{E}\mathbf{z} &= \mathbf{0}, \end{aligned} \tag{2.113}$$

with the absence of fixed nodes $\mathbf{x}^{\mathrm{f}}, \mathbf{y}^{\mathrm{f}}, \mathbf{z}^{\mathrm{f}}$. The nodal coordinates $\mathbf{x}, \mathbf{y}, \mathbf{z}$ cannot be uniquely determined by solving Eq. (2.113), because the force density matrix of a free-standing structure is always rank deficient, or singular, and therefore, it is not invertible.

Example 2.17 Verification of self-equilibrium equations of the two-dimensional free-standing structure as shown in Fig. 2.2 associated with force density matrix.

In this example, we adopt the geometry realization as given in Eq. (2.51) in Example 2.7 for the two-dimensional free-standing structure as shown in Fig. 2.2. Using the prestress mode $\bar{\mathbf{s}}$ as presented in Eq. (2.73) in Example 2.12, the force densities $\mathbf{q}$ are computed from the prestresses $\mathbf{s}$ as

$$
\begin{aligned}
\mathbf{q} = \mathbf{L}^{-1}\mathbf{s} &= \begin{pmatrix} 1 & & & & & & & \\ & 1 & & & \mathbf{O} & & & \\ & & 1 & & & & & \\ & & & 1 & & & & \\ & & & & \sqrt{2} & & & \\ & & & & & \sqrt{2} & & \\ & & \mathbf{O} & & & & \sqrt{2} & \\ & & & & & & & \sqrt{2} \end{pmatrix}^{-1} \frac{t}{\sqrt{12}} \begin{pmatrix} \sqrt{2} \\ \sqrt{2} \\ \sqrt{2} \\ \sqrt{2} \\ -1 \\ -1 \\ -1 \\ -1 \end{pmatrix} \\
&= \frac{t}{\sqrt{24}} \begin{pmatrix} 2 \\ 2 \\ 2 \\ 2 \\ -1 \\ -1 \\ -1 \\ -1 \end{pmatrix},
\end{aligned}
\tag{2.114}
$$

where $t(\neq 0)$ is an arbitrary value, the absolute value of which indicates level of the prestresses; moreover, $\mathbf{O}$ indicates zero entries in the matrix, and the member lengths in Eq. (2.53) have been used.

The force density matrix $\mathbf{E} \in \mathbb{R}^{5\times 5}$ is calculated as follows, by substituting the force densities in Eq. (2.114) into Eq. (2.109):

$$
\mathbf{E} = \frac{t}{\sqrt{24}} \begin{pmatrix} 8 & -2 & -2 & -2 & -2 \\ -2 & 0 & 0 & 1 & 1 \\ -2 & 0 & 0 & 1 & 1 \\ -2 & 1 & 1 & 0 & 0 \\ -2 & 1 & 1 & 0 & 0 \end{pmatrix}. \tag{2.115}
$$

Thus, the self-equilibrium equations in terms of force density matrix **E** associated with nodal coordinates **x** and **y** in x- and y-directions, respectively, are

$$\mathbf{E}\mathbf{x} = \frac{t}{\sqrt{24}} \begin{pmatrix} 8 & -2 & -2 & -2 & -2 \\ -2 & 0 & 0 & 1 & 1 \\ -2 & 0 & 0 & 1 & 1 \\ -2 & 1 & 1 & 0 & 0 \\ -2 & 1 & 1 & 0 & 0 \end{pmatrix} \begin{pmatrix} 0 \\ -1 \\ 1 \\ 0 \\ 0 \end{pmatrix} = \begin{pmatrix} 0 \\ 0 \\ 0 \\ 0 \\ 0 \end{pmatrix},$$
$$\mathbf{E}\mathbf{y} = \mathbf{0}, \qquad (2.116)$$

which show that the self-equilibrium equations are satisfied with the given geometry realization (nodal coordinates) as well as prestresses.

Define rank deficiency $\bar{r}^{\mathrm{E}}$ of force density matrix **E** as

$$\bar{r}^{\mathrm{E}} = n - \mathrm{rank}(\mathbf{E}), \qquad (2.117)$$

i.e., there are in total $\bar{r}^{\mathrm{E}}$ zero eigenvalues in **E**. The eigenvectors corresponding to these zero eigenvalues are denoted by $\boldsymbol{\phi}_i$ $(i = 1, 2, \ldots, \bar{r}^{\mathrm{E}})$, with

$$\boldsymbol{\phi}_i^{\top}\boldsymbol{\phi}_j = \delta_{ij}. \qquad (2.118)$$

As discussed previously, the vector $(\boldsymbol{\phi}_1 =)\mathbf{i}_n$ in Eq. (2.112) is obviously one of these eigenvectors.

The nodal coordinates **x**, **y**, and **z** satisfying Eq. (2.113) can be generally written as a linear combination of the linearly independent eigenvectors $\boldsymbol{\phi}_i$ corresponding to the zero eigenvalues of **E**:

$$\mathbf{x} = \alpha_0^x \mathbf{i}_n + \sum_{i=1}^{\bar{r}^{\mathrm{E}}-1} \alpha_i^x \boldsymbol{\phi}_i,$$
$$\mathbf{y} = \alpha_0^y \mathbf{i}_n + \sum_{i=1}^{\bar{r}^{\mathrm{E}}-1} \alpha_i^y \boldsymbol{\phi}_i,$$
$$\mathbf{z} = \alpha_0^z \mathbf{i}_n + \sum_{i=1}^{\bar{r}^{\mathrm{E}}-1} \alpha_i^z \boldsymbol{\phi}_i, \qquad (2.119)$$

where α_i^x, α_i^y, and α_i^z $(i = 0, 1, \ldots, \bar{r}^{\mathrm{E}} - 1)$ are arbitrary coefficients. To construct a structure in the space with desired dimensions (two or three in practice), certain conditions should be satisfied for nodal coordinates **x**, **y**, and **z**.

Degeneracy of a Tensegrity Structure:

If a structure lies in a space with less dimensions than the specific dimensions d, then the structure is said to be *degenerate* in the d-dimensional space; otherwise, it is *non-degenerate*.

For example, the structure in Fig. 2.2 is non-degenerate in the two-dimensional space, and it is degenerate in the three-dimensional space, because it lies in a two-dimensional space (the plane parallel to the paper).

From the definition of degeneracy, we have the following lemma, on linear independence of the nodal coordinates, for a non-degenerate structure in d-dimensional space.

Lemma 2.1 *If a structure is non-degenerate in a d-dimensional space, its nodal coordinate vectors in each direction are linearly independent.*

Proof Consider the three-dimensional case, with $d = 3$, for instance. Suppose that the coordinate vectors **x**, **y**, and **z** of a three-dimensional structure are linearly dependent. Thus, the following equation holds:

$$\beta^x \mathbf{x} + \beta^y \mathbf{y} + \beta^z \mathbf{z} = \mathbf{0}, \tag{2.120}$$

where the coefficients β^x, β^y, and β^z are not equal to zero simultaneously. Equation (2.120) defines a plane such that the structure is degenerate in the three-dimensional space.

Therefore, the nodal coordinate vectors **x**, **y**, and **z** have to be linearly independent, if the structure is non-degenerate in three-dimensional space.

Two-dimensional case can be proved in the same way, which completes the proof. □

Lemma 2.2 ***Non-degeneracy condition for free-standing structures:***

To guarantee that a free-standing prestressed pin-jointed structure is non-degenerate in d-dimensional space, the following relation should hold for the rank deficiency $\bar{r}^{\mathrm{E}}$ of its force density matrix:

$$\bar{r}^{\mathrm{E}} \geq d + 1. \tag{2.121}$$

Proof From Lemma 2.1, the nodal coordinate vectors of a structure should be linearly independent to ensure its non-degeneracy in the space with specific dimensions.

Because the coordinate vectors $\mathbf{x}$, $\mathbf{y}$, and $\mathbf{z}$ of a three-dimensional structure are in the same format as given in Eq. (2.119), they can be linearly independent only if there are no less than three independent vectors $\boldsymbol{\phi}_i$, except for the common eigenvector $\mathbf{i}_n$, in the null-space of $\mathbf{E}$. Hence, rank deficiency of the force density matrix should be equal to or greater than four for a three-dimensional structure.

For two-dimensional cases, there should be no less than two independent vectors $\boldsymbol{\phi}_i$, except for the common eigenvector $\mathbf{i}_n$, to ensure linear independency of nodal coordinate vectors $\mathbf{x}$ and $\mathbf{y}$. Hence, rank deficiency of the force density matrix should be equal to or greater than three for a two-dimensional structure.

In summary, the lemma has been proved for both two- and three-dimensional free-standing structures. □

Lemma 2.2 can be explained by geometry realization of a structure by using Eq. (2.119). Define $\mathbf{x}_0$, $\mathbf{y}_0$, and $\mathbf{z}_0$ as

$$\begin{aligned} \mathbf{x}_0 &= \alpha_0^x \mathbf{i}_n, \\ \mathbf{y}_0 &= \alpha_0^y \mathbf{i}_n, \\ \mathbf{z}_0 &= \alpha_0^z \mathbf{i}_n. \end{aligned} \tag{2.122}$$

The solutions of Eq. (2.113) can be written in a general form as follows:

$$\begin{pmatrix} \mathbf{x} \\ \mathbf{y} \\ \mathbf{z} \end{pmatrix} = \begin{pmatrix} \mathbf{x}_0 \\ \mathbf{y}_0 \\ \mathbf{z}_0 \end{pmatrix} + \sum_{i=1}^{\bar{r}^{\mathrm{E}}-1} \begin{pmatrix} \alpha_i^x & 0 & 0 \\ 0 & \alpha_i^y & 0 \\ 0 & 0 & \alpha_i^z \end{pmatrix} \begin{pmatrix} \boldsymbol{\phi}_i \\ \boldsymbol{\phi}_i \\ \boldsymbol{\phi}_i \end{pmatrix}, \tag{2.123}$$

where $\boldsymbol{\phi}_i$ is in the null-space of $\mathbf{E}$ such that $\mathbf{E}\boldsymbol{\phi}_i = \mathbf{0}$.

For a tensegrity structure with different rank deficiency $\bar{r}^{\mathrm{E}}$ in its force density matrix $\mathbf{E}$, we have the following discussions:

1. If $\bar{r}^{\mathrm{E}} = 1$, there exists only one non-zero solution, $\mathbf{x}_0$, $\mathbf{y}_0$, and $\mathbf{z}_0$, in each direction, hence, all nodes degenerate into the node $(\alpha_0^x, \alpha_0^y, \alpha_0^z)$. This node is called *base node*.
2. If $\bar{r}^{\mathrm{E}} = 2$, Eq. (2.123) defines a line that passes through the base node.
3. Equation (2.123) defines a two-dimensional space (plane) in the case of $\bar{r}^{\mathrm{E}} = 3$, and a three-dimensional space in the case of $\bar{r}^{\mathrm{E}} = 4$. Both of these solution spaces contain the base node.

Therefore, in order to ensure a non-degenerate tensegrity structure in d-dimensional space, rank deficiency $\bar{r}^{\mathrm{E}}$ of its force density matrix $\mathbf{E}$ should be equal to or larger than $d + 1$.

The condition in Lemma 2.2 is called the *non-degeneracy condition* for free-standing prestressed pin-jointed structures. However, it should be noted that this condition is only a necessary condition, but not a sufficient condition. As have been mentioned in Lemma 2.1, linear independence of the coordinate vectors should also be satisfied to guarantee a non-degenerate configuration.

Example 2.18 Non-degeneracy of the two-dimensional free-standing structure as shown in Fig. 2.2.

The two-dimensional free-standing structure as shown in Fig. 2.2 consists of five (free) nodes; i.e., $d = 2$ and $n = 5$.

Following the assignments of geometry and prestresses given in Example 2.16, numerical calculation shows that the rank of $\mathbf{E}$ is 2, and therefore, its rank deficiency $\bar{r}^{\mathrm{E}}$ is 3, since we have

$$\begin{aligned} \bar{r}^{\mathrm{E}} &= n - \mathrm{rank}(\mathbf{E}) = 5 - 2 = 3 \\ &= d + 1 = 3. \end{aligned} \tag{2.124}$$

This satisfies the non-degeneracy condition for a free-standing prestressed pin-jointed structure in two-dimensional space, which obviously coincides with the fact that it is a two-dimensional structure.

2.6 Remarks

In this chapter, the (self-)equilibrium equations of a (prestressed) pin-jointed structure have been presented in two different ways: those with respect to axial forces (prestresses) associated with the equilibrium matrix, and those with respect to nodal coordinates associated with the force density matrix.

Using rank (deficiency) of the equilibrium matrix, or equivalently the compatibility matrix, detailed information about degrees of static indeterminacy and kinematic indeterminacy of the structure can be achieved. Tensegrity structures are always statically indeterminate, so that they can carry prestresses in the absence of external loads; and they are usually kinematically indeterminate, such that they are unstable in the absence of prestresses. Stability criteria and conditions of tensegrity structures will be discussed in detail in Chap. 4.

Tensegrity structures are free-standing and prestressed pin-jointed structures, which are different from other types of structures, such as trusses carrying no prestress or cable-nets attached to supports. Configurations of the structures carrying prestresses cannot be arbitrarily determined, because the nodes and members have to be in the balance of prestresses. Hence, form-finding is a basic and important problem for design of tensegrity structures.

Moreover, the concept of force density is very useful for form-finding of cable-nets, but it cannot be directly utilized for tensegrity structures, because the force density matrix is singular, and therefore, non-invertible due to the fact that they are free-standing.

The non-degeneracy condition, in terms of rank deficiency of the force density matrix, has to be satisfied for a free-standing structure. The rank deficiency of a three-dimensional structure should be larger than three, while that of a two-dimensional structure should be larger than two. This condition will be used to present a strategy, making use of the idea of force density, for the form-finding of tensegrity structures in Chap. 5.

References

1. Calladine, C. R. (1978). Buckminster Fuller's "Tensegrity" structures and Clerk Maxwell's rules for the construction of stiff frames. *International Journal of Solids and Structures*, *14*(2), 161–172.
2. Calladine, C. R., & Pellegrino, S. (1991). First-order infinitesimal mechanisms. *International Journal of Solids and Structures*, *27*(4), 505–515.
3. Calladine, C. R., & Pellegrino, S. (1992). Further remarks on first-order infinitesimal mechanisms. *International Journal of Solids and Structures*, *29*(17), 2119–2122.
4. Connelly, R. (1982). Rigidity and energy. *Inventiones Mathematicae*, *66*(1), 11–33.
5. Harary, F. (1969). *Graph theory*. Reading, MA: Addison-Wesley.
6. Kaveh, A. (1995). *Structural mechanics: graph and matrix methods*. New York: Research Studies Press.
7. Lanczos, C. (1986). *The variational principles of mechanics* (4th ed.). New York: Dover Publications.
8. Livesley, R. K. (1975). *Matrix methods of structural analysis* (2nd ed.). Oxford: Pergamon Press.
9. Maurin, B., & Motro, R. (1998). The surface stress density method as a form-finding tool for tensile membranes. *Engineering Structures*, *20*(8), 712–719.
10. Maxwell, J. C. (1864). On the calculation of the equilibrium and stiffness of frames. *Philosophical Magazine*, *27*(182), 294–299.
11. Schek, H.-J. (1974). The force density method for form finding and computation of general networks. *Computer Methods in Applied Mechanics and Engineering*, *3*(1), 115–134.

Chapter 3
Self-equilibrium Analysis by Symmetry

Abstract For a tensegrity structure with high level of symmetry, its equilibrium analysis can be significantly simplified by considering the representative nodes only. This makes presentation of analytical conditions possible. In this chapter, we study several classes of symmetric structures, including the X-cross structures with four-fold rotational symmetry, the prismatic as well as star-shaped structures with dihedral symmetry, and the regular truncated tetrahedral structures with tetrahedral symmetry. These symmetric structures will be revisited in Chaps. 6–8 for stability investigation in a more sophisticated way.

Keywords Self-equilibrium analysis · Symmetry-based · Analytical solution · Reduced force density matrix

3.1 Symmetry-based Equilibrium

Symmetry of a structure can be utilized to simplify the self-equilibrium analysis by considering self-equilibrium equations of the representative nodes, instead of the whole structure. A representative node is in the same state of self-equilibrium as some other nodes by symmetry. Furthermore, this process allows us to present its self-equilibrated configuration, including prestresses (or force densities) and geometry (or nodal coordinates), in an analytical manner. Before investigating specific symmetric structures in the coming sections, we first present the basic formulations in this section.

Consider one of the representative nodes as a reference node. The self-equilibrium equations of other representative nodes can be obtained in a similar manner. Let $\mathbf{x}_0 \in \mathbb{R}^3$ denote the coordinate vector of the reference node, which is numbered as node 0; and let $\mathbf{x}_i$ denote the coordinate vector of node i.

Suppose that the reference node is connected to node i_j by member k_j, and there are in total $\hat{m}$ such members. The prestress, or axial force, carried in member k_j is denoted by s_{k_j} $(j = 1, 2, \ldots, \hat{m})$.

The unit direction vector $\mathbf{d}_{k_j}$ of member k_j can be written as

$$\mathbf{d}_{k_j} = \frac{1}{l_{k_j}}(\mathbf{x}_{i_j} - \mathbf{x}_0), \tag{3.1}$$

J.Y. Zhang and M. Ohsaki, *Tensegrity Structures*, Mathematics for Industry 6,
DOI 10.1007/978-4-431-54813-3_3

where $l_{k_j}(=|\mathbf{x}_{i_j}-\mathbf{x}_0|)$ denotes the length of member k_j. Note that we adopt that the unit direction vector $\mathbf{d}_{k_j}$ is directing from the reference node 0 to node i_j.

A tensegrity structure is pin-jointed and carries only axial forces in its members. Moreover, direction of the axial force is identical to that of the member. Thus, the axial force vector $\mathbf{f}_{k_j}$ of member s_{k_j} can be written as

$$\mathbf{f}_{k_j} = q_{k_j}\mathbf{d}_{k_j}, \tag{3.2}$$

for which the definition of force density q_{k_j} of member k_j

$$q_{k_j} = \frac{s_{k_j}}{l_{k_j}} \tag{3.3}$$

and the definition of unit direction vector of member k_j in Eq. (3.1) have been used.

When external load is absent, the reference node should be in the state of self-equilibrium; i.e., all forces applied at the reference node should sum up to zero:

$$\sum_{j=1}^{\hat{m}} \mathbf{f}_{k_j} = \mathbf{0}. \tag{3.4}$$

For some cases, for example the star-shaped structure in Sect. 3.4, the following equivalent self-equilibrium equation of the reference node may be convenient to use:

$$\sum_{j=1}^{\hat{m}} q_{i_j}(\mathbf{x}_{k_j} - \mathbf{x}_0) = \mathbf{0}. \tag{3.5}$$

Example 3.1 Self-equilibrium equation of the reference node 0 as shown in Fig. 3.1.

The reference node 0 as shown in Fig. 3.1 is the only representative node of a symmetric prismatic structure considered later in Sect. 3.3. The reference node 0 is connected to four other nodes 1–4 via members [1]–[4], thus, $\hat{m} = 4$.

Using Eq. (3.4), self-equilibrium equation of the reference node is written as

$$\sum_{j=1}^{4} \mathbf{f}_j = \mathbf{0}, \tag{3.6}$$

where

$$\mathbf{f}_j = q_j\mathbf{d}_j, \quad (j = 1, 2, 3, 4). \tag{3.7}$$

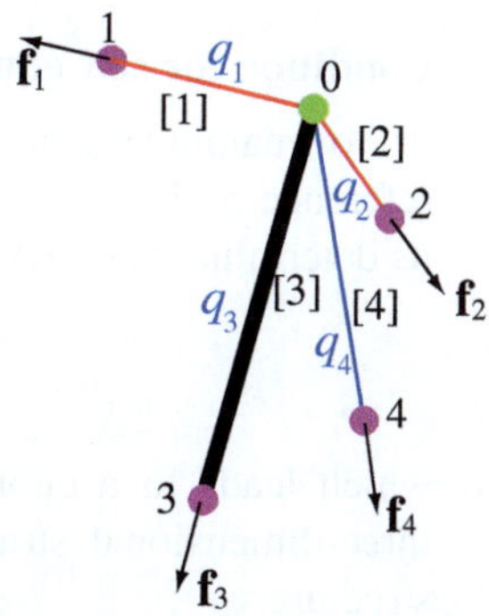

Fig. 3.1 Self-equilibrium of the reference node 0 (of a prismatic tensegrity structure considered in Sect. 3.3)

For a symmetric structure, the nodes 0 and i_j can exchange their positions to each other by a proper symmetry operation. This can be expressed in terms of their coordinate vectors $\mathbf{x}_0$ and $\mathbf{x}_{i_j}$ through the corresponding *transformation matrix* $\mathbf{T}_{i_j} \in \mathbb{R}^{3\times 3}$:

$$\mathbf{x}_{i_j} = \mathbf{T}_{i_j}\mathbf{x}_0. \tag{3.8}$$

Using Eq. (3.8), the unit direction vector $\mathbf{d}_{k_j}$ of member k_j defined in Eq. (3.1) can be written as

$$\mathbf{d}_{k_j} = \frac{1}{l_{k_j}}(\mathbf{T}_{i_j} - \mathbf{I}_3)\mathbf{x}_0, \tag{3.9}$$

where $\mathbf{I}_3$ denotes the 3-by-3 identity matrix. Furthermore, the axial force vector $\mathbf{f}_{k_j}$ of member s_{k_j} becomes

$$\mathbf{f}_{k_j} = q_{k_j}(\mathbf{T}_{k_j} - \mathbf{I}_3)\mathbf{x}_0. \tag{3.10}$$

Hence, the self-equilibrium equation Eq. (3.4) or Eq. (3.5) can be expressed in a matrix form as follows:

$$\hat{\mathbf{E}}\mathbf{x}_0 = \mathbf{0}. \tag{3.11}$$

In Eq. (3.11), $\hat{\mathbf{E}} \in \mathbb{R}^{3\times 3}$ is called the *reduced force density matrix* corresponding to the reference node, and it is defined as

$$\hat{\mathbf{E}} = \sum_{j=1}^{\hat{m}} q_{k_j}(\mathbf{T}_{k_j} - \mathbf{I}_3). \tag{3.12}$$

In the self-equilibrium analysis of a (symmetric) tensegrity structure, we find the force densities q_{k_j} and coordinates of the reference node $\mathbf{x}_0$ such that the self-equilibrium equation Eq. (3.4) or Eq. (3.5) is satisfied.

Condition for self-equilibrium of the reference node:

To guarantee a non-trivial solution for the coordinate vector $\mathbf{x}_0$ of the reference node, the reduced force density matrix $\hat{\mathbf{E}}$ has to be singular; i.e., its determinant $\det(\hat{\mathbf{E}})$ is zero:

$$\det(\hat{\mathbf{E}}) = 0, \tag{3.13}$$

which leads to a cubic equation with respect to the force densities for a three-dimensional structure, and a square equation for a two-dimensional structure.

The procedure for determination of prestresses (force densities) and geometry of a tensegrity structure in the self-equilibrium analysis of a highly symmetric tensegrity structure is summarized as follows:

Procedure for self-equilibrium analysis of a symmetric tensegrity structure:

Step 1: Determine the force densities of the members connected to the reference node by solving Eq. (3.13);
Step 2: Determine coordinates of the reference node satisfying Eq. (3.11);
Step 3: Determine the coordinates of other nodes by applying Eq. (3.8) via the corresponding transformation matrices.

The self-equilibrium analysis procedure can be separated into two independent part: determination of the force densities in Step 1 and that of geometry (nodal coordinates) in Steps 2 and 3. Moreover, it is notable that the force densities have to have full symmetry; on the other hand, as will be demonstrated in Examples 3.2 and 3.3, the geometry realization might be fully or partially symmetric as long as the self-equilibrium equation of the reference node in Eq. (3.11) is satisfied.

Following the above-mentioned procedure, we will investigate the self-equilibrium of several classes of symmetric tensegrity structures. These structures include the X-cross structures with four-fold rotational symmetry, the prismatic and star-shaped structures with dihedral symmetry, and the regular truncated tetrahedral structures with tetrahedral symmetry.

3.2 Symmetric X-cross Structure

The first symmetric structure considered in this chapter is the X-cross structure as shown in Fig. 3.2. This is the simplest tensegrity structure in two-dimensional space. The structure is composed of four nodes and six members; i.e., $n = 4$, $m = 6$.

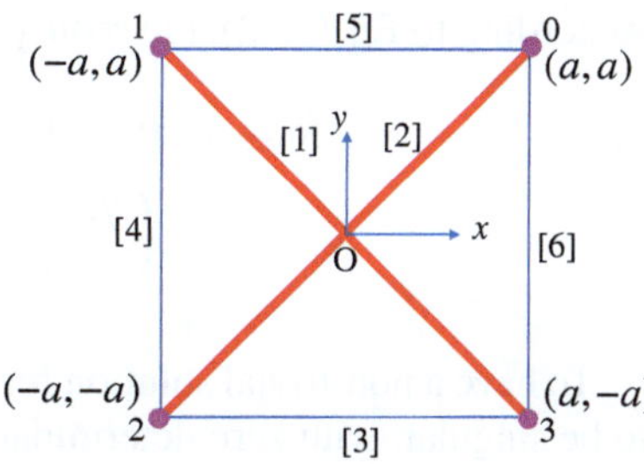

Fig. 3.2 X-cross structure with four-fold rotational symmetry. This is the simplest tensegrity structure in two-dimensional space

Every node is connected by three members—two cables and one strut. Note that the members connecting every node could be two struts and one cable, depending on the signs of the prestresses. The members [1] and [2] do not mechanically contact with each other on the plane they are lying.

With the full four-fold symmetry, any node can be moved to another node by applying an appropriate four-fold rotation about the origin 'O'; and moreover, the four cables have the same length and prestress, and therefore, have the same force density denoted by q_c. Similarly, force density of the two struts is denoted by q_s.

Counter-clockwise rotation about the origin through an angle θ_i can be represented by the transformation matrix $\mathbf{T}_i$ defined as

$$\mathbf{T}_i = \begin{pmatrix} \cos\theta_i & -\sin\theta_i \\ \sin\theta_i & \cos\theta_i \end{pmatrix}. \tag{3.14}$$

For the four-fold rotations, the angles θ_i are defined as

$$\theta_i = \frac{i\pi}{2}, \quad (i = 0, 1, 2, 3). \tag{3.15}$$

Consider, for instance, node 0 as the reference node. It can be moved to itself by applying the transformation matrix $\mathbf{T}_0$, or an identity matrix $\mathbf{I}_2 \in \mathbb{R}^{2\times 2}$:

$$\mathbf{T}_0 = \mathbf{I}_2 = \begin{pmatrix} 1 & 0 \\ 0 & 1 \end{pmatrix}, \tag{3.16}$$

and it can be moved to nodes 1, 2, and 3 by respectively applying the transformation matrices $\mathbf{T}_1$, $\mathbf{T}_2$, and $\mathbf{T}_3$:

$$\mathbf{T}_1 = \begin{pmatrix} 0 & -1 \\ 1 & 0 \end{pmatrix}, \quad \mathbf{T}_2 = \begin{pmatrix} -1 & 0 \\ 0 & -1 \end{pmatrix}, \quad \mathbf{T}_3 = \begin{pmatrix} 0 & 1 \\ -1 & 0 \end{pmatrix}. \tag{3.17}$$

Moreover, node 0 is connected to nodes 1, 2, and 3 by members [5], [2], and [6], respectively. The force densities of these members are

$$q_5 = q_6 = q_c \quad \text{and} \quad q_2 = q_s. \tag{3.18}$$

According to Eq. (3.12), the reduced force density matrix $\hat{\mathbf{E}} \in \mathbb{R}^{2\times 2}$ is written as

$$\begin{aligned}\hat{\mathbf{E}} &= q_5(\mathbf{T}_1 - \mathbf{I}_2) + q_2(\mathbf{T}_2 - \mathbf{I}_2) + q_6(\mathbf{T}_3 - \mathbf{I}_2)\\ &= -2\begin{pmatrix} q_s + q_c & 0 \\ 0 & q_s + q_c \end{pmatrix}.\end{aligned} \tag{3.19}$$

To have a non-trivial solution for nodal coordinates of the reference node 0, $\hat{\mathbf{E}}$ has to be singular, with zero determinant:

$$\begin{aligned}\det(\hat{\mathbf{E}}) &= -2(q_s + q_c)^2\\ &= 0,\end{aligned} \tag{3.20}$$

which leads to the solution of force density of the cables q_c in terms of that q_s of the struts as follows

$$q_c = -q_s. \tag{3.21}$$

It can be observed from Eq. (3.21) that self-equilibrium condition of the X-cross structure is in a simple form when it is described in terms of force densities.

With the symmetric distribution of force densities, the reduced force density matrix $\hat{\mathbf{E}}$ turns out to be a zero matrix:

$$\hat{\mathbf{E}} = \begin{pmatrix} 0 & 0 \\ 0 & 0 \end{pmatrix}. \tag{3.22}$$

Therefore, coordinate vector $\mathbf{x}_0$ of the reference node 0 can be represented by

$$\mathbf{x}_0 = \begin{pmatrix} a \\ b \end{pmatrix}, \tag{3.23}$$

where a and b are arbitrary values. The vector $\mathbf{x}_0$ in Eq. (3.23) obviously satisfies the self-equilibrium equation Eq. (3.11), because $\hat{\mathbf{E}}$ is a zero matrix.

Example 3.2 Fully symmetric geometry realization of the two-dimensional X-cross structure as shown in Fig. 3.2.

Figure 3.2 shows a possible geometry realization with fully (four-fold rotationally) symmetric appearance of the X-cross structure by specifying $a = b$; i.e., the coordinate vector of the reference node 0 is assigned as

$$\mathbf{x}_0 = \begin{pmatrix} a \\ a \end{pmatrix}. \tag{3.24}$$

The coordinates of other nodes are determined as follows by using Eq. (3.8) and the transformation matrices defined in Eq. (3.17):

$$\mathbf{x}_1 = \begin{pmatrix} -a \\ a \end{pmatrix}, \quad \mathbf{x}_2 = \begin{pmatrix} -a \\ -a \end{pmatrix}, \quad \mathbf{x}_3 = \begin{pmatrix} a \\ -a \end{pmatrix}. \tag{3.25}$$

Because the parameters a and b in Eq. (3.23) are mutually independent, and their values can be arbitrarily specified, it is possible to have a partially symmetric geometry realization as demonstrated in Example 3.3, although the force densities are fully symmetric.

Example 3.3 Partially symmetric geometry realization of the two-dimensional X-cross structure as shown in Fig. 3.3.

Assigning, for example, $b = 2a$ for geometry realization of the two-dimensional X-cross structure, its configuration is shown in Fig. 3.3 with the nodal coordinates listed as follows:

$$\mathbf{x}_0 = \begin{pmatrix} a \\ 2a \end{pmatrix}, \quad \mathbf{x}_1 = \begin{pmatrix} -a \\ 2a \end{pmatrix}, \quad \mathbf{x}_2 = \begin{pmatrix} -a \\ -2a \end{pmatrix}, \quad \mathbf{x}_3 = \begin{pmatrix} a \\ -2a \end{pmatrix}. \tag{3.26}$$

It can be observed from Eq. (3.26) or Fig. 3.3 that the geometry realization in this example is of reflectional symmetry about x- and y-axes, which is of less symmetry than the one as shown in Fig. 3.2, although the same set of force densities has been used.

Fig. 3.3 Partially symmetric realization of a two-dimensional X-cross structure

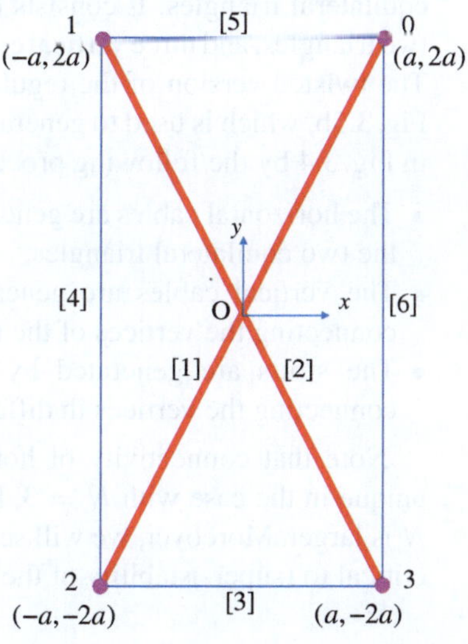

As can be observed in Example 3.3, the symmetry constraints are imposed on the force densities of the members, although we have started from the fully symmetric geometry realization. Therefore, using the highly symmetric force densities, it is possible to derive a geometry realization with the same or lower level of symmetry. This can be explained by applying affine motions (transformations) to the self-equilibrated configuration of a tensegrity structure, while preserving its self-equilibrium state, as discussed in more detail in Appendix B.1.

3.3 Symmetric Prismatic Structures

In this section, we consider the self-equilibrium analysis of a class of prismatic tensegrity structures, which are of dihedral symmetry. They are called *prismatic* tensegrity structures because this kind of structures are generated from twisted prisms, or twisted regular N-gonal dihedrons. A *regular N-gonal dihedron* refers to a polyhedron having two parallel regular polygon faces, each of which has N edges and is equiangular (all angels are equal) as well as equilateral (all edges have the same length). See Example 3.4 for the generation of the prismatic structure $\mathbf{D}_3^{1,1}$ as shown in Fig. 3.4 by using the twisted 3-gonal dihedron as shown in Fig. 3.5b.

Example 3.4 Generation of the simplest symmetric prismatic tensegrity structure as shown in Fig. 3.4 by using the (twisted) regular 3-gonal dihedron as shown in Fig. 3.5.

Figure 3.4 shows the simplest prismatic tensegrity structure, which is also the simplest tensegrity structure in three-dimensional space. The structure is composed of six nodes and twelve members, including six horizontal cables, three vertical cables, and three struts.

The regular 3-gonal dihedron as shown in Fig. 3.5a is made of two parallel equilateral triangles. It consists of six vertices, six (horizontal) edges of the two triangles, and three vertical edges connecting vertices of different triangles. The twisted version of the regular 3-gonal dihedron in Fig. 3.5a is shown in Fig. 3.5b, which is used to generate the prismatic tensegrity structure as shown in Fig. 3.4 by the following processes:

- The horizontal cables are generated by replacing the six horizontal edges of the two equilateral triangles;
- The vertical cables are generated by replacing the three vertical edges connecting the vertices of the two equilateral triangles; and
- The struts are generated by replacing the (additional) diagonal edges connecting the vertices in different triangles.

Note that connectivity of horizontal cables as well as vertical cables is unique in the case with $N = 3$, however, there are many other choices when N is larger. Moreover, we will see that the connectivity of horizontal cables is critical to (super-)stability of the prismatic tensegrity structures in Chap. 6.

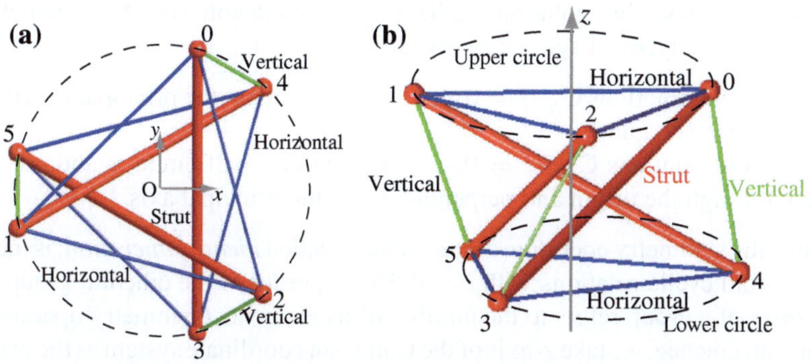

Fig. 3.4 The simplest prismatic tensegrity structure $\mathbf{D}_3^{1,1}$. The structure is of dihedral symmetry $\mathbf{D}_3$, and consists of six nodes, six horizontal cables, three vertical cables, and three struts. **a** Top view, **b** diagonal view

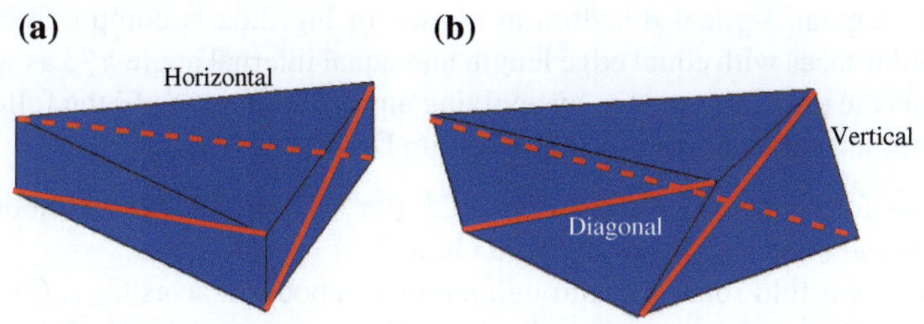

Fig. 3.5 Regular 3-gonal dihedron in (**a**) and its twisted version in (**b**). The horizontal and vertical edges of the twisted regular 3-gonal dihedron are respectively replaced by horizontal and vertical cables of the prismatic structure as shown in Fig. 3.4. Moreover, the diagonals indicated by thick lines are replaced by struts

Since a symmetric prismatic tensegrity structure is generated from a (twisted) regular dihedron, they have the same dihedral symmetry. When a structure is said to have dihedral symmetry, it implies that the structure is physically indistinguishable by the symmetry operations of a dihedral group.

3.3.1 Dihedral Symmetry

A *group* is a set of elements together with an operation that combines any two of its elements to form a third element also in the set while satisfying four specific conditions.[1] A *dihedral group* is the group of symmetry (rotation) operations of a regular dihedron. Dihedral groups are among the simplest examples of finite groups, and they play an important role in group theory, geometry, and chemistry.

---

[1] More details about group and its representation theory can be found in Appendix D.

Using the Schoenflies notation, a dihedral group is denoted by $\mathbf{D}_N$, in which there are two different types of symmetry operations:

- N-fold cyclic rotations C_N^i $(i = 0, 1, \ldots, N-1)$ about the principal axis through an angle $2i\pi/N$, and
- N two-fold rotations $\mathrm{C}_{2,i}$ $(i = 0, \ldots, N-1)$ of a half circle π about the axes going through the origin and perpendicular to the principal axis.

Note that the symmetry operation doing nothing, called *identity operation*, is included in the N-fold cyclic rotations with $i = 0$. Hence, order of the dihedral group $\mathbf{D}_N$ is $2N$. *Order* of a group refers to the number of its elements (symmetry operations).

For convenience, we take z-axis of the Cartesian coordinate system as the principal axis, and regard the point (0, 0, 0) as the origin 'O'.

Example 3.5 Symmetry operations in the dihedral group $\mathbf{D}_3$.

The regular 3-gonal dihedron as shown in Fig. 3.6a is composed of two triangular faces with equal edge length and equal internal angle $\pi/3$ as well. It remains the same appearance by applying any (combination of) the following six symmetry operations in dihedral group $\mathbf{D}_3$:

- Three-fold rotations C_3^0 (identity), C_3^1 $(2\pi/3)$, and C_3^2 $(4\pi/3)$ about the principal z-axis as shown in Fig. 3.6b; and
- Three two-fold rotations (through angle π) about the axes $\mathrm{C}_{2,1}$, $\mathrm{C}_{2,2}$, and $\mathrm{C}_{2,3}$ in the xy-plane or perpendicular to the z-axis, as shown in Fig. 3.6c.

The structure as shown in Fig. 3.4, which is of dihedral symmetry $\mathbf{D}_3$ remains the same appearance by applying the above-mentioned symmetry operations. This is obviously true while observing top view of the structure associated with the rotation axes as shown in Fig. 3.7.

The nodes of a symmetric prismatic tensegrity structure have one-to-one correspondence to the symmetry operations of the corresponding dihedral symmetry. The one-to-one correspondence means that a node can be moved to any other nodes, including the node itself, by only one specific symmetry operation. Thus, the number n of nodes of a symmetric prismatic structure is equal to the order of the corresponding dihedral group $\mathbf{D}_N$; i.e., $n = 2N$.

The nodes of a symmetric prismatic tensegrity structure are symmetrically located on the two (horizontal) circles. The two circles are parallel to xy-plane, and they have the same radius. The nodes of a prismatic structures with dihedral symmetry $\mathbf{D}_N$ are labelled as follows:

- The nodes on the upper circle is numbered as $0, 1, \ldots, N-1$;
- The nodes on the lower circle is numbered as $N, N-1, \ldots, 2N-1$.

Without loss of generality, the numbering is arranged counterclockwise, when the structure is observed from the top.

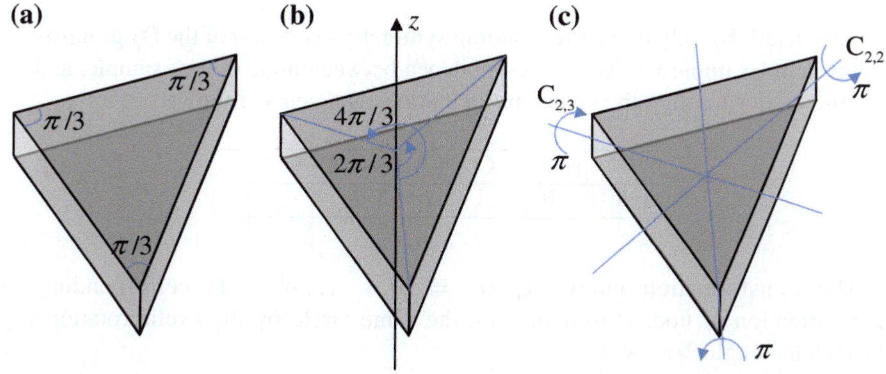

Fig. 3.6 A regular 3-gonal dihedron with *triangular faces* and the symmetry operations in the dihedral group $\mathbf{D}_3$. **a** Identify: $C_3^0(E)$. **b** Three-fold rotations: C_3^1, C_3^2. **c** Two-fold rotations: $C_{2,1}$, $C_{2,2}$, $C_{2,3}$

Fig. 3.7 *Top view* of the prismatic tensegrity structure $\mathbf{D}_3^{1,1}$, illustrating the dihedral symmetry $\mathbf{D}_3$ of the structure

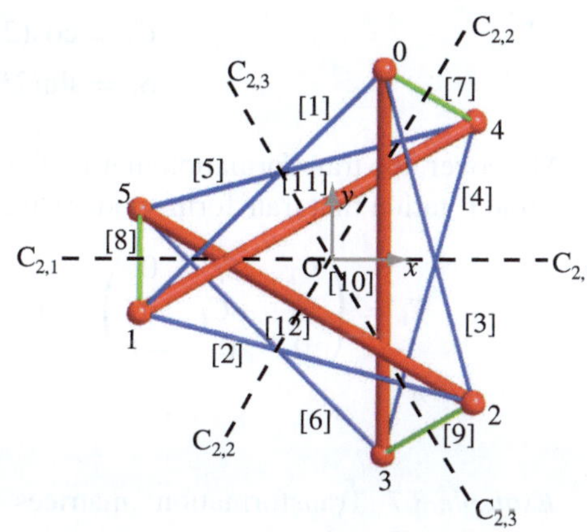

Example 3.6 Nodes of the simplest symmetric prismatic tensegrity $\mathbf{D}_3^{1,1}$ as shown in Fig. 3.4.

The prismatic tensegrity structure, as shown in Fig. 3.4 and denoted by the notation $\mathbf{D}_3^{1,1}$, is of $\mathbf{D}_3$ symmetry. The order of the dihedral group $\mathbf{D}_3$ is six, and therefore, the structure consists of six nodes, which are numbered from 0 to 5. Three of the nodes 0, 1, and 2 are symmetrically located on the upper circle, and the other three nodes 3, 4, and 5 are located on the lower circle.

The nodes have one-to-one correspondence to the symmetry operations of $\mathbf{D}_3$ group; i.e., one of these nodes can be moved to another node, including the node itself, by only one corresponding symmetry operation of the $\mathbf{D}_3$ group as given in Example 3.5. The correspondence between node 0, for example, and other nodes through the symmetry operations is given as follows:

	$E(C_3^0)$	C_3^1	C_3^2	$C_{2,1}$	$C_{2,2}$	$C_{2,3}$
Node 0	0	1	2	3	4	5

(3.27)

The transformation matrix $\mathbf{T}_i$ $(i = 0, 1, \ldots, N-1)$ corresponding to transformation of node 0 to node i on the same circle by the cyclic rotation C_N^i through the angle $2i\pi/N$ is

$$\mathbf{T}_i = \begin{pmatrix} C_i & -S_i & 0 \\ S_i & C_i & 0 \\ 0 & 0 & 1 \end{pmatrix}, \quad (i = 0, 1, \ldots, N-1), \tag{3.28}$$

where C_i and S_i are defined as

$$\begin{aligned} C_i &= \cos(2i\pi/N), \\ S_i &= \sin(2i\pi/N). \end{aligned} \tag{3.29}$$

Moreover, the transformation matrix $\mathbf{T}_i$ $(N \leq i \leq 2N-1)$ corresponding to the two-fold rotation that transforms node 0 to node i on different circles is

$$\mathbf{T}_i = \begin{pmatrix} C_i & S_i & 0 \\ S_i & -C_i & 0 \\ 0 & 0 & -1 \end{pmatrix}, \quad (i = N, N+1, \ldots, 2N-1). \tag{3.30}$$

Example 3.7 Transformation matrices for the structure with dihedral symmetry $\mathbf{D}_3$.

The values of $C_i = \cos(2i\pi/3)$ and $S_i = \sin(2i\pi/3)$ $(i = 0, 1, 2)$ for the dihedral symmetry $\mathbf{D}_3$ are

$$\begin{aligned} &C_0 = C_3 = 1, \quad && S_0 = S_3 = 0, \\ &C_1 = C_4 = -\tfrac{1}{2}, && S_1 = S_4 = \tfrac{\sqrt{3}}{2}, \\ &C_2 = C_5 = -\tfrac{\sqrt{3}}{2}, && S_2 = C_5 = -\tfrac{1}{2}. \end{aligned} \tag{3.31}$$

The transformation matrices $\mathbf{T}_i$ $(i = 0, 1, 2)$ corresponding to the cyclic rotations C_3^i are

$$\mathbf{T}_0 = \begin{pmatrix} 1 & 0 & 0 \\ 0 & 1 & 0 \\ 0 & 0 & 1 \end{pmatrix},$$

$$\mathbf{T}_1 = \frac{1}{2}\begin{pmatrix} -1 & -\sqrt{3} & 0 \\ \sqrt{3} & -1 & 0 \\ 0 & 0 & 2 \end{pmatrix},$$

$$\mathbf{T}_2 = \frac{1}{2}\begin{pmatrix} -\sqrt{3} & 1 & 0 \\ -1 & -\sqrt{3} & 0 \\ 0 & 0 & 2 \end{pmatrix}, \tag{3.32}$$

and the transformation matrices $\mathbf{T}_i$ $(i = 3, 4, 5)$ of the two-fold rotations $C_{2,i}$ are

$$\mathbf{T}_3 = \begin{pmatrix} 1 & 0 & 0 \\ 0 & -1 & 0 \\ 0 & 0 & -1 \end{pmatrix},$$

$$\mathbf{T}_4 = \frac{1}{2}\begin{pmatrix} -1 & \sqrt{3} & 0 \\ \sqrt{3} & 1 & 0 \\ 0 & 0 & -2 \end{pmatrix},$$

$$\mathbf{T}_5 = \frac{1}{2}\begin{pmatrix} -\sqrt{3} & -1 & 0 \\ -1 & \sqrt{3} & 0 \\ 0 & 0 & -2 \end{pmatrix}. \tag{3.33}$$

3.3.2 Connectivity

There exist three types of members in a symmetric prismatic tensegrity structure: horizontal cables, vertical cables, and struts. By fixing connectivity of the struts, connectivity of horizontal cables as well as vertical cables is respectively defined by using the parameters h and v. A symmetric prismatic tensegrity structure with $\mathbf{D}_N$ symmetry can therefore be denoted by $\mathbf{D}_N^{h,v}$.

Connectivity of a symmetric prismatic tensegrity structure $\mathbf{D}_N^{h,v}$:

- **Struts**:
  A strut connects node i $(i = 0, 1, \ldots, N-1)$ on the upper circle to node $N + i$ on the lower circle.
- **Horizontal cables**:
  The horizontal cables lie on the horizontal planes that contain the (node) circles. On the upper plane, a horizontal cable connects node i $(i = 0, 1, \ldots, N-1)$ to node $h + i$, or $h + i - N$ when $h + i \geq N$; on the lower plane, a horizontal cable connects node i $(i = N, N+1, \ldots, 2N-1)$ to node $h + i$, or $h + i - N$ when $h + i \geq 2N$. Note that symmetry implies

that a horizontal cable must also connect node i to node $N - h + i$. Thus, in the following, we restrict $1 \leq h < N/2$.

- **Vertical cables**:
A vertical cable connects node i $(i = 0, 1, \ldots, N-1)$ on the upper circle to node $N + v + i$, or $v + i$ when $v + i \geq N$, on the lower circle. We restrict $1 \leq v < N/2$. Note that choosing $N/2 \leq v < N$ will give essentially the same set of structures, but in the opposite rotational direction.

Accordingly, every node of a symmetric prismatic tensegrity structure is connected by two horizontal cables, one vertical cable, and one strut. Furthermore, a structure $\mathbf{D}_N^{h,v}$ is composed of $4N$ members—N struts, $2N$ horizontal cables, and N vertical cables.

Example 3.8 Connectivity of the structure $\mathbf{D}_3^{1,1}$ in Fig. 3.4.

The symmetric prismatic tensegrity structure denoted by $\mathbf{D}_3^{1,1}$ is shown in Fig. 3.4. According to its notation, the structure is of dihedral symmetry $\mathbf{D}_3$, and the parameters describing connectivity of horizontal cables and vertical cables are $h = 1$ and $v = 1$, respectively. The structure consists of three struts, six horizontal cables, and three vertical cables.

A member $[k]$ $(k = 1, 2, \ldots, 12)$ connected by nodes i and j is denoted by $[k : i, j]$, and the members of structure $\mathbf{D}_3^{1,1}$ as shown in Fig. 3.7 are listed as follows:

Members	Pair of nodes
Horizontal cable	$[1:0,1]$, $[2:1,2]$, $[3:2,0]$ $[4:3,4]$, $[5:4,5]$, $[6:5,3]$
Vertical cable	$[7:0,4]$, $[8:1,5]$, $[9:2,3]$
Strut	$[10:0,3]$, $[11:1,4]$, $[12:2,5]$

(3.34)

The connectivity of horizontal cables and vertical cables is unique for the prismatic structure with $\mathbf{D}_3$ symmetry as shown in Example 3.8, because the parameters h and v defining connectivity of the cables are positive integers, and furthermore, we limit them to be less than $(N/2 = 3/2 =)1.5$. Therefore, $h = v = 1$ is the only choice for the structures with $\mathbf{D}_3$ symmetry. However, the parameters h and v are not unique when N is larger than four, see, for instance, Example 3.9 for the connectivity of horizontal cables of the structures with $\mathbf{D}_8$ symmetry.

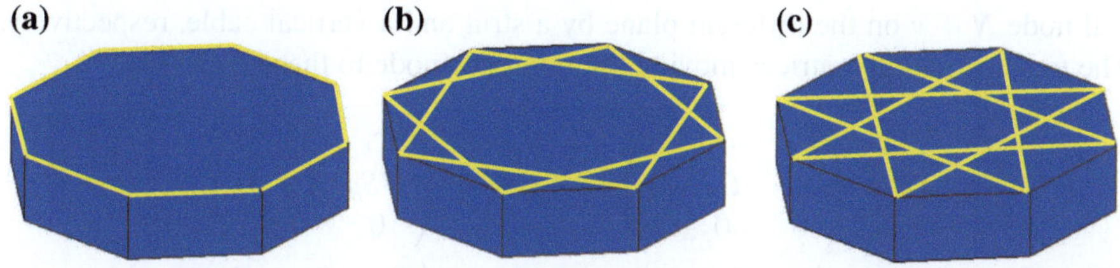

Fig. 3.8 Connectivity of horizontal cables of the prismatic structure with dihedral symmetry $\mathbf{D}_8$. **a** $h = 1$, **b** $h = 2$, **c** $h = 3$

Example 3.9 Connectivity of horizontal cables of the prismatic structure with dihedral symmetry $\mathbf{D}_8$ in Fig. 3.8.

The prismatic structures with dihedral symmetry $\mathbf{D}_8$ are generated from the twisted regular 8-gonal dihedron. For the 8-gonal dihedron, there are in total eight vertices for each polygon surface; i.e., $N = 8$. Because

$$1 \leq h < \frac{N}{2} = 4, \quad 1 \leq v < 4, \tag{3.35}$$

there are three possible choices respectively for connectivity of horizontal and vertical cables:

$$h = 1, 2, 3, \quad v = 1, 2, 3. \tag{3.36}$$

In Fig. 3.8, we illustrate the three possible connectivity patterns of horizontal cables.

3.3.3 *Self-equilibrium Analysis*

Using the dihedral symmetry, we will show in this subsection that there exists a general formulation for self-equilibrium condition of the whole class of symmetric prismatic structures. The force densities of different types of members as well as geometry in terms of coordinates in the state of self-equilibrium can be expressed in a simple and compact form.

Due to symmetry, the members of each type have the same length, and carry the same prestress, and therefore, they have the same force density. Let q_{h}, q_{s}, and q_{v} denote the force densities of the horizontal cables, struts, and vertical cables, respectively.

Consider node 0 as the reference node. Node 0 is connected to node h and node $N - h$ on the same plane by two horizontal cables, and it is connected to node N

and node $N + v$ on the different plane by a strut and a vertical cable, respectively. The transformation matrices moving the reference node to these nodes are

$$\mathbf{T}_h = \begin{pmatrix} C_h & -S_h & 0 \\ S_h & C_h & 0 \\ 0 & 0 & 1 \end{pmatrix}, \quad \mathbf{T}_{N-h} = \begin{pmatrix} C_h & S_h & 0 \\ -S_h & C_h & 0 \\ 0 & 0 & 1 \end{pmatrix},$$
$$\mathbf{T}_N = \begin{pmatrix} 1 & 0 & 0 \\ 0 & -1 & 0 \\ 0 & 0 & -1 \end{pmatrix}, \quad \mathbf{T}_{N+v} = \begin{pmatrix} C_v & S_v & 0 \\ S_v & -C_v & 0 \\ 0 & 0 & -1 \end{pmatrix}. \tag{3.37}$$

In Eq. (3.37), the relations

$$\begin{aligned} C_{N-h} &= \cos\frac{2(N-h)\pi}{N} = \cos(2\pi - \frac{2h\pi}{N}) = \cos(\frac{2h\pi}{N}) \\ &= C_h, \\ S_{N-h} &= \sin\frac{2(N-h)\pi}{N} = \sin(2\pi - \frac{2h\pi}{N}) = -\sin(\frac{2h\pi}{N}) \\ &= -S_h, \end{aligned} \tag{3.38}$$

as well as

$$\begin{aligned} C_N &= \cos\frac{2(N)\pi}{N} = \cos(2\pi) = 1, \\ S_N &= \cos\frac{2(N)\pi}{N} = \cos(2\pi) = 0, \\ C_{N+v} &= \cos\frac{2(N+v)\pi}{N} = \cos(2\pi + \frac{2v\pi}{N}) = \cos(\frac{2h\pi}{N}) \\ &= C_v, \\ S_{N+v} &= \sin\frac{2(N+v)\pi}{N} = \sin(2\pi + \frac{2v\pi}{N}) = \sin(\frac{2v\pi}{N}) \\ &= S_v \end{aligned} \tag{3.39}$$

have been used.

Example 3.10 Transformation matrices for the nodes connecting the reference node 0 of the structure $\mathbf{D}_8^{2,1}$ in Fig. 3.9.

From the notation $\mathbf{D}_8^{2,1}$, the structure as shown in Fig. 3.9a is of $\mathbf{D}_8$ symmetry, and moreover, the connectivity of horizontal cables and vertical cables is respectively defined by $h = 2$ and $v = 1$. According to the definition of connectivity, the reference node 0 is connected to node 2 and node 6 on the same horizontal plane by two horizontal cables, and it is connected to nodes 8 and 9 on different horizontal plane respectively by a strut and a vertical cable, as indicated in the figure.

The transformation matrices moving the reference node to these nodes are

$$\mathbf{T}_2 = \begin{pmatrix} 0 & -1 & 0 \\ 1 & 0 & 0 \\ 0 & 0 & 1 \end{pmatrix}, \quad \mathbf{T}_6 = \begin{pmatrix} 0 & 1 & 0 \\ -1 & 0 & 0 \\ 0 & 0 & 1 \end{pmatrix},$$

$$\mathbf{T}_8 = \begin{pmatrix} 1 & 0 & 0 \\ 0 & -1 & 0 \\ 0 & 0 & -1 \end{pmatrix}, \quad \mathbf{T}_9 = \frac{\sqrt{2}}{2}\begin{pmatrix} 1 & 1 & 0 \\ 1 & -1 & 0 \\ 0 & 0 & -\sqrt{2} \end{pmatrix}, \tag{3.40}$$

where

$$C_h = C_2 = 0, \quad S_h = S_2 = 1, \quad C_v = C_1 = \frac{\sqrt{2}}{2}, \quad S_v = S_1 = \frac{\sqrt{2}}{2} \tag{3.41}$$

have been used.

According to the definition of the reduced force density matrix $\hat{\mathbf{E}}$ defined in Eq. (3.12), we have

$$\begin{aligned} \hat{\mathbf{E}} = \; & 2q_{\mathrm{h}} \begin{pmatrix} C_h - 1 & 0 & 0 \\ 0 & C_h - 1 & 0 \\ 0 & 0 & 0 \end{pmatrix} + q_{\mathrm{s}} \begin{pmatrix} 0 & 0 & 0 \\ 0 & -2 & 0 \\ 0 & 0 & -2 \end{pmatrix} \\ & + q_{\mathrm{v}} \begin{pmatrix} C_v - 1 & S_v & 0 \\ S_v & -C_v - 1 & 0 \\ 0 & 0 & -2 \end{pmatrix}. \end{aligned} \tag{3.42}$$

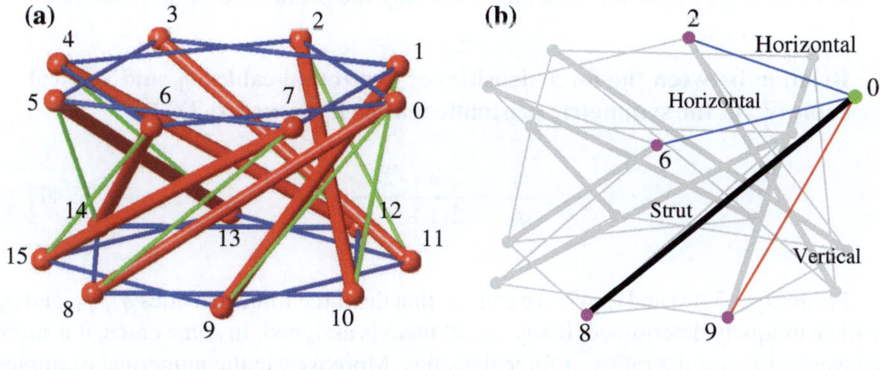

Fig. 3.9 The tensegrity structure $\mathbf{D}_8^{2,1}$ and the nodes connected to its reference node 0. **a** Structure $\mathbf{D}_8^{2,1}$. **b** Reference node 0

It is obvious that $\hat{\mathbf{E}}$ is a block-diagonal matrix, which is constructed by a 2-by-2 and a 1-by-1 sub-matrices on its leading diagonal. To allow a non-trivial solution of Eq. (3.11) for the coordinate vector $\mathbf{x}_0$ of the reference node in three-dimensional space, both of these sub-matrices have to be singular.

For the singularity of the 1-by-1 sub-matrix, we have

$$0 - 2q_{\mathrm{s}} - 2q_{\mathrm{v}} = 0, \tag{3.43}$$

which leads to the following relation:

Relation between the force densities of struts q_{s} and vertical cables q_{v} of the symmetric prismatic tensegrity structure $\mathbf{D}_N^{h,v}$:

$$\frac{q_{\mathrm{s}}}{q_{\mathrm{v}}} = -1 \quad \text{or equivalently} \quad q_{\mathrm{s}} = -q_{\mathrm{v}}. \tag{3.44}$$

Singularity of the 2-by-2 sub-matrix is guaranteed by its zero determinant:

$$[2q_{\mathrm{h}}(C_h - 1) + q_{\mathrm{v}}(C_v - 1)][2q_{\mathrm{h}}(C_h - 1) - 2q_{\mathrm{s}} - q_{\mathrm{v}}(C_v + 1)] - q_{\mathrm{v}}^2 S_v^2 = 0. \tag{3.45}$$

Using the relation $q_{\mathrm{s}} = -q_{\mathrm{v}}$ in Eq. (3.44) and the trigonometric equality $C_v^2 + S_v^2 = 1$, Eq. (3.45) reduces to

$$4\left(\frac{q_{\mathrm{h}}}{q_{\mathrm{v}}}\right)^2 (C_h - 1)^2 + 2C_v - 2 = 0. \tag{3.46}$$

Both of q_{h} and q_{v} should have positive signs because they are force densities of cables which are in tension, and therefore, only the positive solution is adopted:

Relation between the force densities of horizontal cables q_{h} and vertical cables q_{v} of the symmetric prismatic tensegrity structure $\mathbf{D}_N^{h,v}$:

$$\frac{q_{\mathrm{h}}}{q_{\mathrm{v}}} = \frac{\sqrt{2 - 2C_v}}{2(1 - C_h)}. \tag{3.47}$$

From Eqs. (3.44) and (3.47) we can see that the three force densities q_{s}, q_{h}, and q_{v} will be uniquely determined, if any one of them is assigned. In some cases, it is more convenient to use the ratios of force densities. Moreover, in the numerical examples in this chapter, units will be omitted without any loss of generality in discussions on self-equilibrium.

Example 3.11 Force densities of the symmetric prismatic tensegrity structure $\mathbf{D}_3^{1,1}$.

From the notation of the structure $\mathbf{D}_3^{1,1}$; i.e., $N = 3$, $h = 1$, and $v = 1$, we have

$$C_h = C_v = C_1 = \cos\frac{2\pi}{3} = -\frac{1}{2}. \tag{3.48}$$

Moreover, from Eqs. (3.44) and (3.47) we have the following relations for the force densities q_{s}, q_{h}, and q_{v}:

$$\frac{q_{\mathrm{s}}}{q_{\mathrm{v}}} = -1, \quad \frac{q_{\mathrm{h}}}{q_{\mathrm{v}}} = \frac{\sqrt{3}}{3}. \tag{3.49}$$

Assign a positive value to the force density of vertical cables, for example $q_{\mathrm{v}} = 1$, then the force densities for each type of members are calculated as follows by using Eq. (3.49):

$$q_{\mathrm{v}} = 1, \quad q_{\mathrm{s}} = -1, \quad q_{\mathrm{h}} = \frac{\sqrt{3}}{3}. \tag{3.50}$$

Substituting the above results into Eq. (3.42), the reduced force density matrix $\hat{\mathbf{E}}$ is

$$\begin{aligned}\hat{\mathbf{E}} &= \frac{2\sqrt{3}}{3}\begin{pmatrix} -\frac{3}{2} & 0 & 0 \\ 0 & -\frac{3}{2} & 0 \\ 0 & 0 & 0 \end{pmatrix} + (-1)\begin{pmatrix} 0 & 0 & 0 \\ 0 & -2 & 0 \\ 0 & 0 & -2 \end{pmatrix} + \begin{pmatrix} -\frac{3}{2} & \frac{\sqrt{3}}{2} & 0 \\ \frac{\sqrt{3}}{2} & -\frac{1}{2} & 0 \\ 0 & 0 & -2 \end{pmatrix} \\ &= \frac{1}{2}\begin{pmatrix} -2\sqrt{3}-3 & \sqrt{3} & 0 \\ \sqrt{3} & -2\sqrt{3}+3 & 0 \\ 0 & 0 & 0 \end{pmatrix}. \end{aligned} \tag{3.51}$$

It is easy to see that $\hat{\mathbf{E}}$ has two zero eigenvalues, so that there exists a non-trivial coordinate vector $\mathbf{x}_0$ of the reference node satisfying the self-equilibrium equation Eq. (3.11).

When both Eqs. (3.44) and (3.47) hold, the reduced force density matrix $\hat{\mathbf{E}}$ has a nullity of 2, and hence, there exist two independent solutions for coordinates of the reference node. In general, the coordinate vector $\mathbf{x}_0$ of the reference node can be written as follows by using two arbitrary parameters R and H:

Fig. 3.10 Geometry of the structure $\mathbf{D}_3^{1,1}$, defined by the parameters R and H

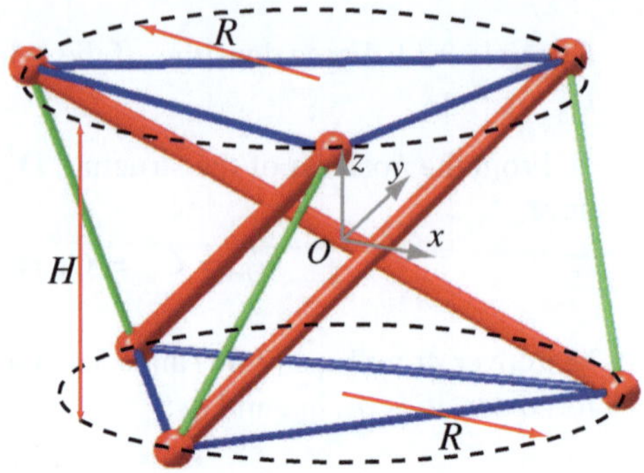

Coordinates of the reference node of the symmetric prismatic tensegrity structure $\mathbf{D}_N^{h,v}$:

$$\mathbf{x}_0 = \frac{R}{R_0}\begin{pmatrix} C_v - 1 + \sqrt{2-2C_v} \\ S_v \\ 0 \end{pmatrix} + \frac{H}{2}\begin{pmatrix} 0 \\ 0 \\ 1 \end{pmatrix}, \tag{3.52}$$

where R_0 is the norm of the first vector on the right-hand side of Eq. (3.52) representing the xy-coordinates, and R and H denote the radius and height of the structure.

From Eq. (3.52), the two planes containing the nodes are at $z = \pm H/2$. Figure 3.10 illustrates the definition of R and H. Moreover, R and H are mutually independent, and they can have arbitrary positive real values. By the application of Eq. (3.8), the coordinates of all the other nodes can be determined by assigning i from 1 to $2N-1$.

It is notable that self-equilibrated configuration of a prismatic tensegrity structure does not depend on connectivity of the horizontal cables; however, we will see that the connectivity pattern of horizontal cables affects its stability in Chap. 6.

Example 3.12 Self-equilibrated configuration of the structure $\mathbf{D}_3^{1,1}$.

From the notation of the structure $\mathbf{D}_3^{1,1}$; i.e., $N = 3$, $h = 1$, and $v = 1$, we have

$$\begin{aligned} C_v &= C_1 = \cos\frac{2\pi}{3} = -\frac{1}{2}, \\ S_v &= S_1 = \sin\frac{2\pi}{3} = \frac{\sqrt{3}}{2}. \end{aligned} \tag{3.53}$$

From Eq. (3.52), the coordinate vector $\mathbf{x}_0$ of the reference node 0 is

$$\mathbf{x}_0 = \frac{R}{R_0}\begin{pmatrix} -\frac{3}{2}+\sqrt{3} \\ \frac{\sqrt{3}}{2} \\ 0 \end{pmatrix} + \frac{H}{2}\begin{pmatrix} 0 \\ 0 \\ 1 \end{pmatrix}, \tag{3.54}$$

where $R_0 = \sqrt{6-3\sqrt{3}}$.

Assign, for example, radius and height of the structure as $R = 1$ and $H = 1$ in Eq. (3.54). The nodal coordinates $\mathbf{x}_i$ $(i = 0, 1, \ldots, 5)$ of the structure can be calculated as follows by using Eq. (3.8) together with the transformation matrices $\mathbf{T}_i$ given in Example 3.7:

$$\mathbf{x}_0 = \mathbf{T}_0\mathbf{x}_0 = \begin{pmatrix} 0.2588 \\ 0.9659 \\ 0.5000 \end{pmatrix}, \quad \mathbf{x}_1 = \mathbf{T}_1\mathbf{x}_0 = \begin{pmatrix} -0.9659 \\ -0.2588 \\ 0.5000 \end{pmatrix},$$
$$\mathbf{x}_2 = \mathbf{T}_2\mathbf{x}_0 = \begin{pmatrix} 0.7071 \\ -0.7071 \\ 0.5000 \end{pmatrix}, \quad \mathbf{x}_3 = \mathbf{T}_3\mathbf{x}_0 = \begin{pmatrix} 0.2588 \\ -0.9659 \\ -0.5000 \end{pmatrix},$$
$$\mathbf{x}_4 = \mathbf{T}_4\mathbf{x}_0 = \begin{pmatrix} 0.7071 \\ 0.7071 \\ -0.5000 \end{pmatrix}, \quad \mathbf{x}_5 = \mathbf{T}_5\mathbf{x}_0 = \begin{pmatrix} -0.9659 \\ 0.2588 \\ -0.5000 \end{pmatrix}. \tag{3.55}$$

The geometry realization of the structure $\mathbf{D}_3^{1,1}$ with the above nodal coordinates is plotted in Fig. 3.4.

3.4 Symmetric Star-shaped Structures

The symmetric star-shaped tensegrity structures considered in this section have similar symmetric configurations to the prismatic structures studied in the previous section—both of these two classes of structures are of dihedral symmetry. However, compared with prismatic structures which have only one type of (representative) nodes, there are two different types of nodes in star-shaped structures. The nodes belonging to different types cannot be moved to each other by the symmetry operations in a dihedral group. In this section, we conduct self-equilibrium analysis of the star-shaped tensegrity structures. The existence of two different types of nodes makes the analysis a little more complicated than that of the prismatic structures.

3.4.1 Connectivity

Compared with prismatic structures, a star-shaped structure of the same symmetry has two more nodes lying on the z-axis going through the centers of the two node circles. These two nodes are called *center nodes*, and other nodes located on the two parallel circles are *boundary nodes*. For example, the two different types of nodes of the simplest star-shaped structure are illustrated in Fig. 3.11.

Therefore, a star-shaped structure that is of dihedral symmetry $\mathbf{D}_N$ has $2N + 2$ nodes, including $2N$ boundary nodes and two center nodes, which are labeled as follows:

- The boundary nodes on the upper and lower circles (planes) are respectively numbered as $\{0, 1, \ldots, N - 1\}$ and $\{N, N + 1, \ldots, 2N - 1\}$;
- The center node on the upper plane is numbered as $2N$, and the center node on the lower plane is $2N + 1$.

Note that the boundary nodes have one-to-one correspondence to the symmetry operations of the corresponding dihedral group $\mathbf{D}_N$, however, the center nodes have one-to-N correspondence to the symmetry operations.

Moreover, there are three different types of members in a symmetric star-shaped tensegrity structure.

- *Radial cables*, which are in tension and connected to the boundary nodes and the center node on the same circle;
- *Vertical cables*, which are in tension and connected to the boundary nodes on different circles;
- *Struts*, which are in compression and connected to the boundary nodes on different circles.

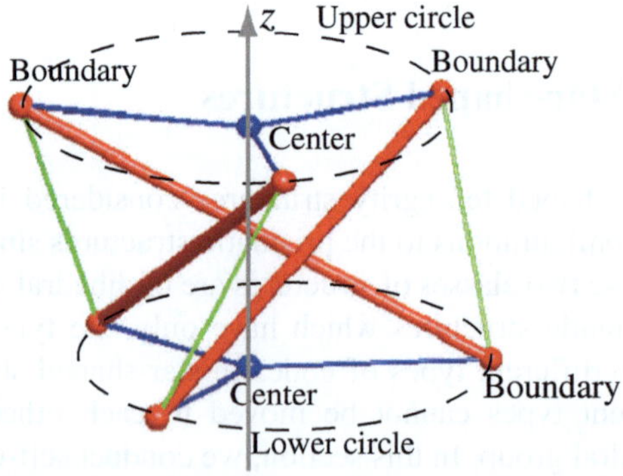

Fig. 3.11 The simplest *star-shaped* tensegrity structure, denoted as $\mathbf{D}_3^1$. This structure is of the same dihedral symmetry $\mathbf{D}_3$ as the simplest prismatic structure $\mathbf{D}_3^{1,1}$ in Fig. 3.4, but there are two additional (center) nodes in the *star-shaped* structure $\mathbf{D}_3^1$

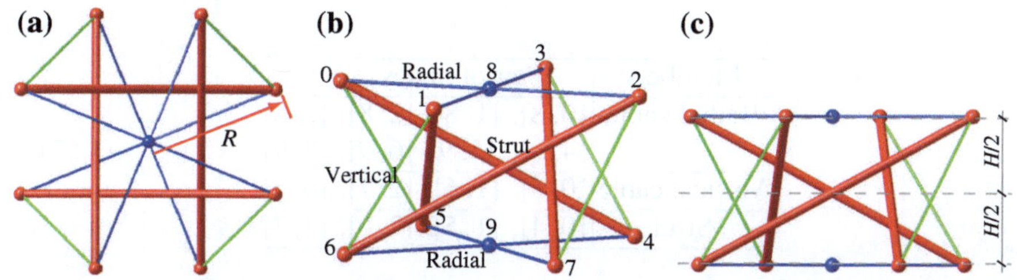

Fig. 3.12 *Star-shaped* tensegrity structure $\mathbf{D}_4^1$. Geometry of the structure is defined by the parameters R and H. **a** Top, **b** diagonal, **c** side

It is obvious that connectivity of the radial cables are fixed. If we fix connectivity of the struts, as presented later, connectivity of a star-shaped structure depends only on that of the vertical cables, for which a parameter v is used. Thus, similarly to the notation $\mathbf{D}_N^{h,v}$ for the symmetric prismatic tensegrity structures, symmetry and connectivity of a symmetric star-shaped tensegrity structure can be generally denoted by $\mathbf{D}_N^v$. Let $[i, j]$ indicate that the member is connected by nodes i and j. The connectivity of the three types of members of the star-shaped structure $\mathbf{D}_N^v$ by the nodes is defined as follows:

Connectivity of the symmetric star-shaped tensegrity structure $\mathbf{D}_N^v$:

$$\begin{aligned} \textbf{Radial cables}: &\ [i, 2N] \text{ and } [N+i, 2N+1], \\ \textbf{Vertical cables}: &\ [i, N+i+v], \qquad (i = 0, 1, \ldots, N-1), \\ \textbf{Struts}: &\ [i, N+i], \end{aligned} \tag{3.56}$$

where we set $N + i + v := i + v$ if $N + i + v \geq 2N$.

Note that value of the parameter v is limited to $1 \leq v < N/2$ in the following discussions; larger values would give the same set of structures, but in the opposite rotational direction, in the same manner as the prismatic structures.

Example 3.13 Connectivity of the symmetric star-shaped structure $\mathbf{D}_4^1$ as shown in Fig. 3.12.

The structure $\mathbf{D}_4^1$ as shown in Fig. 3.12 is of $\mathbf{D}_4$ symmetry, with the connectivity of vertical cables defined by the parameter $v = 1$. It is composed of four struts, four vertical cables, and eight radial cables. The nodes and members of the structure are shown in Fig. 3.12b. The members $[i, j]$ connected by pairs of nodes i and j are listed as follows:

Members	Pair of nodes
Radial cable	[0, 8], [1, 8], [2, 8], [3, 8]
	[4, 9], [5, 9] [6, 9], [7, 9]
Vertical cable	[0, 5], [1, 6], [2, 7], [3, 4]
Strut	[0, 4], [1, 5], [2, 6], [3, 7]

(3.57)

3.4.2 Self-equilibrium Analysis

Analytical conditions for self-equilibrium of a symmetric star-shaped tensegrity structure $\mathbf{D}_N^\nu$ is given in this subsection. As discussed previously, there are two different types of nodes, while the nodes of the same type have the same connectivity—every center node is connected by N radial cables, and every boundary node is connected by one radial cable, one vertical cable, and one strut. Therefore, there are two representative nodes for a star-shaped structure: any center node and any boundary node. Self-equilibrium analysis of the structure is reduced to finding the three force densities associated with coordinates of the two representative nodes satisfying the reduced self-equilibrium equations.

To describe geometry of a star-shaped structure, we use the parameter R for radius of the circles on which the boundary nodes are lying, and the parameter H for distance between the two parallel circles. Figure 3.12 shows the example structure $\mathbf{D}_4^1$ illustrating these two geometry parameters. Because the center nodes are located on z-axis, coordinate vector $\mathbf{x}_c \in \mathbb{R}^3$ of the representative center node, for example the one on the upper plane, is defined as follows:

$$\mathbf{x}_c = \begin{pmatrix} 0 \\ 0 \\ H/2 \end{pmatrix}. \tag{3.58}$$

Moreover, coordinate vector of the representative boundary node is denoted by $\mathbf{x}_0 \in \mathbb{R}^3$. Because the boundary nodes are located on the two circles parallel to the xy-plane, their z-coordinates are $\pm H/2$, as indicated in Fig. 3.12. Hence, coordinate vector $\mathbf{x}_0$ of the representative boundary node can be given by using the xy-coordinate vector $\bar{\mathbf{x}}_0 \in \mathbb{R}^2$ and the height H as follows:

$$\mathbf{x}_0 = \begin{pmatrix} \bar{\mathbf{x}}_0 \\ H/2 \end{pmatrix}. \tag{3.59}$$

The boundary nodes on the same circle are of rotational symmetry. Thus, coordinate vector $\mathbf{x}_i$ of the boundary node i can be computed as follows by using the transformation matrix $\mathbf{T}_i$ and the coordinate vector $\mathbf{x}_0$ of the representative boundary node:

$$\mathbf{x}_i = \mathbf{T}_i \mathbf{x}_0, \tag{3.60}$$

where the transformation matrix $\mathbf{T}_i$ (counterclockwise rotation) is defined as

$$\mathbf{T}_i = \begin{pmatrix} C_i & -S_i & 0 \\ S_i & C_i & 0 \\ 0 & 0 & 1 \end{pmatrix}, \tag{3.61}$$

with C_i and S_i respectively denoting $\cos(2i\pi/N)$ and $\sin(2i\pi/N)$.

Coordinates $\mathbf{x}_c$ of the center node on the same plane can indeed be calculated from the coordinates $\mathbf{x}_0$ of the representative boundary node as

$$\mathbf{x}_c = \mathbf{T}_c \mathbf{x}_0, \tag{3.62}$$

by using a non-unitary transformation matrix $\mathbf{T}_c$ defined as

$$\mathbf{T}_c = \begin{pmatrix} 0 & 0 & 0 \\ 0 & 0 & 0 \\ 0 & 0 & 1 \end{pmatrix}. \tag{3.63}$$

Denote the force densities of the strut, vertical cable, and radial cable by q_s, q_v, and q_r, respectively. Consider first self-equilibrium of the center node. Since the center node is connected by N radial cables, the forces $\mathbf{f}_c$ applied at the center node is

$$\mathbf{f}_c = \sum_{j=1}^{N} \mathbf{f}_{k_j} = \sum_{i=0}^{N-1} q_r(\mathbf{x}_i - \mathbf{x}_c). \tag{3.64}$$

Substituting Eqs. (3.59)–(3.61) into Eq. (3.64), the forces $\mathbf{f}_c$ become

$$\mathbf{f}_c = q_r \begin{pmatrix} \sum_{i=0}^{N-1} C_i & -\sum_{i=0}^{N-1} S_i & 0 \\ \sum_{i=0}^{N-1} S_i & \sum_{i=0}^{N-1} C_i & 0 \\ 0 & 0 & N-1 \end{pmatrix} \begin{pmatrix} \bar{\mathbf{x}}_0 \\ 0 \end{pmatrix}. \tag{3.65}$$

Because the relations

$$\sum_{i=0}^{N-1} C_i = 0, \quad \sum_{i=0}^{N-1} S_i = 0 \tag{3.66}$$

hold, the forces $\mathbf{f}_c$ turn out to be zero:

$$\mathbf{f}_c = \mathbf{0}, \tag{3.67}$$

which means that the center node is always in the state of self-equilibrium irrespective of value of the force density q_r of the radial cables.

Self-equilibrium of the center nodes can also be confirmed by using the reduced force density matrix. Because the boundary nodes and center nodes cannot move to each other by any of the symmetry operations of a dihedral group, the definition of the reduced force density matrix $\hat{\mathbf{E}}_c$ in Eq. (3.12) is modified as follows for the representative center node:

$$\hat{\mathbf{E}}_c = \sum_{i=0}^{N-1} q_r(\mathbf{T}_i - \mathbf{T}_c). \tag{3.68}$$

Substituting transformation matrices $\mathbf{T}_i$ defined in Eq. (3.61) and $\mathbf{T}_c$ defined in Eq. (3.63) into Eq. (3.68), we have

$$\hat{\mathbf{E}}_c = q_r \begin{pmatrix} \sum_{i=0}^{N-1} C_i & -\sum_{i=0}^{N-1} S_i & 0 \\ \sum_{i=0}^{N-1} S_i & \sum_{i=0}^{N-1} C_i & 0 \\ 0 & 0 & 0 \end{pmatrix}, \tag{3.69}$$

which is always singular because $\hat{\mathbf{E}}_c$ can be shown to be a zero matrix by using Eq. (3.66).

Next, we consider the representative boundary node 0 on the upper plane as a reference node. The coordinates of the two boundary nodes on the lower plane, which are connected to the reference node by a strut and a vertical cable, respectively, are denoted by $\mathbf{x}_s$ and $\mathbf{x}_v$. Because the boundary nodes are of the same type, the reference node $\mathbf{x}_0$ can be transformed to the two boundary nodes $\mathbf{x}_s$ and $\mathbf{x}_v$ by the proper two-fold rotations $\mathbf{T}_N$ and $\mathbf{T}_{N+v}$, respectively:

$$\begin{aligned} \mathbf{x}_s &= \mathbf{T}_N \mathbf{x}_0, \\ \mathbf{x}_v &= \mathbf{T}_{N+v} \mathbf{x}_0, \end{aligned} \tag{3.70}$$

where the transformation matrices $\mathbf{T}_N$ and $\mathbf{T}_{N+v}$ are

$$\mathbf{T}_N = \begin{pmatrix} 1 & 0 & 0 \\ 0 & -1 & 0 \\ 0 & 0 & -1 \end{pmatrix}, \quad \mathbf{T}_{N+v} = \begin{pmatrix} C_v & S_v & 0 \\ S_v & -C_v & 0 \\ 0 & 0 & -1 \end{pmatrix}, \tag{3.71}$$

with C_v and S_v denoting $\cos(2v\pi/N)$ and $\sin(2v\pi/N)$, respectively.

The self-equilibrium equation of the reference node 0 can be written as follows by using Eqs. (3.1) and (3.4):

$$\begin{aligned}\sum_{j=1}^{3} \mathbf{f}_{k_j} &= q_s(\mathbf{x}_s - \mathbf{x}_0) + q_v(\mathbf{x}_v - \mathbf{x}_0) + q_r(\mathbf{x}_c - \mathbf{x}_0) \\ &= \mathbf{H}\mathbf{x}_0 + q_r\mathbf{x}_c \\ &= \mathbf{0}_3, \end{aligned} \tag{3.72}$$

where $\mathbf{0}_3 \in \mathbb{R}^3$ is a zero vector and the matrix $\mathbf{H} \in \mathbb{R}^{3\times 3}$ is

$$\mathbf{H} = \begin{pmatrix} \mathbf{H}_1 & \mathbf{0}_2 \\ \mathbf{0}_2^\top & \mathbf{H}_2 \end{pmatrix} \tag{3.73}$$

with $\mathbf{H}_1 \in \mathbb{R}^{2\times 2}$ and $\mathbf{H}_2 \in \mathbb{R}^{1\times 1}$ respectively defined as

$$\mathbf{H}_1 = -\begin{pmatrix} q_v(1 - C_v) + q_r & -q_v S_v \\ -q_v S_v & 2q_s + q_v(C_v + 1) + q_r \end{pmatrix} \tag{3.74}$$

and

$$\mathbf{H}_2 = -(2q_s + 2q_v + q_r). \tag{3.75}$$

Since $\mathbf{H}_1$ and $\mathbf{H}_2$ are independent to each other, Eq. (3.72) can be separated into the following two equations:

$$\begin{aligned} \mathbf{H}_1\bar{\mathbf{x}}_0 &= \mathbf{0}_2, \\ \mathbf{H}_2 H + q_r H &= 0. \end{aligned} \tag{3.76}$$

Substituting Eq. (3.75) into the second equation in Eq. (3.76), we have

Relation between force densities of struts q_s and vertical cables q_v of the symmetric star-shaped tensegrity structure $\mathbf{D}_N^v$:

$$\frac{q_s}{q_v} = -1 \quad \text{or equivalently} \quad q_s = -q_v. \tag{3.77}$$

For the first equation in Eq. (3.76), $\mathbf{H}_1$ should be singular so as to guarantee a non-trivial solution for the *xy*-coordinate vector $\bar{\mathbf{x}}_0(\neq \mathbf{0})$. Equivalently, determinant of $\mathbf{H}_1$ should be zero:

$$\begin{aligned}\det(\mathbf{H}_1) &= q_r^2 - 2q_v^2(1 - C_v) \\ &= 0, \end{aligned} \tag{3.78}$$

in which Eq. (3.77) and $C_v^2 + S_v^2 = 1$ have been used. Therefore, we have the positive solution for the force densities of the cables as

Relation between force densities of radial cables q_r and vertical cables q_v of the symmetric star-shaped tensegrity structure $\mathbf{D}_N^v$:

$$\frac{q_r}{q_v} = \sqrt{2(1 - C_v)}. \tag{3.79}$$

So far, we have derived the relations among the force densities of the radial cables q_r, the vertical cables q_v, and the struts q_s in Eqs. (3.77) and (3.79). If the value of one of the three force densities is specified, the remaining two are consequently determined.

The same set of self-equilibrium equations as Eq. (3.72) can be obtained by using the reduced force density matrix $\hat{\mathbf{E}}_0$ corresponding to the representative boundary node. With slight modification to the definition of reduced force density matrix in Eq. (3.12), $\hat{\mathbf{E}}_0$ is written as

$$\begin{aligned}\hat{\mathbf{E}}_0 &= \sum_{j=1}^{3} q_{k_j}(\mathbf{T}_{k_j} - \mathbf{I}_3)\\ &= q_r(\mathbf{T}_c - \mathbf{I}_3) + q_s(\mathbf{T}_N - \mathbf{I}_3) + q_v(\mathbf{T}_{N+v} - \mathbf{I}_3),\end{aligned} \tag{3.80}$$

where $\mathbf{I}_3 \in \mathbb{R}^{3\times 3}$ is an identify matrix. Substituting the transformation matrix $\mathbf{T}_c$ defined in Eq. (3.63) together with $\mathbf{T}_N$ and $\mathbf{T}_{N+v}$ defined in Eq. (3.71) into Eq. (3.80), we have

$$\hat{\mathbf{E}}_0 = \begin{pmatrix} \mathbf{H}_1 & \mathbf{0}_2 \\ \mathbf{0}_2^\top & \mathbf{H}_2 + q_r \end{pmatrix}. \tag{3.81}$$

For the purpose of having non-trivial solutions for coordinates of the representative boundary node, the independent sub-matrices $(\mathbf{H}_2 + q_r)$ and $\mathbf{H}_1$ of $\hat{\mathbf{E}}_0$ have to be singular respectively leading to the same equations in Eqs. (3.76) and (3.78).

From the first equation in Eq. (3.76), xy-coordinates of the representative boundary node lie in the null-space of $\mathbf{H}_1$. Substituting the force densities in Eqs. (3.77) and (3.79) into Eq. (3.74), the sub-matrix $\mathbf{H}_1$ becomes

$$\mathbf{H}_1 = -q_v \begin{pmatrix} 1 - C_v + \sqrt{2(1 - C_v)} & -S_v \\ -S_v & -(1 - C_v) + \sqrt{2(1 - C_v)} \end{pmatrix}. \tag{3.82}$$

The xy-coordinates $\bar{\mathbf{x}}_0$ of the representative boundary node lie in the null-space of $\mathbf{H}_1$, and its particular solution is

$$\bar{\mathbf{x}}_0 = \begin{pmatrix} C_v - 1 + \sqrt{2 - 2C_v} \\ S_v \end{pmatrix}. \tag{3.83}$$

Accordingly, the general solution for the coordinate vector $\mathbf{x}_0$ of the representative boundary node is summarized as

Coordinates of the representative boundary node of the symmetric star-shaped structure $\mathbf{D}_N^v$:

$$\mathbf{x}_0 = \frac{R}{R_0} \begin{pmatrix} C_v - 1 + \sqrt{2 - 2C_v} \\ S_v \\ 0 \end{pmatrix} + \frac{H}{2} \begin{pmatrix} 0 \\ 0 \\ 1 \end{pmatrix}, \tag{3.84}$$

where R_0 is the norm of the first vector on the right-hand side of the equality representing the xy-coordinates. The parameters R and H are respectively the radius of the parallel circles and height of the structure.

It should be noted that the coordinates of the representative boundary node of a star-shaped structure are the same as those of the prismatic structure, when they have the same $\mathbf{D}_N$ symmetry and the same connectivity v for vertical cables. Using $\mathbf{x}_0$ in Eq. (3.84) and transformation of nodes in Eq. (3.60), coordinates of all other boundary nodes can be uniquely determined.

3.5 Regular Truncated Tetrahedral Structures

In this section, we consider a class of tensegrity structures with more complicated symmetry—tetrahedral symmetry. The tensegrity structures with tetrahedral symmetry were first introduced by Fuller [1], when he intended to make novel tensegrity structures utilizing the well-known geometries.

An example of this kind of structures is shown in Fig. 3.13. The structure consists of 12 nodes and 24 members; i.e., $n = 12, m = 24$. There are six struts in compression and 18 cables in tension. The cables lie along the edges of a truncated tetrahedron as shown in Fig. 3.14b, which is generated by symmetrically cutting off the vertices of a regular tetrahedron as shown in Fig. 3.14a; and the struts are the diagonals connecting the vertices of the truncated tetrahedron.

There are many other tensegrity structures with tetrahedral symmetry, for example the structures achieved by truncating the vertices of a regular tetrahedron using the polyhedral truncation scheme [2]. In this section, we will concentrate only on the structures with nodes having one-to-one correspondence to the symmetry operations of a tetrahedral group.

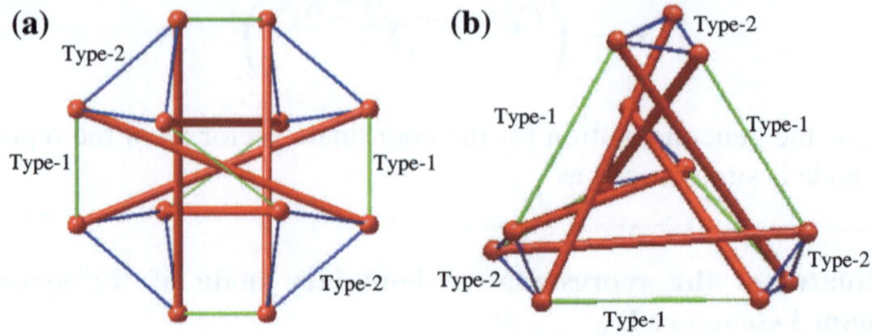

Fig. 3.13 A tensegrity structure with tetrahedral symmetry. The cables lie on the edges, and the struts connect the vertices of a regular truncated tetrahedron as shown in Fig. 3.14. **a** Top view, **b** side view

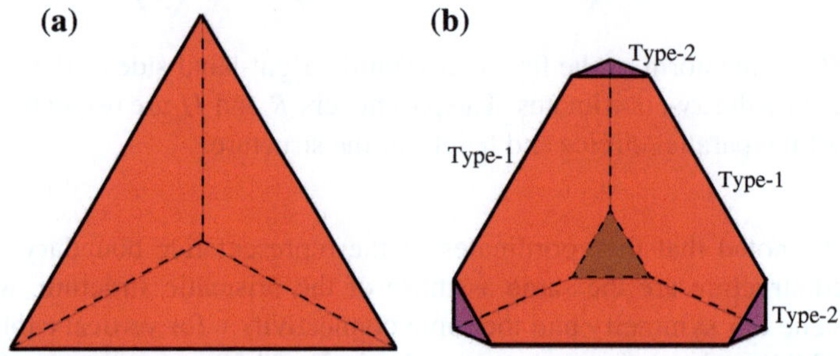

Fig. 3.14 Regular tetrahedron and truncated tetrahedron. The original edges are replaced by Type-1 cables of the structure with tetrahedral symmetry, for example the structure as shown in Fig. 8.1, and the new edges appeared in the truncated tetrahedron are replaced by Type-2 cables. **a** Tetrahedron, **b** truncated tetrahedron

There are in total three different types of members in a regular truncated tetrahedral structure:

- *Type-1 cables*, which are in tension: they are generated by replacing the original edges of the tetrahedron.
- *Type-2 cables*, which are in tension: they are generated by replacing the edges of the components (small tetrahedrons) cut off.
- *Struts*, which are in compression: they are generated by connecting the vertices of the truncated tetrahedron.

Accordingly, a regular truncated tetrahedral structure consists of six struts, six Type-1 cables, and twelve Type-2 cables. A possible geometry realization of this class of structures is shown in Fig. 3.13.

In this section, we show that self-equilibrium analysis of the regular truncated tetrahedral tensegrity structures can be analytically conducted by considering the self-equilibrium of only one node, because the nodes have one-to-one correspondence to the symmetry operations of the tetrahedral group. The self-equilibrium condition in terms of force densities is the same as that obtained in Chap. 8 as well as in Ref. [3], by using the symmetry-adapted form of the force density matrix.

3.5.1 Tetrahedral Symmetry

A tetrahedron is composed of four vertices, four faces, and six edges. Moreover, all edges of a *regular tetrahedron* have the same length. An example of regular tetrahedron is shown in Fig. 3.15a. A regular tetrahedron will have the same appearance by applying any of the following twelve symmetry operations:

- The identity operation E, doing nothing, as indicated in Fig. 3.15a.
- The clockwise three-fold rotations τ_j^1 and τ_j^2 ($j = 1, 2, 3, 4$) of the angles $2\pi/3$ and $4\pi/3$ respectively about the j-axis going through the vertex P_j and the origin (central point 'O') of the tetrahedron as indicated in Fig. 3.15b.
- The two-fold rotations σ_k ($k = 1, 2, 3$) of the angle π about the k-axis going through the middle of the edges as well as the origin as indicated in Fig. 3.15c.

Let σ_1 denote the two-fold rotation, which exchanges positions of vertices 1 and 2, and vertices 3 and 4; i.e.,

$$\sigma_1 : \quad (1, 2)(3, 4), \tag{3.85}$$

where a permutation (i, j, k) indicates vertex movements of i to j, j to k, and k to i. Similarly, the other two two-fold rotations σ_2 and σ_3 are defined as

$$\begin{aligned} \sigma_2 &: \quad (1, 3)(2, 4), \\ \sigma_3 &: \quad (1, 4)(2, 3). \end{aligned} \tag{3.86}$$

The clockwise three-fold rotations about the axis OP_j ($j = 1, 2, 3, 4$), going through the origin O and vertex j, are denoted as τ_j^1 and τ_j^2. For example, for the axis OP_1, we have

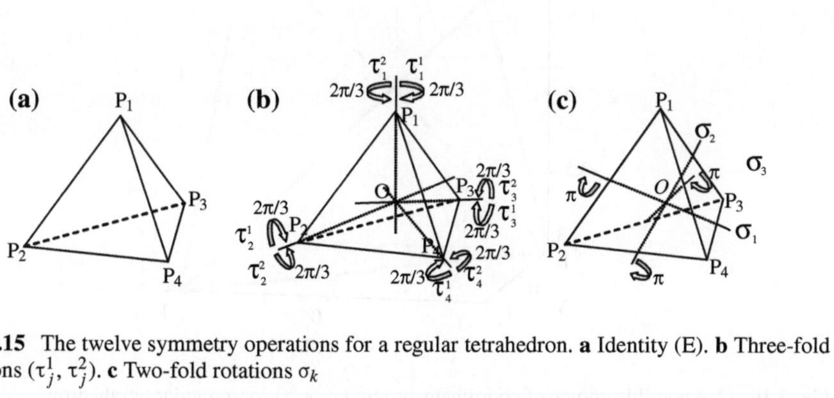

Fig. 3.15 The twelve symmetry operations for a regular tetrahedron. **a** Identity (E). **b** Three-fold rotations (τ_j^1, τ_j^2). **c** Two-fold rotations σ_k

$$\begin{aligned} \tau_1^1 &: \quad (2,3,4), \\ \tau_1^2 &: \quad (2,4,3). \end{aligned} \tag{3.87}$$

For the two- and three-fold symmetry operations of a tetrahedral group, the following relation holds:

$$\tau_{j+1}^i = \sigma_j \tau_1^i \sigma_j, \quad (j = 1,2,3;\ i = 1,2). \tag{3.88}$$

All other three-fold rotations can be generated by using Eqs. (3.85)–(3.88).

There are in total twelve nodes in a regular truncated tetrahedral tensegrity structure, and twelve symmetry operations in a tetrahedral group. One node of the structure can be moved to another node by only one of the (appropriate) symmetry operations. Thus, the nodes of a regular truncated tetrahedral structure have one-to-one correspondence to the symmetry operations of a tetrahedral group.

It should be noted that the transformation matrices corresponding to the symmetry operations of a tetrahedral group are not unique, depending on selection of coordinate system. Figure 3.16 shows a possible choice of Cartesian coordinate system (x, y, z). The coordinates of the four vertices P_j $(j = 1,2,3,4)$ of a regular tetrahedron in this coordinate system are given as

$$\mathrm{P}_1 : \begin{pmatrix} 1 \\ 1 \\ 1 \end{pmatrix}, \quad \mathrm{P}_2 : \begin{pmatrix} 1 \\ -1 \\ -1 \end{pmatrix}, \quad \mathrm{P}_3 : \begin{pmatrix} -1 \\ 1 \\ -1 \end{pmatrix}, \quad \mathrm{P}_4 : \begin{pmatrix} -1 \\ -1 \\ 1 \end{pmatrix}. \tag{3.89}$$

The transformation matrix $\mathbf{T}(\sigma_1)$ corresponding to the symmetry operation σ_1, which exchanges the vertices 1 and 2, and 3 and 4; i.e., $(1,2)(3,4)$, in this coordinate system is

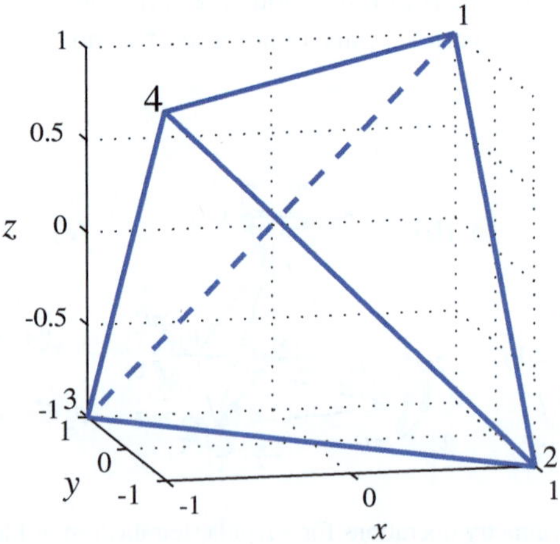

Fig. 3.16 One possible choice of coordinate system (x, y, z) for a regular tetrahedron

$$\mathbf{T}(\sigma_1) = \begin{pmatrix} 1 & 0 & 0 \\ 0 & -1 & 0 \\ 0 & 0 & -1 \end{pmatrix}. \tag{3.90}$$

Example 3.14 An alternative selection of coordinate system for the regular tetrahedron.

Other than the coordinate system adopted for the regular tetrahedron in Eq. (3.89), we can have another possible choice of Cartesian coordinate system $(\bar{x}, \bar{y}, \bar{z})$, with the coordinates of the vertices of the tetrahedron given as

$$\bar{\mathrm{P}}_1 : \begin{pmatrix} 0 \\ 0 \\ 1 \end{pmatrix}, \quad \bar{\mathrm{P}}_2 : \begin{pmatrix} -\frac{2\sqrt{2}}{3} \\ 0 \\ -\frac{1}{3} \end{pmatrix}, \quad \bar{\mathrm{P}}_3 : \begin{pmatrix} \frac{\sqrt{2}}{3} \\ \frac{\sqrt{6}}{3} \\ -\frac{1}{3} \end{pmatrix}, \quad \bar{\mathrm{P}}_4 : \begin{pmatrix} \frac{\sqrt{2}}{3} \\ -\frac{\sqrt{6}}{3} \\ -\frac{1}{3} \end{pmatrix}. \tag{3.91}$$

The transformation matrix $\bar{\mathbf{T}}(\sigma_1)$ corresponding to the symmetry operation $\sigma_1 : (1, 2)(3, 4)$ in this coordinate system is

$$\bar{\mathbf{T}}(\sigma_1) = \begin{pmatrix} \frac{1}{3} & 0 & -\frac{2\sqrt{2}}{3} \\ 0 & -1 & 0 \\ -\frac{2\sqrt{2}}{3} & 0 & -\frac{1}{3} \end{pmatrix}. \tag{3.92}$$

It is obvious that the transformation matrix $\mathbf{T}(\sigma_1)$ in the coordinate system (x, y, z) as shown in Fig. 3.16 is much simpler than the transformation matrix $\bar{\mathbf{T}}(\sigma_1)$ in this alternative coordinate system $(\bar{x}, \bar{y}, \bar{z})$.

In the following discussions, we adopt the coordinate system and transformation matrices corresponding the tetrahedron in the coordinate system (x, y, z) as shown in Fig. 3.16, because this could simplify our further analytical investigations. Note that selection of coordinate system will not lead to any loss of generality in the discussions on self-equilibrium analysis.

The transformation matrices $\mathbf{T}(\sigma_2)$ and $\mathbf{T}(\sigma_3)$ for the two-fold rotations $\sigma_2 = (1\ 3)(2\ 4)$ and $\sigma_3 = (1\ 4)(2\ 3)$ are defined as

$$\mathbf{T}(\sigma_2) = \begin{pmatrix} -1 & 0 & 0 \\ 0 & 1 & 0 \\ 0 & 0 & -1 \end{pmatrix}, \quad \mathbf{T}(\sigma_3) = \begin{pmatrix} -1 & 0 & 0 \\ 0 & -1 & 0 \\ 0 & 0 & 1 \end{pmatrix}, \tag{3.93}$$

and the three-fold rotations about the axis OP_1 are

$$\mathbf{T}(\tau_1^1) = \begin{pmatrix} 0 & 0 & 1 \\ 1 & 0 & 0 \\ 0 & 1 & 0 \end{pmatrix}, \quad \mathbf{T}(\tau_1^2) = \begin{pmatrix} 0 & 1 & 0 \\ 0 & 0 & 1 \\ 1 & 0 & 0 \end{pmatrix}. \tag{3.94}$$

Since the relation $\tau_{j+1}^i = \sigma_j \tau_1^i \sigma_j$ $(i = 1, 2;\ j = 1, 2, 3)$ holds, we have

$$\begin{aligned} \mathbf{T}(\tau_{j+1}^1) &= \mathbf{T}(\sigma_j)\mathbf{T}(\tau_1^1)\mathbf{T}(\sigma_j), \\ \mathbf{T}(\tau_{j+1}^2) &= \mathbf{T}(\sigma_j)\mathbf{T}(\tau_1^2)\mathbf{T}(\sigma_j), \quad (j = 1, 2, 3). \end{aligned} \tag{3.95}$$

Therefore, the transformation matrices for the other three-fold rotations can be generated as

$$\begin{aligned} \mathbf{T}(\tau_2^1) &= \begin{pmatrix} 0 & 0 & -1 \\ -1 & 0 & 0 \\ 0 & 1 & 0 \end{pmatrix}, \quad \mathbf{T}(\tau_2^2) = \begin{pmatrix} 0 & -1 & 0 \\ 0 & 0 & 1 \\ -1 & 0 & 0 \end{pmatrix}, \\ \mathbf{T}(\tau_3^1) &= \begin{pmatrix} 0 & 0 & 1 \\ -1 & 0 & 0 \\ 0 & -1 & 0 \end{pmatrix}, \quad \mathbf{T}(\tau_3^2) = \begin{pmatrix} 0 & -1 & 0 \\ 0 & 0 & -1 \\ 1 & 0 & 0 \end{pmatrix}, \\ \mathbf{T}(\tau_4^1) &= \begin{pmatrix} 0 & 0 & -1 \\ 1 & 0 & 0 \\ 0 & -1 & 0 \end{pmatrix}, \quad \mathbf{T}(\tau_4^2) = \begin{pmatrix} 0 & 1 & 0 \\ 0 & 0 & -1 \\ -1 & 0 & 0 \end{pmatrix}. \end{aligned} \tag{3.96}$$

3.5.2 Self-equilibrium Analysis

The force densities of Type-1 cables, Type-2 cables, and struts are denoted by $q_{\rm v}$, $q_{\rm h}$, and $q_{\rm s}$, respectively. Because the nodes of a regular truncated tetrahedral structure have one-to-one correspondence to the symmetry operations of the tetrahedral group, there exists only one type of node so that we can choose any of them as a reference node. Every node of the structure is connected by one Type-1 cable, two Type-2 cables, and one strut. From Eq. (3.12), the reduced force density matrix $\hat{\mathbf{E}}$ corresponding to the reference node can be written in a general form as follows:

$$\hat{\mathbf{E}} = q_{\rm h}(\mathbf{T}(\tau_h^1) - \mathbf{I}) + q_{\rm h}(\mathbf{T}(\tau_h^2) - \mathbf{I}) + q_{\rm v}(\mathbf{T}(\sigma_v) - \mathbf{I}) + q_{\rm s}(\mathbf{T}(\sigma_s) - \mathbf{I}), \quad (h = 1, 2, 3, 4;\ v, s = 1, 2, 3;\ v \neq s). \tag{3.97}$$

In Eq. (3.97), values of h, v, and s refer to the choice of connectivity pattern. Different values of (h, v, s) will lead to different entries in the transformation matrices, but importantly, they will not alter the eigenvalues of the three-dimensional reduced force density matrix $\hat{\mathbf{E}}$. Thus, different connectivity pattern will not affect self-equilibrium analysis of this class of structures, because the condition of zero determinant, which is product of individual eigenvalues, of the reduced force density matrix is concerned.

Example 3.15 The reduced force density matrix of a truncated tetrahedral structure.

For the case of $h = 1$, $v = 2$, and $s = 3$, we have the reduced force density matrix $\hat{\mathbf{E}}$ as follows by using Eq. (3.97) and the transformation matrices $\mathbf{T}(\tau_1^1)$, $\mathbf{T}(\tau_1^2)$, $\mathbf{T}(\sigma_2)$, and $\mathbf{T}(\sigma_3)$:

$$\hat{\mathbf{E}} = \begin{pmatrix} 2(q_h + q_v + q_s) & -q_h & -q_h \\ -q_h & 2(q_h + q_s) & -q_h \\ -q_h & -q_h & 2(q_h + q_v) \end{pmatrix}. \tag{3.98}$$

For the case of $h = 2$, $v = 1$, and $s = 2$, we have

$$\hat{\mathbf{E}} = \begin{pmatrix} 2(q_h + q_s) & q_h & q_h \\ q_h & 2(q_h + q_v) & -q_h \\ q_h & -q_h & 2(q_h + q_v + q_s) \end{pmatrix}. \tag{3.99}$$

Obviously, the above two versions of reduced force density matrix corresponding to different connectivity patterns defined by (h, v, s) have the same eigenvalues, although arrangement of the diagonal entries is different.

In the self-equilibrium analysis, the reduced force density matrix $\hat{\mathbf{E}}$ should be singular to ensure a non-trivial solution for coordinates of the reference node; i.e., its determinant is zero:

$$\det(\hat{\mathbf{E}}) = 0, \tag{3.100}$$

which leads to

$$3q_h^2 q_v + 3q_h^2 q_s + 2q_h q_v^2 + 2q_v^2 q_s + 2q_h q_s^2 + 2q_v q_s^2 + 6q_h q_v q_s = 0. \tag{3.101}$$

Utilizing the symmetry with respect to q_v and q_s in Eq. (3.101), we set

$$q_v = \psi + \phi \ (> 0) \quad \text{and} \quad q_s = \psi - \phi \ (< 0), \tag{3.102}$$

such that

$$\psi = \frac{1}{2}(q_v + q_s) \quad \text{and} \quad \phi = \frac{1}{2}(q_v - q_s) > 0. \tag{3.103}$$

Furthermore, by letting

$$\begin{aligned} a &= 3(q_v + q_s) = 6\psi, \\ b &= 2(q_v^2 + q_s^2 + 3q_v q_s) = 2(5\psi^2 - \phi^2), \\ c &= 2q_v q_s (q_v + q_s) = 4\psi(\psi^2 - \phi^2), \end{aligned} \tag{3.104}$$

Eq. (3.101) reduces to

$$aq_{\mathrm{h}}^2 + bq_{\mathrm{h}} + c = 0. \tag{3.105}$$

Regarding the value of $q_{\mathrm{v}} + q_{\mathrm{s}}$, there are two possibilities:

1. When $q_{\mathrm{v}} + q_{\mathrm{s}} = 2\psi = 0$, such that $a = c = 0$, the equation has only one solution:

$$q_{\mathrm{h}} = 0. \tag{3.106}$$

   This is a trivial solution, and will not be considered any more in the following discussions.
2. When $q_{\mathrm{v}} + q_{\mathrm{s}} = 2\psi \neq 0$, the two solutions for q_{h} are

$$q_{\mathrm{h}_1} = \frac{-b + \sqrt{b^2 - 4ac}}{2a} = \frac{\phi^2 - 5\psi^2 + \sqrt{\psi^4 + 14\psi^2\phi^2 + \phi^4}}{6\psi}, \tag{3.107}$$

   and

$$q_{\mathrm{h}_2} = \frac{-b - \sqrt{b^2 - 4ac}}{2a} = \frac{\phi^2 - 5\psi^2 - \sqrt{\psi^4 + 14\psi^2\phi^2 + \phi^4}}{6\psi}. \tag{3.108}$$

   Because $\psi \neq 0$ and $\phi > 0$, we have

$$\psi^4 + 14\psi^2\phi^2 + \phi^4 > 0, \tag{3.109}$$

such that the solutions q_{h_1} and q_{h_2} of Eq. (3.105) have real values.

Furthermore, the following inequality is satisfied

$$\begin{aligned}(\psi^4 + 14\psi^2\phi^2 + \phi^4) - (\phi^2 - 5\psi^2)^2 &= -24\psi^2(\psi + \phi)(\psi - \phi) \\ &= -96\psi^2 q_{\mathrm{v}} q_{\mathrm{s}} \\ &> 0,\end{aligned} \tag{3.110}$$

because $q_{\mathrm{v}} > 0$ holds for Type-1 cables, and $q_{\mathrm{s}} < 0$ holds for strut. From Eqs. (3.109) and (3.110), we have

$$(\phi^2 - 5\psi^2) + \sqrt{\psi^4 + 14\psi^2\phi^2 + \phi^4} > 0. \tag{3.111}$$

Hence, the first solution q_{h_1} for the force density of Type-2 cable can be positive if and only if

$$\psi = \frac{1}{2}(q_{\mathrm{v}} + q_{\mathrm{s}}) > 0; \tag{3.112}$$

and similarly, the second solution q_{h_2} for the force density of Type-2 cable can be positive if and only if

$$\psi = \frac{1}{2}(q_{\mathrm{v}} + q_{\mathrm{s}}) < 0, \tag{3.113}$$

because

$$(\phi^2 - 5\psi^2) - \sqrt{\psi^4 + 14\psi^2\phi^2 + \phi^4} < 0. \tag{3.114}$$

The solutions can be rewritten with respect to ratios to one of the force density, for instance, q_v of the Type-1 cables:

Force densities of a regular truncated tetrahedral structure:

There are two solutions for relations between the ratios q_h/q_v and q_s/q_v of the force densities of Type-1 cables q_v, Type-2 cables q_h, and struts q_s.

- The first solution q_{h_1}/q_v is given as

$$\frac{q_{h_1}}{q_v} = \frac{\phi^2 - 5\psi^2 + \sqrt{\psi^4 + 14\psi^2\phi^2 + \phi^4}}{6\psi}, \tag{3.115}$$

  where the following inequality should hold for a positive solution for force densities of cables:

$$\psi > 0 \quad \text{or} \quad \frac{q_s}{q_v} > -1. \tag{3.116}$$

- The second solution q_{h_2}/q_v is given as

$$\frac{q_{h_2}}{q_v} = \frac{\phi^2 - 5\psi^2 - \sqrt{\psi^4 + 14\psi^2\phi^2 + \phi^4}}{6\psi}, \tag{3.117}$$

  where the following inequality should hold for a positive solution for force densities of cables:

$$\psi < 0 \quad \text{or} \quad \frac{q_s}{q_v} < -1. \tag{3.118}$$

In the above solutions, we have set $\psi = (q_s + q_v)/2$ and $\phi = (q_v - q_s)/2$.

Example 3.16 Illustration of the force densities of a regular truncated tetrahedral tensegrity structure.

The relations $q_v > 0$ for Type-1 cables, $q_h > 0$ for Type-2 cables, and $q_s < 0$ for struts should be satisfied so as to ensure that cables carry tensions (positive prestresses) and struts carry compressions (negative prestresses). Thus, the following inequalities hold

$$\frac{q_h}{q_v} > 0 \quad \text{and} \quad \frac{q_s}{q_v} < 0. \tag{3.119}$$

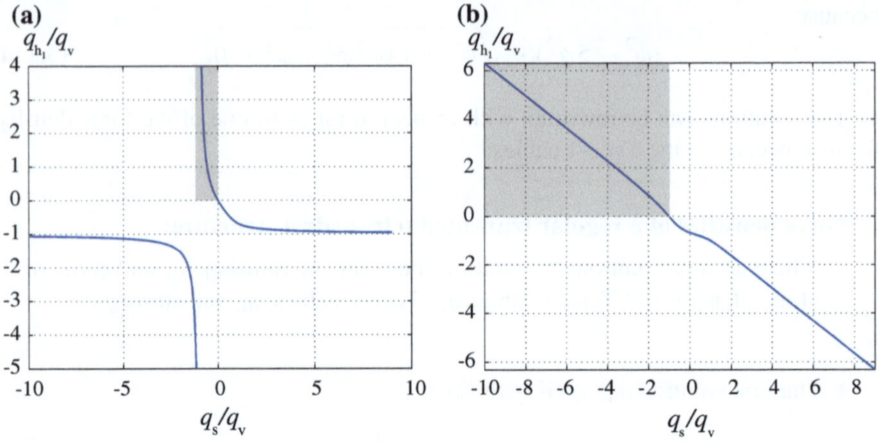

Fig. 3.17 Two possible solutions for force densities of tensegrity structures with tetrahedral symmetry. Only the solutions in the shaded regions are feasible, because the cables should have positive force densities, while the struts have negative force densities. **a** Solution in Eq. (3.115), **b** solution in Eq. (3.117)

Moreover, the conditions $q_s/q_v > -1$ and $q_s/q_v < -1$ should be satisfied for positive q_h/q_v as in Eqs. (3.115) and (3.117), respectively. Therefore, only those falling into the shaded regions in Fig. 3.17a, b are feasible solutions.

The coordinates of the reference node lie in the null-space of the reduced force density matrix, and those of the other nodes can be determined by symmetry operations via transformation matrices defined in Eqs. (3.90), (3.93), (3.94), and (3.96), corresponding to the selected coordinate system. Because nodal coordinates of the structure depend on the selection of coordinate system, analytical solutions are not generally possible. Here, we do not attempt to present the analytical solutions for the nodal coordinates, instead, we present several possible self-equilibrated configurations in a numerical manner. In the following examples, we omit units without any loss of generality.

Example 3.17 Self-equilibrated configuration of the regular truncated tetrahedral structures with different force density ratios.

As discussed previously, the connectivity of the members, defined by the parameters h, v, and s, do not have any effect on the self-equilibrium condition of the structure. In the following examples, we adopt the connectivity pattern of $(h, v, s) = (1, 2, 3)$.

- $q_s/q_v = -0.5$ (>−1):
  From Eqs. (3.115) and (3.117), the solutions for the force densities are

$$\frac{q_{h_1}}{q_v} = 0.7676 \quad \text{and} \quad \frac{q_{h_2}}{q_v} = -0.4343, \tag{3.120}$$

  and their corresponding self-equilibrated configurations are shown in Figs. 3.13 and 3.18, respectively. Note that only the first solution q_{h_1}/q_v is feasible, satisfying the constraints assigned on the type of members: positive force densities for cables and negative for struts.
- $q_s/q_v = -0.2$ (>−1):
  The feasible solution is $q_{h_1}/q_v = 0.3658$ and its self-equilibrated configuration is shown in Fig. 3.19a.
- $q_s/q_v = -0.8$ (>−1):
  The feasible solution is $q_{h_1}/q_v = 2.7288$ and its self-equilibrated configuration is shown in Fig. 3.19b.
- $q_s/q_v = -2.0$ (<−1):
  The feasible solution is $q_{h_2}/q_v = 0.8685$ and its self-equilibrated configuration is shown in Fig. 3.20a.

If we do not strictly follow the definition of the types of members as indicated in Fig. 3.13, we can also have self-equilibrated configurations, for example:

- $q_s/q_v = 1.0$ (>0):
  The first solution is $q_{h_1}/q_v = -1.0$ and its self-equilibrated configuration is shown in Fig. 3.20b.

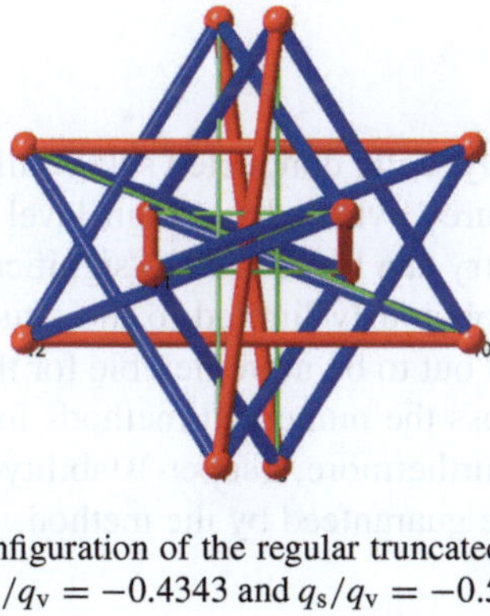

Fig. 3.18 Self-equilibrated configuration of the regular truncated tetrahedral tensegrity structure with the force density ratios $q_{h_2}/q_v = -0.4343$ and $q_s/q_v = -0.5$. The struts contact at their ends

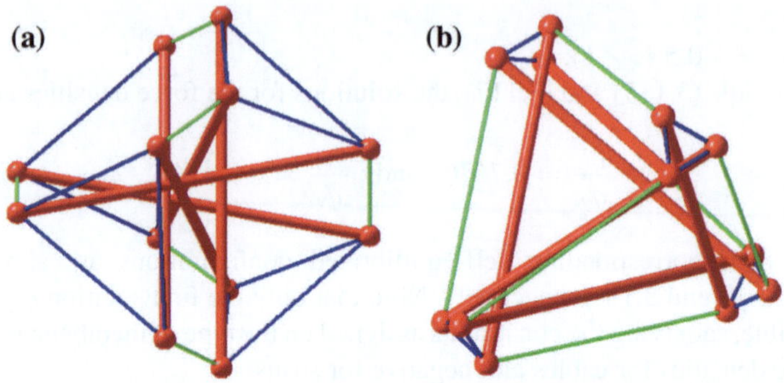

Fig. 3.19 Two regular truncated tetrahedral tensegrity structures that have similar appearances to the structure in Fig. 3.13, while the lengths of cables are different due to different force densities. **a** $q_{h_1}/q_v = 0.3658$, $q_s/q_v = -0.2$, **b** $q_{h_1}/q_v = 2.7288$, $q_s/q_v = -0.8$

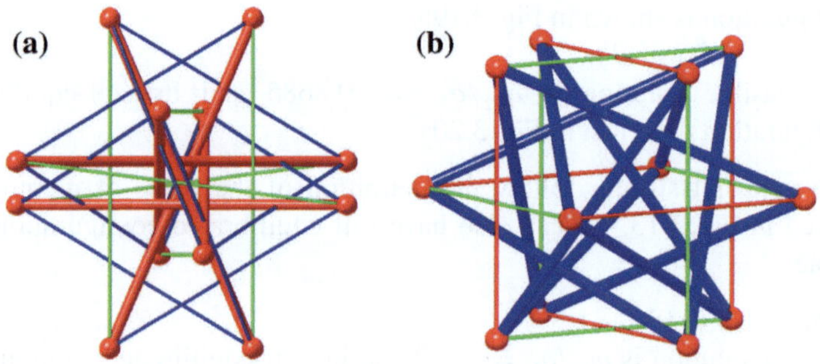

Fig. 3.20 Two regular truncated tetrahedral tensegrity structures. These structures are unstable, as will be discussed in Chap. 8. **a** $q_s/q_v = -2.0$, $q_{h_2}/q_v = 0.8685$, **b** $q_s/q_v = 1.0$, $q_{h_1}/q_v = -1.0$

3.6 Remarks

In this chapter, we have analytically conducted self-equilibrium analysis for several classes of tensegrity structures, which have high level of symmetry. It has been shown that the high symmetry can be utilized to significantly simplify the analysis. However, this approach is obviously limited to the structure with high symmetry, and numerical methods turn out to be more flexible for those with less symmetry.

In Chap. 5, we will discuss the numerical methods for self-equilibrium analysis of tensegrity structures. Furthermore, (super-)stability of their self-equilibrated configurations might also be guaranteed by the method.

Furthermore, stability of these structures have not been discussed in this chapter. Some of these structures consist of a large number of (infinitesimal) mechanisms, hence, they are not stable in the absence of prestresses according to the discussions in Chap. 4.

Example 3.18 Numbers of infinitesimal mechanisms of the symmetric prismatic structure $\mathbf{D}_3^{1,1}$ in Fig. 3.10 and the symmetric star-shaped structures $\mathbf{D}_3^{1}$ in Fig. 3.11.

Both of the structures in Figs. 3.10 and 3.11 are of dihedral symmetry, and they have only one prestress mode $n^{\mathrm{s}} = 1$. By using the modified Maxwell's rule in Eq. (2.61) for free-standing prestressed pin-jointed structures, number n^{m} of infinitesimal mechanisms of these structures can be calculated as follows:

- Prismatic structure $\mathbf{D}_3^{1,1}$ in Fig. 3.10: The structure is composed of six nodes and twelve members; i.e., $n = 6$, $m = 12$, thus, we have

$$\begin{aligned} n^{\mathrm{m}} &= n^{\mathrm{s}} - m + dn - n^{\mathrm{b}} \\ &= 1 - 12 + 3 \times 6 - 6 \\ &= 1, \end{aligned} \tag{3.121}$$

  where the number n^{b} of rigid-body motions is six for a three-dimensional case ($d = 3$).
- Star-shaped structure $\mathbf{D}_3^{1}$ in Fig. 3.11a: The structure is composed of eight nodes and twelve members; i.e., $n = 8$, $m = 12$, thus, we have

$$\begin{aligned} n^{\mathrm{m}} &= n^{\mathrm{s}} - m + dn - n^{\mathrm{b}} \\ &= 1 - 12 + 3 \times 8 - 6 \\ &= 7. \end{aligned} \tag{3.122}$$

In spite of existence of so many mechanisms, the prismatic structure $\mathbf{D}_3^{1,1}$ and the star-shaped structure $\mathbf{D}_3^{1}$ in Fig. 3.11a are actually super-stable owing to prestresses that stabilize the infinitesimal mechanisms. However, this is not always the case. For example, the regular truncated tetrahedral structures as shown in Figs. 3.18 and 3.20 are indeed unstable, with negative eigenvalues in their force density matrices, or equivalently the geometrical stiffness matrices, as well as their tangent stiffness matrices. Super-stability conditions for these structures will be presented in Chaps. 6–8.

References

1. Fuller, R. B. (1962). Tensile-integrity structures. U.S. Patent No. 3,063,521, November 1962.
2. Li, Y., Feng, X.-Q., Cao, Y.-P., & Gao, H. J. (2010). A Monte Carlo form-finding method for large scale regular and irregular tensegrity structures. *International Journal of Solids and Structures*, *47*(14–15), 1888–1898.
3. Raj, R. P., & Guest, S. D. (2006). Using symmetry for tensegrity form-finding. *Journal of International Association for Shell and Spatial Structures*, *47*(3), 1–8.

Chapter 4
Stability

Abstract In this chapter, we present formulations of the stiffness matrices, including the tangent, linear, and geometrical stiffnesses for a prestressed pin-jointed structure. Moreover, three different stability criteria—stability, prestress-stability, and super-stability—are presented for its stability investigation. It is demonstrated that super-stability is the most robust criterion, and therefore, it is usually preferable in the design of tensegrity structures. Furthermore, to guarantee a super-stable structure, we present the necessary conditions and sufficient conditions, which will be extensively used in the coming chapters.

Keywords Stiffness · Stability · Prestress-stability · Super-stability · Necessary conditions and sufficient conditions

4.1 Stability and Potential Energy

A structure deforms; i.e., its geometry changes, more or less when it is subjected to external loads. If it returns to its original configuration when the external loads are released, then it is said to be *stable*; otherwise, it is *unstable*. Obviously, the structures that have finite mechanisms, for example the structures in Figs. 2.6c and 2.7, are unstable because deformations will not be recovered after release of external loads.

To understand stability of a structure, it might be more comprehensible to consider a mechanical system consisting of a ball subjected to gravity.

4.1.1 Equilibrium and Stability of a Ball Under Gravity

Consider a ball, as shown in Fig. 4.1, which is placed on a curved surface consisting of hills and valleys. The ball is subjected to gravity, hence, it always intends to move 'down' to a lower position. This is because the ball tends to possess smaller potential energy, while the potential energy is a linear function of height of the ball in this system. Therefore, the curve in Fig. 4.1 can also be regarded as the graph of potential energy of the ball under gravity.

J.Y. Zhang and M. Ohsaki, *Tensegrity Structures*, Mathematics for Industry 6,
DOI 10.1007/978-4-431-54813-3_4

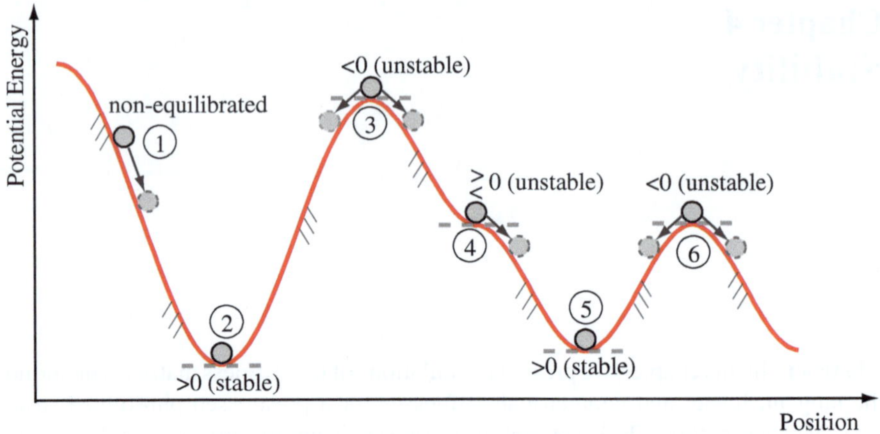

Fig. 4.1 Stability of a ball in view of potential energy. A ball is in equilibrium when it has stationary value of potential energy; i.e., when it is at the hilltop, valley, or saddle point. Moreover, it is stable only if the potential energy is of locally strict minimum

Obviously, the ball will not move if no disturbance (position change) is applied, when the curve of potential energy is flat. In this case, the ball is said to be *in equilibrium*. Mathematically speaking, the ball is in equilibrium when it is at the position with zero gradient of potential energy. Therefore, the ball can be in equilibrium only if it is at the hilltop, valley, or saddle (inflection) point of the curve of potential energy.

On the other hand, if the ball returns to its original position in equilibrium after small disturbance in *any direction*, then it is said to be *in stability*. It is obvious that the ball must be at the valley when it is in stability. Mathematically speaking, the potential energy is of *locally strict minimum*, when the ball is in stability. Locally strict minimum means that the potential energy will increase in any direction of the neighboring region of the current equilibrium position.

It should be noted that stability does imply equilibrium, however, equilibrium does not necessarily imply stability. Depending on position of the ball and its corresponding potential energy, we have the following cases:

- *Non-equilibrated*: When the ball is at the position, which is neither a hilltop nor a valley, e.g., the position ① in Fig. 4.1, it intends to move to the position with smaller potential energy even no disturbance is applied. Hence, the ball at this kind of positions is neither in equilibrium nor in stability.
- *Equilibrated but unstable*: When the ball is at the hilltop, e.g., ③ or ⑥, it does not move if it is not disturbed, thus, it is in equilibrium. However, any small disturbance will make it move to a lower position with smaller potential energy, hence, it is not in stability. When the ball is at the saddle point, e.g., ④ in Fig. 4.1, it is also in equilibrium, however, the potential energy decreases in some specific directions. Hence, the ball at this kind of positions is also unstable, although it is in equilibrium.

- *(Equilibrated and) stable*: When the ball is at the valley, e.g., ② or ⑤ in Fig. 4.1, it stays steady when there is no disturbance, because the curve is flat at the position; i.e., the gradient at the position is zero. Therefore, it is in equilibrium. Furthermore, it tends to return to its original position after small disturbances in any direction, because the current position is of the minimum potential energy. Hence, the ball at this kind of positions is also in stability.
- *Multi-stable*: If the disturbance is large enough, the ball may move from one stable position to another, e.g., from ② to ⑤ or its reverse. Such system is called *bi-stable* system if there are two stable positions, or *multi-stable* system if there are more than two stable positions. In Chap. 7, we will see an example of multi-stable star-shaped tensegrity structure.

Similarly to the ball under gravity, stability of an elastic structure can also be verified by investigating local minimum of its total potential energy, or strain energy when external loads are absent. This comes from the fact that a ball under gravity has direct correspondence to an elastic structure subjected to external loads in view of energy:

- Position of the ball corresponds to geometry of a structure, and small disturbance applied to the ball corresponds to small enforced deformation of the structure;
- Potential energy of the ball corresponds to the total potential energy of the structure.

Because equilibrium and stability of an elastic structure can be investigated in view of energy in the same manner as the ball under gravity, the next subsection presents formulation of the total potential energy of an elastic structure.

4.1.2 Total Potential Energy

A tensegrity structure is usually subjected to large deformation, because it is flexible especially in the directions of mechanisms. However, the resulting strains in the members are usually very small. Therefore, it is reasonable for us to consider large deformation but small strain for a tensegrity structure. In general, we consider a structure consisting of n (free) nodes and m members in three-dimensional space; i.e., $d = 3$.

The (engineering) strain ε_k in member k due to member extension is defined as

$$\varepsilon_k = \frac{l_k - l_k^0}{l_k^0}, \tag{4.1}$$

where l_k^0 is the length of member k in the unstressed state, and l_k is the member length in the stressed state. The strains are the same anywhere along a member, because the members of a tensegrity structure carry only axial forces.

Let E_k and A_k denote Young's modulus and the cross-sectional area of member k, respectively. Assuming that the members are made of linear elastic materials, the stress-strain relation is linear in the form of

$$\sigma_k = E_k \varepsilon_k, \tag{4.2}$$

where σ_k denotes the normal stress in the member. From Eq. (4.2), the axial force s_k of member k can be computed as follows:

$$s_k = E_k \varepsilon_k A_k, \tag{4.3}$$

where change in cross-section area of the member in the stressed state is ignored, since we have assumed that the stain is small enough. Substituting Eq. (4.1) into Eq. (4.3), we have

$$s_k = E_k A_k \frac{l_k - l_k^0}{l_k^0}. \tag{4.4}$$

The total potential energy Π of a system is defined as follows using the strain energy Π_{E} stored in the structure and the work Π_{W} done by external loads:

$$\Pi = \Pi_{\mathrm{E}} - \Pi_{\mathrm{W}}. \tag{4.5}$$

The strain energy Π_{E} of a structure with m members are given as

$$\begin{aligned} \Pi_{\mathrm{E}} &= \sum_{k=1}^{m} \frac{1}{2} \int_0^{l_k^0} A_k \sigma_k \varepsilon_k dx = \sum_{k=1}^{m} \frac{1}{2} s_k \varepsilon_k l_k^0 = \frac{1}{2} \sum_{k=1}^{m} E_k A_k \varepsilon_k^2 l_k^0 \\ &= \frac{1}{2} \sum_{k=1}^{m} \frac{E_k A_k (l_k - l_k^0)^2}{l_k^0}, \end{aligned} \tag{4.6}$$

because stress σ_k and strain ε_k are the same throughout each member.

Let X_j $(j = 1, 2, \ldots, 3n)$ denote the generalized nodal coordinates, representing nodal coordinates in x-, y-, and z-directions of a structure. The generalized nodal coordinate vector $\mathbf{X} \in \mathbb{R}^{3n}$ is defined as

$$\begin{aligned} \mathbf{X} &= (x_1, x_2, \ldots, x_i, \ldots, x_n, y_1, y_2, \ldots, y_i, \ldots, y_n, z_1, z_2, \ldots, z_i, \ldots, z_n)^\top \\ &= \left(\mathbf{x}^\top, \mathbf{y}^\top, \mathbf{z}^\top\right)^\top. \end{aligned} \tag{4.7}$$

In Eq. (4.6), Π_{E} is a function of $\mathbf{X}$, because member length l_k is a (nonlinear) function of $\mathbf{X}$.

Let ΔX_j denote generalized displacements. The external work Π_{W}, or equivalently its increment $\Delta\Pi_{\mathrm{W}}$ from self-equilibrium state, done by the external loads $\mathbf{p} = (p_1, p_2, \ldots, p_{3n})^\top$ is given as

$$\Delta\Pi_{\mathrm{W}} = \Pi_{\mathrm{W}} = \sum_{j=1}^{3n} p_j \Delta X_j. \tag{4.8}$$

The increment of the strain energy $\Delta\Pi_{\mathrm{E}}$ due to deformations $\Delta\mathbf{X}$ of the system (structure) is written as follows, by applying Taylor's expansion:

$$\Delta\Pi_{\mathrm{E}} = \delta\Pi_{\mathrm{E}} + \delta^2\Pi_{\mathrm{E}} + \cdots, \tag{4.9}$$

where the higher-order terms have been ignored, and the first-order $\delta\Pi_{\mathrm{E}}$ and second-order $\delta^2\Pi_{\mathrm{E}}$ increments of strain energy Π_{E} respectively are

$$\begin{aligned} \delta\Pi_{\mathrm{E}} &= \sum_{j=1}^{3n} \frac{\partial\Pi_{\mathrm{E}}}{\partial X_j}\Delta X_j, \\ \delta^2\Pi_{\mathrm{E}} &= \frac{1}{2}\sum_{i=1}^{3n}\sum_{j=1}^{3n} \frac{\partial^2\Pi_{\mathrm{E}}}{\partial X_i \partial X_j}\Delta X_i \Delta X_j. \end{aligned} \tag{4.10}$$

From Eqs. (4.8)–(4.10), increment $\Delta\Pi$ of the total potential energy can be written as

$$\begin{aligned} \Delta\Pi &= \Delta\Pi_{\mathrm{E}} - \Delta\Pi_{\mathrm{W}} \\ &= \sum_{j=1}^{3n}\left(\frac{\partial\Pi_{\mathrm{E}}}{\partial X_j} - p_j\right)\Delta X_j + \frac{1}{2}\sum_{i=1}^{3n}\sum_{j=1}^{3n}\frac{\partial^2\Pi_{\mathrm{E}}}{\partial X_i \partial X_j}\Delta X_i \Delta X_j + \cdots, \end{aligned} \tag{4.11}$$

from which the first-order $\Delta\Pi_1$ and second-order $\Delta\Pi_2$ of $\Delta\Pi$ is further summarized as

$$\begin{aligned} \Delta\Pi_1 &= \delta\Pi_{\mathrm{E}} - \Delta\Pi_{\mathrm{W}} = \sum_{j=1}^{3n}\left(\frac{\partial\Pi_{\mathrm{E}}}{\partial X_j} - p_j\right)\Delta X_j, \\ \Delta\Pi_2 &= \delta^2\Pi_{\mathrm{E}} = \frac{1}{2}\sum_{i=1}^{3n}\sum_{j=1}^{3n}\frac{\partial^2\Pi_{\mathrm{E}}}{\partial X_i \partial X_j}\Delta X_i \Delta X_j. \end{aligned} \tag{4.12}$$

4.2 Equilibrium and Stiffness

Equilibrium of a structure is related to the first-order increment of the total potential energy, and its stability is related to the second-order increment. In this section, we present the equilibrium matrix as well as the stiffness matrices, including the tangent, linear, and geometrical stiffnesses, by investigation of increment of the total potential energy.

4.2.1 Equilibrium Equations

Equilibrium equations for a prestressed pin-jointed structure have already been presented in Chap. 2 in two different ways: those by force balance of the free-nodes as well as those by the principle of virtual work. In this subsection, we present the same equations in the third way—by using the stationary condition of the total potential energy of the structure.

From Eq. (4.6), the first partial derivative of Π_{E} with respect to X_j is given as

$$\begin{aligned}\frac{\partial \Pi_{\mathrm{E}}}{\partial X_j} &= \sum_{k=1}^{m} \frac{E_k A_k (l_k - l_k^0)}{l_k^0} \frac{\partial l_k}{\partial X_j} = \sum_{k=1}^{m} s_k \frac{\partial l_k}{\partial X_j} \\ &= \left(\frac{\partial l_1}{\partial X_j}, \frac{\partial l_2}{\partial X_j}, \ldots, \frac{\partial l_k}{\partial X_j}, \ldots, \frac{\partial l_m}{\partial X_j} \right) \begin{pmatrix} s_1 \\ s_2 \\ \vdots \\ s_k \\ \vdots \\ s_m \end{pmatrix} \\ &= \mathbf{A}\mathbf{s},\end{aligned} \tag{4.13}$$

where

$$\mathbf{s} = (s_1,\ s_2,\ \ldots,\ s_k,\ \ldots,\ s_m)^\top, \tag{4.14}$$

and

$$\mathbf{A} = \begin{pmatrix} \frac{\partial l_1}{\partial X_1} & \frac{\partial l_2}{\partial X_1} & \cdots & \frac{\partial l_k}{\partial X_1} & \cdots & \frac{\partial l_m}{\partial X_1} \\ \frac{\partial l_1}{\partial X_2} & \frac{\partial l_2}{\partial X_2} & \cdots & \frac{\partial l_k}{\partial X_2} & \cdots & \frac{\partial l_m}{\partial X_2} \\ \vdots & \vdots & \vdots & \vdots & \vdots & \vdots \\ \frac{\partial l_1}{\partial X_j} & \frac{\partial l_2}{\partial X_j} & \cdots & \frac{\partial l_k}{\partial X_j} & \cdots & \frac{\partial l_m}{\partial X_j} \\ \vdots & \vdots & \vdots & \vdots & \vdots & \vdots \\ \frac{\partial l_1}{\partial X_{3n}} & \frac{\partial l_2}{\partial X_{3n}} & \cdots & \frac{\partial l_k}{\partial X_{3n}} & \cdots & \frac{\partial l_m}{\partial X_{3n}} \end{pmatrix}. \tag{4.15}$$

Note that the jth row $\mathbf{A}_j$ of the matrix $\mathbf{A} \in \mathbb{R}^{3n \times m}$ is

$$\mathbf{A}_j = \frac{\partial \mathbf{l}^\top}{\partial X_j}, \tag{4.16}$$

where $\mathbf{l}$ is the member length vector.

Consequently, $\delta\Pi_{\mathrm{E}}$ can be simply written in a matrix form as

$$\delta\Pi_{\mathrm{E}} = \Delta\mathbf{X}^{\top}\mathbf{A}\mathbf{s}, \tag{4.17}$$

where

$$\Delta\mathbf{X} = \left(\Delta X_1, \Delta X_2, \ldots, \Delta X_j, \ldots, \Delta X_{3n}\right)^{\top}. \tag{4.18}$$

Using the definition of $\mathbf{l}$ by the nodal coordinate difference vectors $\mathbf{u}$, $\mathbf{v}$, and $\mathbf{w}$ in each direction in Eq. (2.20), its first derivative with respect to X_j is given as

$$\frac{\partial\mathbf{L}}{\partial X_j}\mathbf{l}+\mathbf{L}\frac{\partial\mathbf{l}}{\partial X_j} = \frac{\partial\mathbf{U}}{\partial X_j}\mathbf{u}+\mathbf{U}\frac{\partial\mathbf{u}}{\partial X_j}+\frac{\partial\mathbf{V}}{\partial X_j}\mathbf{v}+\mathbf{V}\frac{\partial\mathbf{v}}{\partial X_j}+\frac{\partial\mathbf{W}}{\partial X_j}\mathbf{w}+\mathbf{W}\frac{\partial\mathbf{w}}{\partial X_j}, \tag{4.19}$$

which can be rearranged as

$$\mathbf{L}\frac{\partial\mathbf{l}}{\partial X_j} = \mathbf{U}\frac{\partial\mathbf{u}}{\partial X_j}+\mathbf{V}\frac{\partial\mathbf{v}}{\partial X_j}+\mathbf{W}\frac{\partial\mathbf{w}}{\partial X_j}, \tag{4.20}$$

because $\mathbf{L}$, $\mathbf{U}$, $\mathbf{V}$, and $\mathbf{W}$ are the diagonal versions of $\mathbf{l}$, $\mathbf{u}$, $\mathbf{v}$, and $\mathbf{w}$ such that

$$\begin{aligned} &\frac{\partial\mathbf{L}}{\partial X_j}\mathbf{l} = \mathbf{L}\frac{\partial\mathbf{l}}{\partial X_j}, \quad \frac{\partial\mathbf{U}}{\partial X_j}\mathbf{u} = \mathbf{U}\frac{\partial\mathbf{u}}{\partial X_j}, \\ &\frac{\partial\mathbf{V}}{\partial X_j}\mathbf{v} = \mathbf{V}\frac{\partial\mathbf{v}}{\partial X_j}, \quad \frac{\partial\mathbf{W}}{\partial X_j}\mathbf{w} = \mathbf{W}\frac{\partial\mathbf{w}}{\partial X_j}. \end{aligned} \tag{4.21}$$

From Eq. (4.20), we have

$$\frac{\partial\mathbf{l}}{\partial X_j} = \mathbf{L}^{-1}\mathbf{U}\frac{\partial\mathbf{u}}{\partial X_j}+\mathbf{L}^{-1}\mathbf{V}\frac{\partial\mathbf{v}}{\partial X_j}+\mathbf{L}^{-1}\mathbf{W}\frac{\partial\mathbf{w}}{\partial X_j}, \tag{4.22}$$

which can be further rewritten as

$$\frac{\partial\mathbf{l}^{\top}}{\partial X_j} = \frac{\partial\mathbf{u}^{\top}}{\partial X_j}\mathbf{U}\mathbf{L}^{-1}+\frac{\partial\mathbf{v}^{\top}}{\partial X_j}\mathbf{V}\mathbf{L}^{-1}+\frac{\partial\mathbf{w}^{\top}}{\partial X_j}\mathbf{W}\mathbf{L}^{-1}. \tag{4.23}$$

From the definition of coordinate difference vectors in Eq. (2.10), we have

$$\begin{aligned} \frac{\partial\mathbf{u}^{\top}}{\partial X_j} &= \frac{\partial(\mathbf{C}\mathbf{x})^{\top}}{\partial X_j} = \frac{\partial\mathbf{x}^{\top}}{\partial X_j}\mathbf{C}^{\top}, \\ \frac{\partial\mathbf{v}^{\top}}{\partial X_j} &= \frac{\partial\mathbf{y}^{\top}}{\partial X_j}\mathbf{C}^{\top}, \\ \frac{\partial\mathbf{w}^{\top}}{\partial X_j} &= \frac{\partial\mathbf{z}^{\top}}{\partial X_j}\mathbf{C}^{\top}, \end{aligned} \tag{4.24}$$

since the connectivity matrix $\mathbf{C}$ is a constant matrix.

If X_j represents nodal coordinates x_i in x-direction; i.e., $X_j = x_i$, then Eq. (4.24) becomes

$$\frac{\partial \mathbf{u}^\top}{\partial x_i} = (\mathbf{C}^\top)_i, \quad \frac{\partial \mathbf{v}^\top}{\partial x_i} = \mathbf{0}, \quad \frac{\partial \mathbf{w}^\top}{\partial x_i} = \mathbf{0}, \tag{4.25}$$

where $(\mathbf{C}^\top)_i$ denotes the ith row of $\mathbf{C}^\top$. Similar equations can be obtained for coordinates in y- and z-directions, $X_j = y_i$ and $X_j = z_i$:

$$\begin{aligned} &\frac{\partial \mathbf{u}^\top}{\partial y_i} = \mathbf{0}, \quad \frac{\partial \mathbf{v}^\top}{\partial y_i} = (\mathbf{C}^\top)_i, \quad \frac{\partial \mathbf{w}^\top}{\partial y_i} = \mathbf{0}, \\ &\frac{\partial \mathbf{u}^\top}{\partial z_i} = \mathbf{0}, \quad \frac{\partial \mathbf{v}^\top}{\partial z_i} = \mathbf{0}, \quad \frac{\partial \mathbf{w}^\top}{\partial z_i} = (\mathbf{C}^\top)_i. \end{aligned} \tag{4.26}$$

Therefore, Eq. (4.23) is simplified as

$$\begin{aligned} \frac{\partial \mathbf{l}^\top}{\partial X_i} &= (\mathbf{C}^\top)_i \mathbf{U}\mathbf{L}^{-1}, \quad \text{if } X_j = x_i; \\ \frac{\partial \mathbf{l}^\top}{\partial X_i} &= (\mathbf{C}^\top)_i \mathbf{V}\mathbf{L}^{-1}, \quad \text{if } X_j = y_i; \\ \frac{\partial \mathbf{l}^\top}{\partial X_i} &= (\mathbf{C}^\top)_i \mathbf{W}\mathbf{L}^{-1}, \quad \text{if } X_j = z_i. \end{aligned} \tag{4.27}$$

Substituting Eq. (4.27) into Eq. (4.16) and assembling them through $j = 1, 2, \ldots, 3n$, we have

$$\mathbf{A} = \begin{pmatrix} \mathbf{C}^\top \mathbf{U}\mathbf{L}^{-1} \\ \mathbf{C}^\top \mathbf{V}\mathbf{L}^{-1} \\ \mathbf{C}^\top \mathbf{W}\mathbf{L}^{-1} \end{pmatrix}, \tag{4.28}$$

which turns out to be the same as the equilibrium matrix $\mathbf{D}$ defined in Eq. (2.34) in Chap. 2; i.e.,

$$\mathbf{A} = \mathbf{D}. \tag{4.29}$$

In the following, we will use the notation $\mathbf{D}$ for the equilibrium matrix.

The first-order term $\Delta\Pi_1$ of $\Delta\Pi$ with respect to nodal coordinate X_j in Eq. (4.11) can be summarized in a matrix form as

$$\begin{aligned} \Delta\Pi_1 &= \sum_{j=1}^{3n} \left(\frac{\partial \Pi_{\mathrm{E}}}{\partial X_j} - p_j \right) \Delta X_j \\ &= \Delta\mathbf{X}^\top \left(\frac{\partial \Pi_{\mathrm{E}}}{\partial \mathbf{X}} - \mathbf{p} \right), \end{aligned} \tag{4.30}$$

where the jth entry of $\mathbf{p} \in \mathbb{R}^{3n}$ is p_j, and $\mathbf{p}$ consists of the x-, y-, and z-components $\mathbf{p}^x, \mathbf{p}^y, \mathbf{p}^z$ $(\in \mathbb{R}^n)$ as follows:

$$\mathbf{p} = \left((\mathbf{p}^x)^\top, (\mathbf{p}^y)^\top, (\mathbf{p}^z)^\top \right)^\top. \tag{4.31}$$

When the structure is in equilibrium subjected to external loads $\mathbf{p}$, the total potential energy should be stationary, and thus, the following equation representing its first-order increment should hold for any non-zero ΔX_j or $\Delta \mathbf{X}$

$$\Delta \Pi_1 = \Delta \mathbf{X}^\top \left(\frac{\partial \Pi_\mathrm{E}}{\partial \mathbf{X}} - \mathbf{p} \right) = 0. \tag{4.32}$$

The only possibility for Eq. (4.32) to be true is that

$$\frac{\partial \Pi_\mathrm{E}}{\partial \mathbf{X}} - \mathbf{p} = \mathbf{D}\mathbf{s} - \mathbf{p} = \mathbf{0}, \tag{4.33}$$

for which the first partial derivative of Π_E with respect to $\mathbf{X}$ given in Eq. (4.13) and the relation $\mathbf{A} = \mathbf{D}$ have been used. Equation (4.33) can be rearranged as

$$\mathbf{D}\mathbf{s} = \mathbf{p}, \tag{4.34}$$

which is in fact the equilibrium equation as presented previously in Eq. (2.33).

4.2.2 Stiffness Matrices

Stability of a structure is directly related to positiveness of the second-order increment $\Delta \Pi_2$ of the total potential energy [5], because $\Delta \Pi_1 = 0$ holds according to the stationary condition. As will be discussed in the next section, positiveness of $\Delta \Pi_2$ can be verified by investigation of positive definiteness of the tangent stiffness matrix, which is the second-order derivative of Π.

From Eq. (4.12), the second-order increment $\Delta \Pi_2$ of the total potential energy due to small deformations $\Delta \mathbf{X}$ is

$$\begin{aligned} \Delta \Pi_2 &= \frac{1}{2} \sum_{i=1}^{3n} \sum_{j=1}^{3n} \frac{\partial^2 \Pi}{\partial X_i \partial X_j} \Delta X_i \Delta X_j \\ &= \frac{1}{2} \Delta \mathbf{X}^\top \mathbf{K} \Delta \mathbf{X}, \end{aligned} \tag{4.35}$$

where $\mathbf{K} \in \mathbb{R}^{3n \times 3n}$ is called Hessian of the total potential energy, or *tangent stiffness matrix* in the field of structural engineering. The (i, j)th entry $K_{(i,j)}$ of $\mathbf{K}$ is defined as

$$K_{(i,j)} = \frac{\partial^2 \Pi}{\partial X_i \partial X_j}, \tag{4.36}$$

which can be written in a matrix form as

$$\mathbf{K} = \frac{\partial}{\partial \mathbf{X}} \left(\frac{\partial \Pi}{\partial \mathbf{X}} \right)^{\top}. \tag{4.37}$$

For further formulation of the tangent stiffness matrix, it is convenient to use the force density matrix **E** defined in Chap. 2 associated with the nodal coordinate vectors **x**, **y**, and **z**, instead of using the equilibrium matrix **D**. First, we write

$$\frac{\partial \Pi_{\mathrm{E}}}{\partial \mathbf{X}} = \mathbf{D}\mathbf{s} = \begin{pmatrix} \mathbf{E}\mathbf{x} \\ \mathbf{E}\mathbf{y} \\ \mathbf{E}\mathbf{z} \end{pmatrix}, \tag{4.38}$$

which comes from Eq. (2.106) with vanishing $\mathbf{E}^{\mathrm{f}}$ corresponding to fixed nodes.

Substituting Eq. (4.38) into Eq. (4.37) gives

$$\mathbf{K} = \left(\frac{\partial(\mathbf{x}^{\top}\mathbf{E})}{\partial \mathbf{X}}, \frac{\partial(\mathbf{y}^{\top}\mathbf{E})}{\partial \mathbf{X}}, \frac{\partial(\mathbf{z}^{\top}\mathbf{E})}{\partial \mathbf{X}} \right) = \begin{pmatrix} \dfrac{\partial(\mathbf{x}^{\top}\mathbf{E})}{\partial \mathbf{x}}, & \dfrac{\partial(\mathbf{y}^{\top}\mathbf{E})}{\partial \mathbf{x}}, & \dfrac{\partial(\mathbf{z}^{\top}\mathbf{E})}{\partial \mathbf{x}} \\ \dfrac{\partial(\mathbf{x}^{\top}\mathbf{E})}{\partial \mathbf{y}}, & \dfrac{\partial(\mathbf{y}^{\top}\mathbf{E})}{\partial \mathbf{y}}, & \dfrac{\partial(\mathbf{z}^{\top}\mathbf{E})}{\partial \mathbf{y}} \\ \dfrac{\partial(\mathbf{x}^{\top}\mathbf{E})}{\partial \mathbf{z}}, & \dfrac{\partial(\mathbf{y}^{\top}\mathbf{E})}{\partial \mathbf{z}}, & \dfrac{\partial(\mathbf{z}^{\top}\mathbf{E})}{\partial \mathbf{z}} \end{pmatrix}, \tag{4.39}$$

where the symmetry properties of **E**; i.e., $\mathbf{E}^{\top} = \mathbf{E}$, has been used.

To obtain the entries in **K**, we consider $\frac{\partial(\mathbf{x}^{\top}\mathbf{E})}{\partial \mathbf{x}}$ in x-direction, for instance, as follows:

$$\frac{\partial(\mathbf{x}^{\top}\mathbf{E})}{\partial \mathbf{x}} = \sum_{i=1}^{n} \frac{\partial \mathbf{E}}{\partial x_i} x_i + \frac{\partial \mathbf{x}^{\top}}{\partial \mathbf{x}} \mathbf{E} = \sum_{i=1}^{n} \frac{\partial \mathbf{E}}{\partial x_i} x_i + \mathbf{E}. \tag{4.40}$$

Using the definition of force density in Eqs. (2.97) and (4.4) for the normal stress s_k, we have

$$q_k = \frac{s_k}{l_k} = E_k A_k \left(\frac{1}{l_k^0} - \frac{1}{l_k} \right). \tag{4.41}$$

Recall that **Q** is the diagonal version of force density vector **q**; i.e., $\mathbf{Q} = \mathrm{diag}(\mathbf{q})$. By using definition of the force density matrix $\mathbf{E} = \mathbf{C}^{\top}\mathbf{Q}\mathbf{C}$ in Eq. (2.104), we have

$$\frac{\partial \mathbf{E}}{\partial x_i} = \mathbf{C}^\top \frac{\partial \mathbf{Q}}{\partial x_i}\mathbf{C}, \tag{4.42}$$

because $\mathbf{C}$ is a constant matrix defining connectivity of the nodes by the members.

Let $\mathbf{L}_0 \in \mathbb{R}^{m\times m}$ denote the diagonal member length matrix, of which the kth diagonal entry is the member length l_k^0 at unstressed state. $E_k A_k$ of member k is the (k,k)th entry of the matrix $\hat{\mathbf{K}} \in \mathbb{R}^{m\times m}$, which is also a diagonal matrix. Hence, we have the following equation for all the force densities:

$$\mathbf{Q} = \hat{\mathbf{K}}(\mathbf{L}_0^{-1} - \mathbf{L}^{-1}). \tag{4.43}$$

The member length matrix $\mathbf{L}_0$ in the unstressed state is constant. Moreover, the matrix $\hat{\mathbf{K}}$ is also constant, since the members are assumed to be linearly elastic such that Young's modulus E_k is constant and changes of the cross-sectional areas A_k are neglected within small strains. Thus, the partial derivative of $\mathbf{Q}$ with respect to x_i can be written as follows from Eq. (4.43):

$$\frac{\partial \mathbf{Q}}{\partial x_i} = \hat{\mathbf{K}}(\mathbf{L}^{-1})^2 \frac{\partial \mathbf{L}}{\partial x_i}. \tag{4.44}$$

From Eq. (2.19), partial derivative of $\mathbf{L}$ with respect to x_i leads to

$$\frac{\partial \mathbf{L}}{\partial x_i} = \mathbf{L}^{-1}\left(\mathbf{U}\frac{\partial \mathbf{U}}{\partial x_i} + \mathbf{V}\frac{\partial \mathbf{V}}{\partial x_i} + \mathbf{W}\frac{\partial \mathbf{W}}{\partial x_i}\right). \tag{4.45}$$

From Eq. (2.10), we have

$$\frac{\partial \mathbf{V}}{\partial x_i} = \mathbf{O}_n, \quad \frac{\partial \mathbf{W}}{\partial x_i} = \mathbf{O}_n, \tag{4.46}$$

where $\mathbf{O}_n \in \mathbb{R}^{n\times n}$ is a zero matrix. Equation (4.46) holds because $\mathbf{V}$ and $\mathbf{W}$ are not functions of the x-coordinate x_i. Let $(\mathbf{C}^\top)_i$ denote the ith row of $\mathbf{C}^\top$; i.e., transpose of the ith column of $\mathbf{C}$. For the first derivative of $\mathbf{U}$, we have

$$\begin{aligned}\frac{\partial \mathbf{U}}{\partial x_i} &= \mathrm{diag}\left(\frac{\partial (\mathbf{Cx})}{\partial x_i}\right)\\ &= \mathrm{diag}\left(\mathbf{C}\frac{\partial \mathbf{x}}{\partial x_i}\right)\\ &= \mathrm{diag}\left((\mathbf{C}^\top)_i\right).\end{aligned} \tag{4.47}$$

From Eqs. (4.42)–(4.47), we obtain

$$\frac{\partial \mathbf{E}}{\partial x_i} = \mathbf{C}^\top \hat{\mathbf{K}}(\mathbf{L}^{-1})^3 \mathbf{U}\mathrm{diag}\left((\mathbf{C}^\top)_i\right)\mathbf{C}. \tag{4.48}$$

Furthermore, we have

$$\sum_{i=1}^{n} \frac{\partial \mathbf{E}}{\partial x_i} x_i = \mathbf{C}^\top \hat{\mathbf{K}} (\mathbf{L}^{-1})^3 \mathbf{U} \left[\sum_{i=1}^{n} \mathrm{diag}\left(x_i (\mathbf{C}^\top)_i \right) \right] \mathbf{C} \\ = \mathbf{C}^\top \hat{\mathbf{K}} (\mathbf{L}^{-1})^3 \mathbf{U}^2 \mathbf{C}, \tag{4.49}$$

because

$$\sum_{i=1}^{n} \mathrm{diag}\left(x_i (\mathbf{C}^\top)_i \right) = \mathrm{diag}\left(\sum_{i=1}^{n} x_i (\mathbf{C}^\top)_i \right) \\ = \mathrm{diag}\left(\mathbf{x}^\top \mathbf{C}^\top \right) = \mathrm{diag}(\mathbf{u}^\top) \\ = \mathbf{U}. \tag{4.50}$$

Substituting Eq. (4.49) into Eq. (4.40) gives

$$\frac{\partial (\mathbf{x}^\top \mathbf{E})}{\partial \mathbf{x}} = \mathbf{C}^\top \hat{\mathbf{K}} (\mathbf{L}^{-1})^3 \mathbf{U}^2 \mathbf{C} + \mathbf{E} \\ = \mathbf{D}^x \hat{\mathbf{K}} \mathbf{L}^{-1} \mathbf{D}^{x\top} + \mathbf{E} \\ = \mathbf{D}^x \bar{\mathbf{K}} \mathbf{D}^{x\top} + \mathbf{E}, \tag{4.51}$$

where the diagonal matrix $\bar{\mathbf{K}}(=\hat{\mathbf{K}}\mathbf{L}^{-1})$ is called the *member stiffness matrix*, and $\bar{\mathbf{K}}\mathbf{L}^{-1}\mathbf{U} = \mathbf{U}\mathbf{L}^{-1}\bar{\mathbf{K}}$ by multiplication properties of diagonal matrices as well as $\mathbf{D}^x = \mathbf{C}^\top \mathbf{U} \mathbf{L}^{-1}$ have been used. The diagonal entry $E_k A_k / l_k$ of $\bar{\mathbf{K}}$ is the stiffness of the corresponding members k.

Similarly, we have

$$\frac{\partial (\mathbf{y}^\top \mathbf{E})}{\partial \mathbf{x}} = \mathbf{D}^y \bar{\mathbf{K}} \mathbf{D}^{x\top}, \quad \frac{\partial (\mathbf{z}^\top \mathbf{E})}{\partial \mathbf{x}} = \mathbf{D}^z \bar{\mathbf{K}} \mathbf{D}^{x\top}, \tag{4.52}$$

respectively, where $\mathbf{D}^y = \mathbf{C}^\top \mathbf{V} \mathbf{L}^{-1}$ and $\mathbf{D}^z = \mathbf{C}^\top \mathbf{W} \mathbf{L}^{-1}$ are defined in Eq. (2.34). Moreover, the partial derivatives of the y- and z-components can be derived in a similar manner as follows:

$$\frac{\partial (\mathbf{x}^\top \mathbf{E})}{\partial \mathbf{y}} = \mathbf{D}^x \bar{\mathbf{K}} \mathbf{D}^{y\top}, \quad \frac{\partial (\mathbf{x}^\top \mathbf{E})}{\partial \mathbf{z}} = \mathbf{D}^x \bar{\mathbf{K}} \mathbf{D}^{z\top}, \\ \frac{\partial (\mathbf{y}^\top \mathbf{E})}{\partial \mathbf{y}} = \mathbf{D}^y \bar{\mathbf{K}} \mathbf{D}^{y\top} + \mathbf{E}, \quad \frac{\partial (\mathbf{y}^\top \mathbf{E})}{\partial \mathbf{z}} = \mathbf{D}^y \bar{\mathbf{K}} \mathbf{D}^{z\top}, \tag{4.53} \\ \frac{\partial (\mathbf{z}^\top \mathbf{E})}{\partial \mathbf{y}} = \mathbf{D}^z \bar{\mathbf{K}} \mathbf{D}^{y\top}, \quad \frac{\partial (\mathbf{z}^\top \mathbf{E})}{\partial \mathbf{z}} = \mathbf{D}^z \bar{\mathbf{K}} \mathbf{D}^{z\top} + \mathbf{E}.$$

Therefore, the tangent stiffness matrix $\mathbf{K}$ is given as

$$
\begin{aligned}
\mathbf{K} &= \begin{pmatrix} \mathbf{D}^x\bar{\mathbf{K}}\mathbf{D}^{x\top} + \mathbf{E} & \mathbf{D}^x\bar{\mathbf{K}}\mathbf{D}^{y\top} & \mathbf{D}^x\bar{\mathbf{K}}\mathbf{D}^{z\top} \\ \mathbf{D}^y\bar{\mathbf{K}}\mathbf{D}^{x\top} & \mathbf{D}^y\bar{\mathbf{K}}\mathbf{D}^{y\top} + \mathbf{E} & \mathbf{D}^y\bar{\mathbf{K}}\mathbf{D}^{z\top} \\ \mathbf{D}^z\bar{\mathbf{K}}\mathbf{D}^{x\top} & \mathbf{D}^z\bar{\mathbf{K}}\mathbf{D}^{y\top} & \mathbf{D}^z\bar{\mathbf{K}}\mathbf{D}^{z\top} + \mathbf{E} \end{pmatrix} \\
&= \begin{pmatrix} \mathbf{D}^x \\ \mathbf{D}^y \\ \mathbf{D}^z \end{pmatrix} \bar{\mathbf{K}} \left(\mathbf{D}^{x\top} \ \mathbf{D}^{y\top} \ \mathbf{D}^{z\top} \right) + \begin{pmatrix} \mathbf{E} & \mathbf{O}_n & \mathbf{O}_n \\ \mathbf{O}_n & \mathbf{E} & \mathbf{O}_n \\ \mathbf{O}_n & \mathbf{O}_n & \mathbf{E} \end{pmatrix} \\
&= \mathbf{D}\bar{\mathbf{K}}\mathbf{D}^\top + \mathbf{I}_3 \otimes \mathbf{E},
\end{aligned} \tag{4.54}
$$

where $\mathbf{I}_3 \in \mathbb{R}^{3\times 3}$ is an identity matrix. Moreover, the notation $\otimes$ in Eq. (4.54) denotes tensor product, which will be given in detail in Appendix A. The first term on the right-hand side of Eq. (4.54) is the *linear stiffness matrix* $\mathbf{K}_\mathrm{E}$ dependent on geometry realization of the structure as well as member stiffness; and the second term is the *geometrical stiffness matrix* $\mathbf{K}_\mathrm{G}$ related to prestresses in the members through the force density matrix. Therefore, we have the formulations for the stiffness matrices as follows:

Linear, geometrical, and tangent stiffness matrices of a prestressed pin-jointed structure:

$$
\begin{aligned}
\mathbf{K} &= \mathbf{K}_\mathrm{E} + \mathbf{K}_\mathrm{G}, \\
\mathbf{K}_\mathrm{E} &= \mathbf{D}\bar{\mathbf{K}}\mathbf{D}^\top, \\
\mathbf{K}_\mathrm{G} &= \mathbf{I}_3 \otimes \mathbf{E}.
\end{aligned} \tag{4.55}
$$

It is obvious that the stiffness matrices $\mathbf{K}_\mathrm{E}$, $\mathbf{K}_\mathrm{G}$, and $\mathbf{K}$ are all symmetric, because $\bar{\mathbf{K}}$ and $\mathbf{E}$ are symmetric. We will clarify later in the this chapter that these stiffness matrices play key roles in stability investigation of a prestressed pin-jointed structure.

In this formulation, the stressed equilibrium state is considered as the reference state, which is generally applicable to any type of pin-jointed structures in the field of elastic systems with small strains. If the structure has no prestress, the geometrical stiffness matrix vanishes, and we have $l_k = l_k^0$ because no member is deformed.

Example 4.1 Stiffness of the two-dimensional free-standing structure as shown in Fig. 4.2.

The structure as shown in Fig. 4.2 consists of five nodes and eight members; i.e., $n = 5$ and $m = 8$. Suppose for simplicity that Young's modulus E_k and cross-sectional areas A_k of all members are the same; i.e., $E_k = E$ and $A_k = A$ $(k = 1, 2, \ldots, 8)$, and moreover, we follow the geometry realization and prestress mode given in Example 2.16.

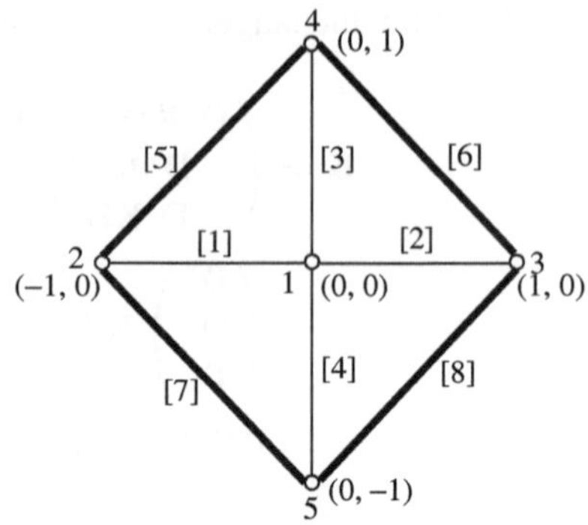

Fig. 4.2 A two-dimensional free-standing structure first studied in Chap. 2. The structure consists of five free nodes 1–5 and eight members [1]–[8]. The structure is statically indeterminate with one prestress mode, and it is kinematically determinate without any mechanism

According to Eq. (4.55), the linear stiffness matrix $\mathbf{K}_{\mathrm{E}} \in \mathbb{R}^{10\times 10}$ of the structure is

$$\mathbf{K}_{\mathrm{E}} = \frac{AE}{2}\begin{pmatrix} 4 & -2 & -2 & 0 & 0 & 0 & 0 & 0 & 0 & 0 \\ -2 & 4 & 0 & -1 & -1 & 0 & 0 & 0 & -1 & 1 \\ -2 & 0 & 4 & -1 & -1 & 0 & 0 & 0 & 1 & -1 \\ 0 & -1 & -1 & 2 & 0 & 0 & -1 & 1 & 0 & 0 \\ 0 & -1 & -1 & 0 & 2 & 0 & 1 & -1 & 0 & 0 \\ 0 & 0 & 0 & 0 & 0 & 4 & 0 & 0 & -2 & -2 \\ 0 & 0 & 0 & -1 & 1 & 0 & 2 & 0 & -1 & -1 \\ 0 & 0 & 0 & 1 & -1 & 0 & 0 & 2 & -1 & -1 \\ 0 & -1 & 1 & 0 & 0 & -2 & -1 & -1 & 4 & 0 \\ 0 & 1 & -1 & 0 & 0 & -2 & -1 & -1 & 0 & 4 \end{pmatrix}, \tag{4.56}$$

and the geometrical stiffness matrix $\mathbf{K}_{\mathrm{G}} \in \mathbb{R}^{10\times 10}$ is

$$\mathbf{K}_{\mathrm{G}} = \frac{t}{\sqrt{2}}\begin{pmatrix} \begin{matrix} 8 & -2 & -2 & -2 & -2 \\ -2 & 0 & 0 & 1 & 1 \\ -2 & 0 & 0 & 1 & 1 \\ -2 & 1 & 1 & 0 & 0 \\ -2 & 1 & 1 & 0 & 0 \end{matrix} & \mathbf{O}_5 \\ \mathbf{O}_5 & \begin{matrix} 8 & -2 & -2 & -2 & -2 \\ -2 & 0 & 0 & 1 & 1 \\ -2 & 0 & 0 & 1 & 1 \\ -2 & 1 & 1 & 0 & 0 \\ -2 & 1 & 1 & 0 & 0 \end{matrix} \end{pmatrix}, \tag{4.57}$$

where t is an arbitrary value, and $\mathbf{O}_5 \in \mathbb{R}^{5\times 5}$ is a zero matrix.

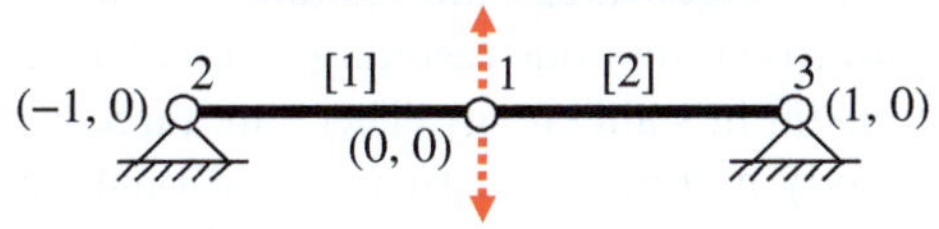

Fig. 4.3 A statically and kinematically indeterminate two-dimensional structure with fixed nodes. Its stability depends on the prestresses: it is stable when its members carry tension; and it is unstable when its members carry compression

In $\mathbf{K}_E$, there exist three zero eigenvalues, corresponding to the three rigid-body motions in two-dimensional space. By contrast, there exist six zero eigenvalues in $\mathbf{K}_G$—three of them correspond to rigid-body motions, and the other three correspond to the non-trivial affine motions as discussed in Appendix B.1.

More detailed investigations on its stability by studying the stiffness matrices will be given in Example 4.3.

Example 4.2 Stiffness of the two-dimensional structure with fixed nodes as shown in Fig. 4.3.

The two-dimensional structure as shown in Fig. 4.3 consists of one free node, tow fixed nodes, and two members; i.e., $n = 1$, $n^{\mathrm{f}} = 2$, and $m = 2$.

Young's modulus and cross-sectional areas of the two members are denoted by E and A, respectively, for simplicity. Using geometry realization and prestress mode given in Example 2.14, the linear stiffness matrix $\mathbf{K}_E \in \mathbb{R}^{2\times 2}$ of the structure is

$$\mathbf{K}_E = \mathbf{D}\bar{\mathbf{K}}\mathbf{D}^\top = AE\begin{pmatrix} 2 & 0 \\ 0 & 0 \end{pmatrix}, \tag{4.58}$$

and the geometrical stiffness matrix $\mathbf{K}_G \in \mathbb{R}^{2\times 2}$ is

$$\mathbf{K}_G = \begin{pmatrix} \mathbf{E} & \mathbf{O} \\ \mathbf{O} & \mathbf{E} \end{pmatrix} = \sqrt{2}t\begin{pmatrix} 1 & 0 \\ 0 & 1 \end{pmatrix}, \tag{4.59}$$

where t is an arbitrary value, and $\mathbf{O} \in \mathbb{R}^{1\times 1}$ is a zero matrix.

4.3 Stability Criteria

In this section, three criteria used for stability investigation of tensegrity structures are presented: stability, prestress-stability, and super-stability. Note that in the following discussions, we assume that the rigid-body motions are appropriately constrained, or the corresponding zero eigenvalues are excluded from stability investigation. Rigid-body motions are trivial motions resulting in no change of member lengths, and therefore, no increment in stain energy. There are three zero eigenvalues corresponding to the rigid-body motions for a two-dimensional free-standing structure, and there are six for a three-dimensional structure. More details can be found in Appendix B.1.

4.3.1 Stability

As discussed previously, a structure is stable if it returns to its original (equilibrium) configuration after release of small enforced deformations (disturbance). In the viewpoint of energy, a stable structure has the locally minimum total potential energy at the equilibrium state, such that any enforced deformations applied to the structure would lead to increase of the total potential energy. This is called the *principle of minimum total potential energy*, and it is a fundamental concept used not only in engineering, but also in many other different disciplines, such as physics, chemistry, and biology.

Definition 3.1: Stability (minimum total potential energy)
A structure is stable, if its total potential energy is at locally strict minimum.

Using the definition of stability of a structure, the following lemma shows that the second-order increment $\Delta\Pi_2$ of the total potential energy of a stable structure should be positive, when its higher-order terms are neglected.

Lemma 4.1 *If a structure is stable,*[1] *then its second-order increment of the total potential energy must be positive, while subjected to any small disturbance to its original equilibrium configuration.*

Proof Increment $\Delta\Pi$ of the total potential energy subjected to infinitesimal disturbances was presented in Eq. (4.11). Because the first-order increment $\Delta\Pi_1$ of the total potential energy defined in Eq. (4.12) is zero when the structure is in the state of equilibrium, $\Delta\Pi$ (approximately) reduces to

$$\Delta\Pi \approx \Delta\Pi_2, \tag{4.60}$$

when higher-order terms are omitted.

---

[1] In this book, stability is investigated up to the second-order term of increment of the total potential energy. Some 'unstable' structures in this book may be actually stable if higher-order terms are included, see, for example, Example 4.4.

Therefore, to guarantee positive increase of the total potential energy; i.e., $\Delta\Pi > 0$, its second-order increment $\Delta\Pi_2$ corresponding to any small displacements has to be positive.

When a structure is stable, from its definition, its total potential energy is at locally strict minimum; or equivalently, any (small) change of configuration results in increase of the total potential energy. Consequently, *a structure is stable if and only if its second-order increment of the total potential energy is positive.* □

For the purpose of stability investigation of a structure, the following lemma is more convenient to use, which is derived from Lemma 4.1.

Lemma 4.2 *A structure is stable, if any of the following equivalent conditions is satisfied after constraining the rigid-body motions:*

1. *The quadratic form Q_{K} of the tangent stiffness matrix* **K** *with respect to any small non-trivial displacements* $\mathbf{d}(\neq \mathbf{0})$ *is positive:*

$$Q_{\mathrm{K}} = \mathbf{d}^{\top}\mathbf{K}\mathbf{d} > 0; \tag{4.61}$$

2. *The tangent stiffness matrix* **K** *is positive definite;*
3. *All eigenvalues of the tangent stiffness matrix* **K** *are positive.*

Proof Let **d** denote small non-trivial displacements as

$$\mathbf{d} = \Delta\mathbf{X}, \tag{4.62}$$

where $\Delta\mathbf{X}$ is the generalized nodal displacement vector.

Using Eq. (4.35) for the second-order increment $\Delta\Pi_2$ of the total potential energy, the following relation holds for a stable structure as has been proved in Lemma 3.1

$$\begin{aligned} Q_{\mathrm{K}} &= \mathbf{d}^{\top}\mathbf{K}\mathbf{d} = \Delta\Pi_2 \\ &> 0, \end{aligned} \tag{4.63}$$

where **K** is the tangent stiffness matrix as given in Eq. (4.54). Thus, the first argument has been proved.

From the definition of positive definiteness of a matrix [3], Eq. (4.63) implies that the tangent stiffness matrix is positive definite. Consequently, the second argument of the lemma has also been proved.

Let λ_i^{K} and $\boldsymbol{\phi}_i^{\mathrm{K}}$ respectively denote the ith eigenvalue and eigenvector of the tangent stiffness matrix **K**; i.e.,

$$\begin{aligned} \mathbf{K}\boldsymbol{\phi}_i^{\mathrm{K}} &= \lambda_i^{\mathrm{K}}\boldsymbol{\phi}_i^{\mathrm{K}}, \\ (\boldsymbol{\phi}_i^{\mathrm{K}})^{\top}\boldsymbol{\phi}_j^{\mathrm{K}} &= \begin{cases} 1 & \text{for } i = j, \\ 0 & \text{for } i \neq j. \end{cases} \end{aligned} \tag{4.64}$$

Because $\boldsymbol{\phi}_i^{\mathrm{K}}$ span the whole space of non-trivial displacements of the structure, the small displacements $\mathbf{d}$ can be written as linear combination of $\boldsymbol{\phi}_i^{\mathrm{K}}$ as follows, with the arbitrary coefficients α_i:

$$\mathbf{d} = \sum_{i=1}^{n^{\mathrm{DOF}}} \alpha_i \boldsymbol{\phi}_i^{\mathrm{K}}, \tag{4.65}$$

where n^{DOF} refers to the degree of freedom of the free-standing d-dimensional structure excluding its rigid-body motions; thus, it is defined as

$$n^{\mathrm{DOF}} = dn - \frac{d^2 + d}{2}. \tag{4.66}$$

Substituting $\mathbf{d}$ in Eq. (4.65) into Eq. (4.63) and using Eq. (4.64), we obtain

$$\begin{aligned} Q_{\mathrm{K}} = \mathbf{d}^{\top}\mathbf{K}\mathbf{d} &= \left(\sum_{i=1}^{n^{\mathrm{DOF}}} \alpha_i (\boldsymbol{\phi}_i^{\mathrm{K}})^{\top}\right) \mathbf{K} \left(\sum_{j=1}^{n^{\mathrm{DOF}}} \alpha_j \boldsymbol{\phi}_j^{\mathrm{K}}\right) \\ &= \sum_{i=1}^{n^{\mathrm{DOF}}} \sum_{j=1}^{n^{\mathrm{DOF}}} \alpha_i \alpha_j (\boldsymbol{\phi}_i^{\mathrm{K}})^{\top} \mathbf{K} \boldsymbol{\phi}_j^{\mathrm{K}} = \sum_{i=1}^{n^{\mathrm{DOF}}} \sum_{j=1}^{n^{\mathrm{DOF}}} \alpha_i \alpha_j \lambda_j^{\mathrm{K}} (\boldsymbol{\phi}_i^{\mathrm{K}})^{\top} \boldsymbol{\phi}_j^{\mathrm{K}} \\ &= \sum_{i=1}^{n^{\mathrm{DOF}}} \alpha_i^2 \lambda_i^{\mathrm{K}}. \end{aligned} \tag{4.67}$$

The coefficients α_i are arbitrary, and not all of them are zero because the displacements are assumed to be non-trivial. Moreover, the inequality $\alpha_i^2 \geq 0$ is always true. Therefore, to guarantee that Q_{K} is positive, all λ_i^{K} have to be positive, which completes the proof for the third argument. □

A structure exhibiting finite mechanisms is obviously unstable, because its deformation in the direction of the finite mechanisms will not lead to any change of member forces as well as member lengths, and therefore, will not change stain energy of the structure. Stability of the structure exhibiting infinitesimal mechanisms strongly depends on distribution of the prestresses, which will be demonstrated later in Example 4.4.

Using Lemma 4.2, stability of a structure can be easily confirmed by checking signs of the eigenvalues of its tangent stiffness matrix.

Example 4.3 Stability investigation of the two-dimensional kinematically determinate free-standing structure as shown in Fig. 4.2.

In Example 2.11, we showed that this structure is statically indeterminate with single prestress mode as presented in Eq. (2.73). Moreover, it is kinematically determinate, such that there exists no (infinitesimal) mechanism.

Using its linear and geometrical stiffness matrices $\mathbf{K}_\mathrm{E}$ and $\mathbf{K}_\mathrm{G}$ derived in Example 4.1, the ten eigenvalues λ_i^E of $\mathbf{K}_\mathrm{E}$ are

$$\{\lambda_i^\mathrm{E}\} = AE\{0.0,\ 0.0,\ 0.0,\ 1.0,\ 1.3820,\ 1.3820,\ 2.0,\ 3.0,\ 3.6180,\ 3.6180\}, \tag{4.68}$$

where A is the cross-sectional area of the members, and E is Young's modulus of the material; and the ten eigenvalues λ_i^G of $\mathbf{K}_\mathrm{G}$ are

$$\{\lambda_i^\mathrm{G}\} = t\{-1.4142,\ -1.4142,\ 0.0,\ 0.0,\ 0.0,\ 0.0,\ 0.0,\ 0.0,\ 7.0711,\ 7.0711\}, \tag{4.69}$$

where $t(\neq 0)$ is an arbitrary value, denoting scale of prestresses introduced to the structure.

Note that there are three zero eigenvalues in $\mathbf{K}_\mathrm{E}$ and six zero eigenvalues in $\mathbf{K}_\mathrm{G}$. All of them have explicit physical meanings:

- Three of the zero eigenvalues in $\mathbf{K}_\mathrm{E}$ and $\mathbf{K}_\mathrm{G}$ correspond to the three rigid-body motions in two-dimensional space;
- The remaining three zero eigenvalues in $\mathbf{K}_\mathrm{G}$ correspond to the non-trivial affine motions, which are discussed in detail in Appendix B.1.

In practical structures, the magnitude of Young's modulus E is more than 1,000 times of the magnitude $|t|$ of prestresses, for instance for steel[2] to avoid material yielding. Hence, eigenvalues of the tangent stiffness matrix $\mathbf{K}$ are usually dominated by those of $\mathbf{K}_\mathrm{E}$.

Suppose that t is 1 % of AE, then eigenvalues λ_i^K of the tangent stiffness matrix $\mathbf{K}$ are

$$\{\lambda_i^\mathrm{K}\} = AE\{0.0,\ 0.0,\ 0.0,\ 1.0,\ 1.4094,\ 1.4094,\ 2.0,\ 3.0,\ 3.6471,\ 3.6471\}, \tag{4.70}$$

which are almost the same as those λ_i^E of $\mathbf{K}_\mathrm{E}$, since the level of prestresses is very low, and therefore, $\mathbf{K}_\mathrm{G}$ has only little influence on the tangent stiffness matrix in this case.

Since the three zero eigenvalues of $\mathbf{K}$ correspond to rigid-body motions, and the remaining eigenvalues of $\mathbf{K}$ are positive, the structure in Fig. 4.2 is stable from Lemma 4.2 if the rigid-body motions are appropriately constrained.

From Example 4.3, we can see that kinematically determinate structures are usually stable. This is because the linear stiffness matrix is positive definite when the rigid-body motions are constrained, and furthermore, it is dominant in the tangent stiffness matrix for practical structures, for which the level of prestresses is small enough so as to avoid material failure.

---

[2] For steel material, for instance, its Young's modulus is appropriately 205 GPa; i.e., $E = 205 \times 10^9\ \mathrm{N/m^2}$.

It may also be immediately noticed that kinematically indeterminate (pin-jointed) structures, are 'unstable' if they carry no prestress. This is because the tangent stiffness matrix is equal to the linear stiffness matrix, when prestress is absent; and moreover, the tangent stiffness matrix is positive semi-definite, with zero eigenvalues corresponding to (infinitesimal) mechanisms.

Example 4.4 Stability investigation of the statically and kinematically indeterminate pin-jointed structure as shown in Fig. 4.3.

The static and kinematic determinacy of the structure in Fig. 4.3 was studied in Example 2.14 in Chap. 2. Using the same settings as in Example 2.14, the tangent stiffness matrix $\mathbf{K}$ of the structure is

$$\begin{aligned} \mathbf{K} &= \mathbf{K}_{\mathrm{E}} + \mathbf{K}_{\mathrm{G}} \\ &= \begin{pmatrix} 2AE + t\sqrt{2} & 0 \\ 0 & t\sqrt{2} \end{pmatrix}, \end{aligned} \tag{4.71}$$

where A is the cross-sectional area, E is Young's modulus, and t is an arbitrary value. It is obvious that the two eigenvalues of $\mathbf{K}$ are

$$\begin{aligned} \lambda_1^{\mathrm{K}} &= 2AE + t\sqrt{2}, \\ \lambda_2^{\mathrm{K}} &= t\sqrt{2}. \end{aligned} \tag{4.72}$$

The parameter AE representing material properties must be positive, and thus, $\lambda_1^{\mathrm{K}} > 0$ is true for $AE > -\sqrt{2}t/2$, which is always the case for practical structures as discussed in Example 4.3. Therefore, stability of the structure depends strongly on the sign of the parameter t:

- When $t > 0$; i.e., tensile forces are introduced into the two members, the structure is stable because both of the two eigenvalues λ_1^{K} and λ_2^{K} of the tangent stiffness matrix are positive;
- When $t = 0$; i.e., prestresses are absent, the structure is 'unstable',[3] because there exists one zero eigenvalue in the tangent stiffness matrix;
- When $t < 0$; i.e., compressive forces are introduced into the two members, the structure is unstable, because the second eigenvalue λ_2^{K} is negative.

As can be observed from Example 4.4, prestresses may be important, and sometimes vital, to stability of a structure. This is true especially for kinematically indeterminate structures.

[3] The structure is indeed stable, if higher-order terms of increment of the total potential energy are taken into consideration. However, we investigate stability of a structure by considering only up to the second-order term in this book, which is sufficient for most cases.

4.3.2 Prestress-stability

Stability of a kinematically indeterminate structure is usually investigated by checking positive definiteness of its tangent stiffness matrix, in which the cross-sectional areas and material properties in the linear stiffness matrix are necessary. This might be cumbersome, especially in preliminary studies. Rather than considering any specific material, we might be interested only in whether a structure with a specific self-equilibrated configuration is stable or not. For such cases, another stability criterion, called *prestress-stability*, introduced in this subsection is more convenient to use.

With the assumption that the member stiffness of the structure is large enough, stability investigation of the structure reduces to verifying its stability only in the directions of infinitesimal mechanisms. The definition of prestress-stability is given as follows:

Definition 3.2 Prestress-stability:
If a prestressed pin-jointed structure is stable in the state of self-equilibrium in the directions of infinitesimal mechanisms, then it is said to be prestress-stable.

Because the tangent stiffness matrix $\mathbf{K}$ can be written as sum of the linear $\mathbf{K}_\mathrm{E}$ and geometrical $\mathbf{K}_\mathrm{G}$ stiffness matrices as in Eq. (4.55), its quadratic form Q_K with respect to a small displacement vector $\mathbf{d}$ can also be written as sum of the quadratic forms Q_E and Q_G of $\mathbf{K}_\mathrm{E}$ and $\mathbf{K}_\mathrm{G}$, respectively:

$$Q_\mathrm{K} = Q_\mathrm{E} + Q_\mathrm{G}, \tag{4.73}$$

where

$$\begin{aligned} Q_\mathrm{E} &= \mathbf{d}^\top \mathbf{K}_\mathrm{E} \mathbf{d}, \\ Q_\mathrm{G} &= \mathbf{d}^\top \mathbf{K}_\mathrm{G} \mathbf{d}. \end{aligned} \tag{4.74}$$

If the structure is kinematically indeterminate, then there exists a mechanism $\mathbf{d}_\mathrm{m} \in \mathbb{R}^{3n}$ that results in $\mathbf{K}_\mathrm{E}\mathbf{d}_\mathrm{m} = \mathbf{0}$, and therefore, $Q_\mathrm{E} = 0$. Hence, it turns out that the quadratic form Q_K of the tangent stiffness matrix $\mathbf{K}$ with respect to a mechanism $\mathbf{d}_\mathrm{m}$ reduces to Q_m which is equal to the quadratic form Q_G of the geometrical stiffness matrix $\mathbf{K}_\mathrm{G}$:

$$Q_\mathrm{K} = 0 + Q_\mathrm{G} = Q_\mathrm{m}, \tag{4.75}$$

where

$$Q_\mathrm{m} = \mathbf{d}_\mathrm{m}^\top \mathbf{K}_\mathrm{G} \mathbf{d}_\mathrm{m}. \tag{4.76}$$

From the definition of stability, we know that the structure is possible to be stable if $Q_\mathrm{K} = Q_\mathrm{m} > 0$. However, it should be noted that $Q_\mathrm{m} > 0$ is only a necessary, but not a sufficient, condition for a stable structure. In Example 4.6 we will see an example that high level of prestresses might result in an unstable structure, even though it is prestress-stable.

Stability of a kinematically indeterminate structure with zero Q_{E} in the directions of mechanisms depends on the sign of Q_{m}, which can be explained intuitively in Fig. 4.4. There are three cases concerning the sign of Q_{m}:

- If $Q_{\mathrm{m}} > 0$ for all mechanisms as in Fig. 4.4a, then the structure is prestress-stable because $Q_{\mathrm{K}} > 0$ holds;
- If $Q_{\mathrm{m}} = 0$ as in Fig. 4.4b, then the structure is 'unstable', when only up to the second-order increment of the total potential energy is considered, because $Q_{\mathrm{K}} = 0$ holds;
- If $Q_{\mathrm{m}} < 0$ as in Fig. 4.4c, then the structure is unstable because $Q_{\mathrm{K}} < 0$ holds.

Denote the mechanisms, which lie in the null-space of the transpose of the equilibrium matrix; i.e., compatibility matrix $\mathbf{D}^{\top}$, or equivalently in the null-space of the linear stiffness matrix, as the columns of the *mechanism matrix* $\mathbf{M} \in \mathbb{R}^{3n \times n^{\mathrm{m}}}$. Remind that n^{m} is the degree of kinematical indeterminacy. From Eqs. (2.82) and (4.54), the quadratic form $\mathbf{Q}_{\mathrm{K}}$ of the tangent stiffness matrix $\mathbf{K}$ with respect to the mechanisms $\mathbf{M}$ turns out to be equal to $\mathbf{Q}_{\mathrm{m}} \in \mathbb{R}^{n^{\mathrm{m}} \times n^{\mathrm{m}}}$ of the geometrical stiffness matrix $\mathbf{K}_{\mathrm{G}}$:

$$\mathbf{Q}_{\mathrm{K}} = \mathbf{Q}_{\mathrm{m}}, \tag{4.77}$$

where

$$\mathbf{Q}_{\mathrm{K}} = \mathbf{M}^{\top}\mathbf{K}\mathbf{M}, \quad \mathbf{Q}_{\mathrm{m}} = \mathbf{M}^{\top}\mathbf{K}_{\mathrm{G}}\mathbf{M}. \tag{4.78}$$

Lemma 4.3 *If a kinematically indeterminate prestressed pin-jointed structure is prestress-stable, then the quadratic form* $\mathbf{Q}_{\mathrm{m}}$ *defined in Eq.* (4.78) *of the geometrical stiffness matrix with respect to the mechanisms is positive definite.*

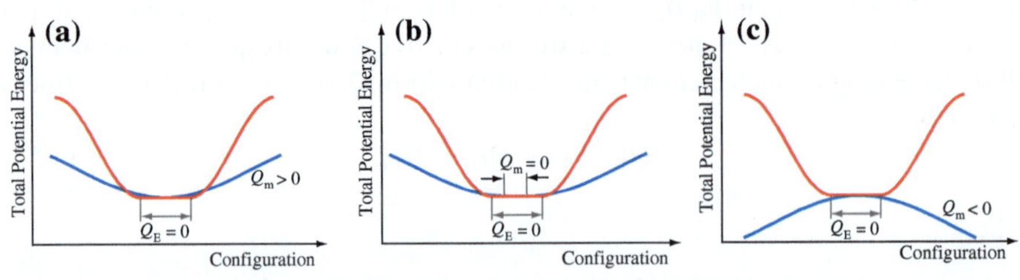

Fig. 4.4 Total potential energy and stability of a kinematically indeterminate structure. The linear stiffness of the kinematically indeterminate structures is positive semi-definite with zero eigenvalues; and positive definiteness of the tangent stiffness matrix depends on the contribution by the geometrical stiffness matrix. **a** Stable ($Q_{\mathrm{m}} > 0$), **b** unstable ($Q_{\mathrm{m}} = 0$), **c** unstable ($Q_{\mathrm{m}} < 0$)

Proof The infinitesimal mechanisms $\mathbf{d}_\mathrm{m}$, lying in the null-space of $\mathbf{D}^\top$ or linear stiffness matrix $\mathbf{K}_\mathrm{E}$, can be written in a matrix form by using the mechanism matrix $\mathbf{M}$ associated with the non-trivial arbitrary coefficient vector $\boldsymbol{\alpha} \in \mathbb{R}^{n^\mathrm{m}}$:

$$\mathbf{d}_\mathrm{m} = \mathbf{M}\boldsymbol{\alpha}. \tag{4.79}$$

When a structure is stable, the quadratic form Q_K defined in Eq. (4.73) with respect to arbitrary (non-trivial) displacements must be positive. These displacements certainly include the infinitesimal mechanisms $\mathbf{d}_\mathrm{m}$, for which we have the following relation from Eq. (4.75):

$$\begin{aligned} Q_\mathrm{K} = Q_\mathrm{m} &= \boldsymbol{\alpha}^\top \mathbf{M}^\top \mathbf{K}_\mathrm{G} \mathbf{M}\boldsymbol{\alpha} = \mathbf{d}_\mathrm{m}^\top \mathbf{K}_\mathrm{G} \mathbf{d}_\mathrm{m} \\ &= \boldsymbol{\alpha}^\top \mathbf{Q}_\mathrm{m} \boldsymbol{\alpha}. \end{aligned} \tag{4.80}$$

Because the coefficient vector $\boldsymbol{\alpha}$ is non-trivial and arbitrary, $\mathbf{Q}_\mathrm{m}$ needs to be positive definite to ensure that Q_m is positive, which completes the proof of the lemma. □

Lemma 4.3 is a necessary condition, but *not* a sufficient condition, for stability of kinematically indeterminate structures. In comparison, we have the following sufficient condition for prestress-stability of a structure.

Lemma 4.4 *If the quadratic form* $\mathbf{Q}_\mathrm{m}$ *defined in Eq. (4.77) of the geometrical stiffness matrix with respect to the mechanisms is positive definite, then the prestressed pin-jointed structure is prestress-stable; moreover, its stability is guaranteed if the axial stiffness of members is large enough compared to the geometrical stiffness due to prestresses.*

Proof Let λ_i^E and $\boldsymbol{\phi}_i^\mathrm{E}$ respectively denote the ith eigenvalue and its corresponding eigenvector of the linear stiffness matrix $\mathbf{K}_\mathrm{E}$; i.e.,

$$\begin{aligned} \mathbf{K}_\mathrm{E}\boldsymbol{\phi}_i^\mathrm{E} &= \lambda_i^\mathrm{E}\boldsymbol{\phi}_i^\mathrm{E}, \\ (\boldsymbol{\phi}_i^\mathrm{E})^\top \boldsymbol{\phi}_j^\mathrm{E} &= \begin{cases} 1 & \text{for } i = j, \\ 0 & \text{for } i \neq j. \end{cases} \end{aligned} \tag{4.81}$$

Let a denote an arbitrary positive value. Suppose that $\mathbf{K} = a\mathbf{K}_\mathrm{E} + \mathbf{K}_\mathrm{G}$, so that the value of a represents the level of the axial stiffness of the members by considering $\mathbf{K}_\mathrm{E} = \mathbf{D}(a\bar{\mathbf{K}})\mathbf{D}^\top$. The eigenvalues of $a\mathbf{K}_\mathrm{E}$ are $a\lambda_i^\mathrm{E}$ with the eigenvectors $\boldsymbol{\phi}_i^\mathrm{E}$ unchanged, because

$$(a\mathbf{K}_\mathrm{E})\boldsymbol{\phi}_i^\mathrm{E} = (a\lambda_i^\mathrm{E})\boldsymbol{\phi}_i^\mathrm{E}. \tag{4.82}$$

Define displacements $\mathbf{d}$ as a linear combination of the eigenvectors $\boldsymbol{\phi}_i^{\mathrm{E}}$, with the arbitrary coefficients α_i:

$$\mathbf{d} = \sum_i \alpha_i \boldsymbol{\phi}_i^{\mathrm{E}}. \tag{4.83}$$

Accordingly, quadratic form Q_{K} of the tangent stiffness matrix $\mathbf{K}$ with respect to $\mathbf{d}$ becomes

$$\begin{aligned} Q_{\mathrm{K}} &= Q_{\mathrm{E}} + Q_{\mathrm{G}} = \mathbf{d}^{\top}\mathbf{K}\mathbf{d} = \mathbf{d}^{\top}\mathbf{K}_{\mathrm{E}}\mathbf{d} + \mathbf{d}^{\top}\mathbf{K}_{\mathrm{G}}\mathbf{d} \\ &= a\sum_i \alpha_i^2 \lambda_i^{\mathrm{E}} + \mathbf{d}\mathbf{K}_{\mathrm{G}}\mathbf{d}. \end{aligned} \tag{4.84}$$

To investigate prestress-stability of a structure, the axial stiffness of the members is set to infinity; i.e., the parameter a tends to infinity as $a \to +\infty$, or a is sufficiently large, while the level of prestresses is relatively low. Concerning the components of the displacements $\mathbf{d}$, we have the following two cases:

- If the displacements $\mathbf{d}$ contain components of any eigenvectors $\boldsymbol{\phi}_i$ corresponding to non-zero (positive) eigenvalues of $\mathbf{K}_{\mathrm{E}}$, then

$$Q_{\mathrm{K}} \approx Q_{\mathrm{E}} = a\sum_i \alpha_i^2 \lambda_i^{\mathrm{E}} \to +\infty, \tag{4.85}$$

  since Q_{G} is small compared to Q_{E}.
- If the displacements $\mathbf{d}(= \mathbf{d}_{\mathrm{m}})$ contain only the components of eigenvectors corresponding to zero eigenvalues; i.e., they are a linear combination of the infinitesimal mechanisms, then we have the following equation using Eqs. (4.79) and (4.80):

$$\begin{aligned} Q_{\mathrm{K}} &= Q_{\mathrm{m}} = Q_{\mathrm{G}} = \mathbf{d}_{\mathrm{m}}^{\top}\mathbf{K}_{\mathrm{G}}\mathbf{d}_{\mathrm{m}} \\ &= \boldsymbol{\alpha}^{\top}\mathbf{Q}_{\mathrm{m}}\boldsymbol{\alpha}. \end{aligned} \tag{4.86}$$

  When $\mathbf{Q}_{\mathrm{m}}$ is positive definite, the structure is stable from Lemma 4.3; if not, it is unstable. Therefore, stability of the structure depends on characteristics of $\mathbf{Q}_{\mathrm{m}}$ in this case.

Based on the above-mentioned discussions, the lemma has been proved for the case where axial stiffness of the members is assumed to be infinite or large enough. □

Example 4.5 Prestress-stability investigation of the two dimensional kinematically indeterminate structure as shown in Fig. 4.3.

From Examples 2.14 and 4.2, the geometrical stiffness matrix $\mathbf{K}_{\mathrm{G}}$ and mechanism matrix (vector) $\mathbf{M}$ of the two dimensional kinematically indeterminate structure as shown in Fig. 4.3 are

$$\mathbf{K}_{\rm G} = t\sqrt{2}\begin{pmatrix} 1 & 0 \\ 0 & 1 \end{pmatrix}, \quad \mathbf{M} = \begin{pmatrix} 0 \\ 1 \end{pmatrix}, \tag{4.87}$$

where the parameter t is an arbitrary value.

Therefore, quadratic form $\mathbf{Q}_{\rm m}$ of $\mathbf{K}_{\rm G}$ with respect to $\mathbf{M}$ turns out to be a scalar:

$$\mathbf{Q}_{\rm m} = \mathbf{M}^\top \mathbf{K}_{\rm G} \mathbf{M} = t\sqrt{2}, \tag{4.88}$$

and prestress-stability of the structure is dependent on sign of the parameter t:

- When $t > 0$; i.e., the members are in tension, the structure is prestress-stable, because $\mathbf{Q}_{\rm m}$ is positive (definite);
- When $t \leq 0$; i.e., prestresses are absent or the members are in compression, the structure is not prestress-stable.

Obviously, the stability criterion with minimum total potential energy is stronger than the prestress-stability, especially when there exit negative eigenvalues in the force density matrix or equivalently in the geometrical stiffness matrix. When a structure is prestress-stable with negative eigenvalues in the geometrical stiffness matrix, it can still be stable with the positive definite tangent stiffness matrix when the level of prestresses is low enough. See, for example, the following Example 4.6.

Furthermore, for the special case where there exists no infinitesimal mechanism in the structure; i.e., the structure is kinematically determinate, the quadratic form of the geometrical stiffness matrix vanishes, and the structure is also classified as a prestress-stable structure for clarity.

Example 4.6 Prestress-stability and stability of the two-dimensional structure as shown in Fig. 4.2 with high level of prestresses.

As has been discussed in Examples 2.11 and 4.3, the two-dimensional structure in Fig. 4.2 is kinematically determinate consisting of no infinitesimal mechanism. Thus, the structure is prestress-stable. However, as will be shown below, the structure might be unstable if the level of prestresses is high enough.

We set the parameter t as two times of AE; i.e., $t = 2AE$, which might not be possible for practical materials. The eigenvalues $\lambda_i^{\rm K}$ of the tangent stiffness matrix $\mathbf{K}$ are

$$\{\lambda_i^{\rm K}\} = AE\{-0.4018, -0.4018, 0.0, 0.0, 0.0, 1.0, 2.0, 3.0, 16.7155, 16.7155\}. \tag{4.89}$$

Therefore, the structure is unstable, because there exist negative eigenvalues in $\mathbf{K}$, although it has the same configuration as that in Example 4.3.

From the above lemmas and examples, we learn that stability of a prestressed pin-jointed structure implies prestress-stability, however, prestress-stability does not necessarily ensure stability, depending on the level of prestresses or member stiffnesses. Nevertheless, investigation of prestress-stability is sufficient in most cases, since the level of prestresses is usually low to avoid material failure, such as yielding, and/or member failure, such as buckling. In the stability investigation of prestressed pin-jointed structures, the criterion of prestress-stability is also more preferable than stability investigation using the tangent stiffness matrix, because the material properties are not needed.

4.3.3 Super-stability

As discussed previously, prestress-stability is convenient to use for stability investigation of a prestressed pin-jointed structure, because there is no need to consider specific materials; but it has also been pointed out that a prestress-stable structure might be unstable, if the level of prestresses is too high. In this subsection, we introduce a new stability criterion, called *super-stability* [2], which is much stronger than the conventional stability criterion. The definition of super-stability is given as follows:

Definition 3.3 Super-stability:
If a prestressed pin-jointed structure is always stable in the state of self-equilibrium, in the sense of having locally strict minimum of the total potential energy, irrespective of material properties as well as level of prestresses,[4] then it is super-stable.

For a super-stable structure, we have the following two necessary conditions concerning positive-definiteness of the force density matrix and the quadratic form $\mathbf{Q}_{\mathrm{m}}$.

Lemma 4.5 *If a free-standing prestressed pin-jointed structure is super-stable, then the following two conditions are satisfied:*

1. *The quadratic form* $\mathbf{Q}_{\mathrm{m}}$ *of the geometrical stiffness matrix* $\mathbf{K}_{\mathrm{G}}$ *with respect to the mechanisms is positive definite;*
2. *Its geometrical stiffness matrix* $\mathbf{K}_{\mathrm{G}}$*, or equivalently the force density matrix* **E***, is positive semi-definite.*

[4] Signs of the prestresses should not be changed, while the magnitude of the prestresses can be arbitrarily scaled in proportion satisfying the self-equilibrium equations.

Proof A super-stable structure is stable, and therefore, is prestress-stable from its definition. Proof of the first argument concerned about positive definiteness of $\mathbf{Q}_{\mathrm{m}}$ can refer to Lemma 4.3 for prestress-stable structures, when the displacements are a linear combination of the infinitesimal mechanisms.

As has been discussed in the non-degeneracy condition for a d-dimensional free-standing structure in Lemma 2.2, there exist at least $d+1$ zero eigenvalues in its force density matrix $\mathbf{E}$, or $(d^2+d)/2$ zero eigenvalues in its geometrical stiffness matrix $\mathbf{K}_{\mathrm{G}}$.

Let λ_i^{G} and $\boldsymbol{\phi}_i^{\mathrm{G}}$ respectively denote the ith eigenvalue and its corresponding eigenvector of the geometrical stiffness matrix $\mathbf{K}_{\mathrm{G}}$; i.e.,

$$\begin{aligned} \mathbf{K}_{\mathrm{G}}\boldsymbol{\phi}_i^{\mathrm{G}} &= \lambda_i^{\mathrm{G}}\boldsymbol{\phi}_i^{\mathrm{G}}, \\ (\boldsymbol{\phi}_i^{\mathrm{G}})^{\top}\boldsymbol{\phi}_j^{\mathrm{G}} &= \begin{cases} 1 & \text{for } i=j, \\ 0 & \text{for } i \neq j. \end{cases} \end{aligned} \tag{4.90}$$

Define displacements $\mathbf{d}$ as a linear combination of the eigenvectors $\boldsymbol{\phi}_i^{\mathrm{G}}$, with the arbitrary coefficients α_i:

$$\mathbf{d} = \sum_i \alpha_i \boldsymbol{\phi}_i^{\mathrm{G}}. \tag{4.91}$$

Let an arbitrary *positive* parameter t to change the level of prestresses proportionally. The tangent stiffness matrix $\mathbf{K}$ becomes

$$\mathbf{K} = \mathbf{K}_{\mathrm{E}} + t\mathbf{K}_{\mathrm{G}}. \tag{4.92}$$

Note that the eigenvalues of $t\mathbf{K}_{\mathrm{G}}$ are $t\lambda_i^{\mathrm{G}}$, with the same eigenvectors $\boldsymbol{\phi}_i^{\mathrm{G}}$.

The quadratic form Q_{K} of the tangent stiffness matrix $\mathbf{K}$ with respect to $\mathbf{d}$ is positive when the structure is stable:

$$\begin{aligned} Q_{\mathrm{K}} &= Q_{\mathrm{E}} + Q_{\mathrm{G}} \\ &= \mathbf{d}\mathbf{K}_{\mathrm{E}}\mathbf{d} + t\sum_i \alpha_i^2\lambda_i^{\mathrm{G}} \\ &> 0. \end{aligned} \tag{4.93}$$

If there exists a negative eigenvalue λ_i^{G} in the geometrical stiffness matrix, then the quadratic form Q_{K} can become negative while the positive parameter t is large enough compared to the axial stiffnesses of the members. This conflicts with the assumption that the structure is super-stable, which is always stable irrespective of level of prestresses, and therefore, the geometrical stiffness matrix should not have negative eigenvalues; i.e., it should be positive semi-definite. Consequently, the second argument of the lemma has been proved. □

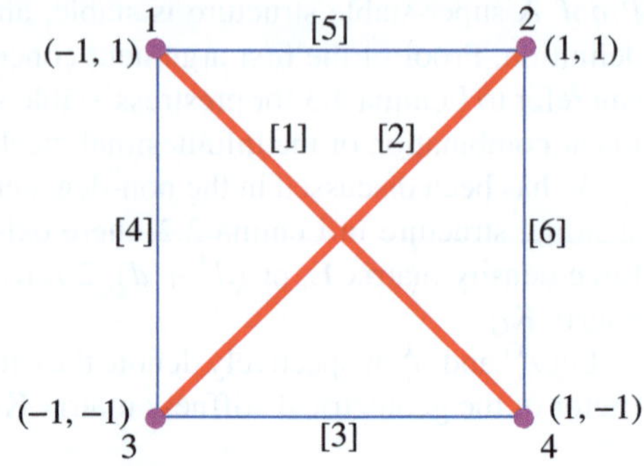

Fig. 4.5 A super-stable two-dimensional free-standing structure. Members [1] and [2] do not mechanically contact with each other

Example 4.7 Super-stability investigation of a two-dimensional free-standing structure as shown in Fig. 4.5.

The free-standing two-dimensional structure as shown in Fig. 4.5 is composed of (n=)4 free nodes and (m=)6 members. Using the nodal coordinates and connectivity of members as indicated in the figure; i.e.,

$$\mathbf{x} = \begin{pmatrix} -1 \\ 1 \\ -1 \\ 1 \end{pmatrix}, \quad \mathbf{y} = \begin{pmatrix} 1 \\ 1 \\ -1 \\ -1 \end{pmatrix}, \tag{4.94}$$

the equilibrium matrix $\mathbf{D} \in \mathbb{R}^{8\times 6}$ is calculated as

$$\mathbf{D} = \frac{1}{2} \begin{pmatrix} -\sqrt{2} & 0 & 0 & 0 & -2 & 0 \\ 0 & \sqrt{2} & 0 & 0 & 2 & 0 \\ 0 & \sqrt{2} & -2 & 0 & 0 & 0 \\ \sqrt{2} & 0 & 2 & 0 & 0 & 0 \\ \sqrt{2} & 0 & 0 & 2 & 0 & 0 \\ 0 & \sqrt{2} & 0 & 0 & 0 & 2 \\ 0 & -\sqrt{2} & 0 & -2 & 0 & 0 \\ -\sqrt{2} & 0 & 0 & 0 & 0 & -2 \end{pmatrix}. \tag{4.95}$$

Numerical investigation shows that rank r^{D} of $\mathbf{D}$ is 5, such that the numbers n^{s} and n^{m} of prestress modes and mechanisms respectively are

$$\begin{aligned} n^{\mathrm{s}} &= m - r^{\mathrm{D}} = 6 - 5 = 1, \\ n^{\mathrm{m}} &= dn - r^{\mathrm{D}} - n^{\mathrm{b}} = 2 \times 4 - 5 - 3 = 0. \end{aligned} \tag{4.96}$$

Therefore, the structure is statically indeterminate and kinematically determinate. The prestress vector $\mathbf{s}$ and its corresponding force density vector $\mathbf{q}$ are calculated as follows:

$$\mathbf{s} = 2t \begin{pmatrix} -\sqrt{2} \\ -\sqrt{2} \\ 1 \\ 1 \\ 1 \\ 1 \end{pmatrix}, \quad \mathbf{q} = t \begin{pmatrix} -1 \\ -1 \\ 1 \\ 1 \\ 1 \\ 1 \end{pmatrix}, \tag{4.97}$$

where t is an arbitrary *positive* value so as to guarantee that the struts in thick lines carry negative prestresses and the cables in thin lines carry positive prestresses as indicated in the figure.

The force density matrix $\mathbf{E} \in \mathbb{R}^{4\times4}$ of the structure is

$$\mathbf{E} = t \begin{pmatrix} 1 & -1 & -1 & 1 \\ -1 & 1 & 1 & -1 \\ -1 & 1 & 1 & -1 \\ 1 & -1 & -1 & 1 \end{pmatrix}, \tag{4.98}$$

the eigenvalues of which are

$$\{\lambda_i^{\mathrm{E}}\} = t\{0.0,\ 0.0,\ 0.0,\ 4.0\}. \tag{4.99}$$

The non-degeneracy condition for a free-standing structure in two-dimensional ($d = 2$) space as presented in Lemma 2.2 is satisfied, because there are three zero eigenvalues in $\mathbf{E}$.

Since there exists no infinitesimal mechanism, and moreover, the force density matrix as well as the geometrical stiffness matrix is positive semi-definite if $t > 0$, the structure is super-stable.

It should be noted that positive semi-definiteness of the geometrical stiffness matrix, or equivalently, the force density matrix cannot guarantee a super-stable structure. For example, the geometrical stiffness matrix of the three-dimensional tensegrity structure as shown in Fig. 4.6 is positive semi-definite, however, it is not super-stable, and in fact, it is unstable.

Example 4.8 Super-stability investigation of the three-dimensional free-standing structure as shown in Fig. 4.6.

The structure as shown in Fig. 4.6 is constructed by assembling two copies of the two-dimensional structure in Fig. 4.5. The structure consists of six nodes and eleven members. Members [1]–[6] and [6]–[11] lie on two different planes I and II, respectively. Planes I and II, which are not parallel to each other, share the

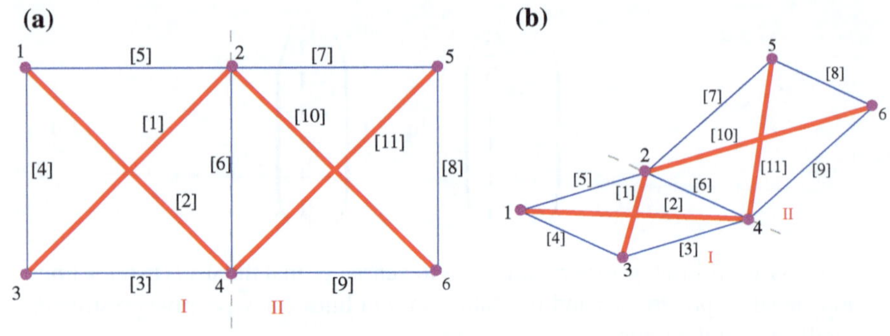

Fig. 4.6 An example of unstable tensegrity structure. The structure has positive semi-definite force density matrix which is one of the (necessary) conditions for super-stability; however, it is not stable in three-dimensional space because there exists a finite mechanism: the plane I or II can be rotated along member [6] without affecting the member lengths in the other plane. **a** Top view, **b** diagonal view

same member [6]. Moreover, we have shown in Example 4.7 that the structure in Fig. 4.5 is super-stable.

It is obvious that the members on the two different planes are mechanically independent, except for the common member [6]:

- The structure has two independent prestress modes. The prestresses of the members on one plane does not affect those on the other plane except the common member [6].
- The structure is unstable, because the two planes can relatively rotate freely about the intersecting member [6]; i.e., there exists a finite mechanism.

Utilizing the results in Example 4.7 for each sub-structure on the two planes, the force density vector $\mathbf{q} \in \mathbb{R}^{11}$ of the structure is

$$\mathbf{q} = a \begin{pmatrix} -1 \\ -1 \\ 1 \\ 1 \\ 1 \\ 1 \\ 0 \\ 0 \\ 0 \\ 0 \\ 0 \end{pmatrix} + b \begin{pmatrix} 0 \\ 0 \\ 0 \\ 0 \\ 0 \\ 1 \\ 1 \\ 1 \\ 1 \\ -1 \\ -1 \end{pmatrix}, \tag{4.100}$$

where $a(>0)$ and $b(>0)$ are arbitrary positive values to ensure that the prestresses are properly assigned to the members.

For instance, if $a = 1$ and $b = 1$, then the eigenvalues of the force density matrix **E** are calculated to be

$$\{\lambda_i^{\mathrm{E}}\} = \{0.0,\ 0.0,\ 0.0,\ 0.0,\ 2.0,\ 6.0\}. \qquad (4.101)$$

E, and therefore, the geometrical stiffness matrix $\mathbf{K}_{\mathrm{G}}$ is positive semi-definite for the structure carrying the prestresses of proper signs; i.e., the members in thin lines are cables carrying tension and the members in thick lines are struts in compression. The four zero eigenvalues in **E** satisfy the non-degeneracy condition for a free-standing prestressed pin-jointed structure in the three-dimensional space.

However, the structure is not super-stable, and even more, it is unstable because there exists a finite mechanism as discussed previously.

Investigation of the structure later in Example 4.10 will also show that it is unstable, because it does not satisfy certain necessary condition for stability.

From Example 4.8, we may learn that the two conditions in Lemma 4.5 are only the necessary conditions, but not sufficient conditions, for super-stability of a prestressed pin-jointed structure. One more condition is needed to ensure super-stability, which will be presented in Sect. 4.4.

4.3.4 Remarks

So far, we have presented the definitions, properties, and conditions of three stability criteria—stability, prestress-stability, and super-stability—for stability investigation of (free-standing) prestressed pin-jointed structures. Among these, super-stability is the strongest stability criterion: a super-stable structure is always stable, in the sense of having minimum strain energy, irrespective of materials and level of prestresses, as long as the members carry proper signs of prestresses.

By contrast, prestress-stability is the weakest stability criterion among these three: a prestress-stable structure might be stable or unstable in the sense of having local minimum strain energy in the state of self-equilibrium; a (prestress-)stable structure may become unstable when the level of prestress is increased significantly. Despite of this, prestress-stability is still useful, especially in the early stage of designing a structure, because there is no need to consider specific materials; and moreover, the level of prestresses is usually low enough, so that a prestress-stable structure is usually stable in practice.

When signs of the prestresses are reversed, a self-equilibrated structure is still in the state of self-equilibrium; however, the super-stability and prestress-stability are also reversed: a super-stable (or prestress-stable) structure is unstable if signs of the prestresses are reversed. The only exception is the special case that the structure is kinematically determinate: a kinematically determinate structure carrying prestresses

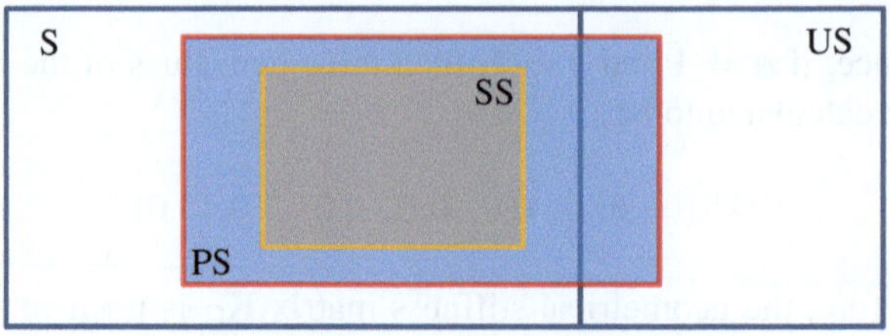

Fig. 4.7 Relationships among the stability criteria—stability, prestress-stability, and super-stability—for prestressed pin-jointed structures in the state of self-equilibrium. Note that level of prestresses is variable, and signs of prestresses are not changed in these discussions. Moreover, the structures carrying no prestress, for example, trusses are not considered here. A super-stable structure is stable, however, a prestress-stable structure might not be stable when the level of prestress is varied

is classified as prestress-stable structure; it is still prestress-stable when signs of the prestresses are reversed. This comes from the fact that the quadratic form of the geometrical stiffness does not exist, since there is no infinitesimal mechanism for these structures.

The relations among these three stability criteria are illustrated in Fig. 4.7, in which only the structures carrying prestresses are considered. In the design of a prestressed pin-jointed structure, super-stability is obviously preferable, as long as it is available. Thus, we will mainly focus on finding super-stable tensegrity structures in the coming chapters.

However, as has been demonstrated in Example 4.8, positive semi-definiteness of the force density matrix, or the geometrical stiffness matrix, is the necessary condition but not the sufficient condition for a super-stable structure. The conditions that sufficiently guarantee a super-stable structure will be presented in the next section.

4.4 Necessary and Sufficient Conditions for Super-stability

In Chap. 2, we proved that the force density matrix of a three-dimensional free-standing structure should have at least four zero eigenvalues, so that its geometry realization does not degenerate into the space with lower dimensions. Therefore, its geometrical stiffness matrix has at least twelve zero eigenvalues. In Appendix B.1, it is proved that six of these zero eigenvalues correspond to the rigid-body motions, and the other six correspond to non-trivial affine motions. An *affine motion* is a motion that preserves colinearity and ratios of distances; i.e., all points lying on a line are transformed to points on a line, and ratios of the distances between any pairs of the points on the line are preserved [4]. The non-trivial affine motions are used in this section to present the sufficient conditions for super-stability of a structure.

The tangent stiffness matrix $\mathbf{K}$ is the sum of the linear $\mathbf{K}_\mathrm{E}$ and geometrical $\mathbf{K}_\mathrm{G}$ stiffness matrices. Moreover, the linear stiffness matrix is always positive (semi-)definite. Hence, if there exist some prestress modes that guarantee $\mathbf{K}_\mathrm{G}$'s positive semi-definiteness, the structure is highly possible to be stable, in the sense of having positive definite tangent stiffness matrix. However, as has been shown in Example 4.8, there may exist a non-trivial motion $\mathbf{d}$, excluding the rigid-body motions, that results in zero increment of the total potential energy satisfying

$$\mathbf{d}^\top \mathbf{K}_\mathrm{E} \mathbf{d} = \mathbf{d}^\top \mathbf{K}_\mathrm{G} \mathbf{d} = 0. \tag{4.102}$$

In this section, we make use of the non-trivial affine motions to present the sufficient conditions for super-stability of tensegrity structures.

4.4.1 Geometry Matrix

Because the non-trivial affine motions $\mathbf{d}_\mathrm{a}^x$, $\mathbf{d}_\mathrm{a}^y$, $\mathbf{d}_\mathrm{a}^z$ (dilation motions) and $\mathbf{d}_\mathrm{a}^{xy}$, $\mathbf{d}_\mathrm{a}^{xz}$, $\mathbf{d}_\mathrm{a}^{yz}$ (shear motions) defined in Eqs. (B.26) and (B.27) are linearly independent from Lemma B.2, any non-trivial affine motion $\mathbf{d}_\mathrm{a}$ can be written as a linear combination of the six non-trivial affine motions:

$$\mathbf{d}_\mathrm{a} = \alpha_x \mathbf{d}_\mathrm{a}^x + \alpha_y \mathbf{d}_\mathrm{a}^y + \alpha_z \mathbf{d}_\mathrm{a}^z + \alpha_{xy} \mathbf{d}_\mathrm{a}^{xy} + \alpha_{xz} \mathbf{d}_\mathrm{a}^{xz} + \alpha_{yz} \mathbf{d}_\mathrm{a}^{yz}, \tag{4.103}$$

where $\alpha_{i(j)}$ $(i, j \in \{x, y, z\})$ are arbitrary coefficients. Moreover, as proved in Lemma B.1, the following equation is always satisfied:

$$\mathbf{K}_\mathrm{G} \mathbf{d}_\mathrm{a} = \mathbf{0}. \tag{4.104}$$

Hence, the quadratic form Q_K of $\mathbf{K}$ with respect to $\mathbf{d}_\mathrm{a}$ is reduced to

$$\begin{aligned} Q_\mathrm{K} &= (\mathbf{d}_\mathrm{a})^\top \mathbf{K} \mathbf{d}_\mathrm{a} = (\mathbf{d}_\mathrm{a})^\top \mathbf{K}_\mathrm{E} \mathbf{d}_\mathrm{a} \\ &= (\mathbf{D}^\top \mathbf{d}_\mathrm{a})^\top \bar{\mathbf{K}} (\mathbf{D}^\top \mathbf{d}^\mathrm{a}), \end{aligned} \tag{4.105}$$

where $\mathbf{D}$ is the equilibrium matrix defined in Eq. (2.34).

Because $\mathbf{K}_\mathrm{E}$ is positive semi-definite, Q_K in Eq. (4.105) cannot be negative. The only possibility for the structure with positive semi-definite force density matrix being unstable is that $Q_\mathrm{K} = 0$. Moreover, the equality $Q_\mathrm{K} = 0$ holds if and only if

$$\mathbf{D}^\top \mathbf{d}_\mathrm{a} = \mathbf{0}, \tag{4.106}$$

because the diagonal matrix $\bar{\mathbf{K}}$ carrying diagonal entries $A_k E_k / l_k$ is positive definite, where A_k, E_k, and l_k are cross sectional area, Young's modulus, and length of member k, respectively. From the member extensions due to nodal displacements as in Eq. (2.82), Eq. (4.106) indicates that member lengths of the structure are not changed by the non-trivial affine motions $\mathbf{d}_{\mathrm{a}}$.

For the shear affine motion $\mathbf{d}_{\mathrm{a}}^{xy}$, for example, we have the following relation from Eq. (2.34):

$$\begin{aligned} \mathbf{D}^\top \mathbf{d}_{\mathrm{a}}^{xy} &= \left(\mathbf{D}^{x\top},\, \mathbf{D}^{y\top},\, \mathbf{D}^{z\top}\right) \begin{pmatrix} \mathbf{y} \\ \mathbf{x} \\ \mathbf{0} \end{pmatrix} \\ &= \mathbf{D}^{x\top}\mathbf{y} + \mathbf{D}^{y\top}\mathbf{x} = \mathbf{L}^{-1}(\mathbf{UCy} + \mathbf{VCx}) \\ &= 2\mathbf{L}^{-1}\mathbf{Uv}. \end{aligned} \tag{4.107}$$

Similarly, we have

$$\begin{aligned} &\mathbf{D}^\top \mathbf{d}_x^{\mathrm{a}} = \mathbf{L}^{-1}\mathbf{Uu}, \quad \mathbf{D}^\top \mathbf{d}_y^{\mathrm{a}} = \mathbf{L}^{-1}\mathbf{Vv}, \quad \mathbf{D}^\top \mathbf{d}_z^{\mathrm{a}} = \mathbf{L}^{-1}\mathbf{Ww}, \\ &\mathbf{D}^\top \mathbf{d}_{\mathrm{a}}^{xz} = 2\mathbf{L}^{-1}\mathbf{Uw},\, \mathbf{D}^\top \mathbf{d}_{\mathrm{a}}^{yz} = 2\mathbf{L}^{-1}\mathbf{Vw}. \end{aligned} \tag{4.108}$$

Substituting Eq. (4.103) into Eq. (2.82) and using Eqs. (4.107) and (4.108), we obtain

$$\mathbf{D}^\top \mathbf{d}_{\mathrm{a}} = \mathbf{L}^{-1}\mathbf{G}\boldsymbol{\alpha} = \mathbf{0}, \tag{4.109}$$

where $\boldsymbol{\alpha} = (\alpha_x, \alpha_y, \alpha_z, 2\alpha_{xy}, 2\alpha_{xz}, 2\alpha_{yz})^\top$ and

$$\mathbf{G} = \left(\mathbf{Uu},\, \mathbf{Vv},\, \mathbf{Ww},\, \mathbf{Uv},\, \mathbf{Uw},\, \mathbf{Vw}\right). \tag{4.110}$$

The matrix $\mathbf{G} \in \mathbb{R}^{m\times(d^2+d)/2}$ is called the *geometry matrix*, because it is related only to the geometry of the structure.

For the two-dimensional tensegrity structures, the geometry matrix $\mathbf{G} \in \mathbb{R}^{m\times 3}$ becomes

$$\mathbf{G} = \left(\mathbf{Uu},\, \mathbf{Vv},\, \mathbf{Uv}\right). \tag{4.111}$$

Because the inverse matrix $\mathbf{L}^{-1}$ of the member length matrix $\mathbf{L}$ is positive definite, Eq. (4.109) has a non-trivial solution $\boldsymbol{\alpha} \neq \mathbf{0}$, if and only if the rank of $\mathbf{G}$ is less than $(d^2+d)/2$, because the inequality $m > (d^2+d)/2$ always holds for a prestressed pin-jointed structure.[5] From this discussion, we have the following necessary condition for stability of tensegrity structures based on rank of the geometry matrix $\mathbf{G}$:

[5] A two-dimensional free-standing pin-jointed structure has at least three members (having the triangular appearance), and a three-dimensional free-standing structure has six (having the tetrahedral appearance). However, these simplest structures cannot carry any prestress in their members. To ensure the possibility of carrying prestresses, the number m of member is always larger than $(d^2+d)/2$ for a d-dimensional free-standing prestressed pin-jointed structure.

Lemma 4.6 *If a d-dimensional tensegrity structure is stable, then the rank of its geometry matrix* **G** *defined in Eq.* (4.110) *or Eq.* (4.111) *is full (column) rank; i.e.,*

$$\text{rank}(\mathbf{G}) = \frac{d^2 + d}{2}. \tag{4.112}$$

Proof The space spanned by the non-trivial affine motions is a sub-space of the null-space of the geometrical stiffness matrix. If the rank of **G** is less than $(d^2+d)/2$, then there exist non-trivial motions in this sub-space that make the quadratic form Q_K equal to zero from Eqs. (4.105) and (4.109). Therefore, the structure is unstable. Hence, the lemma has been proved. □

Note that if a d-dimensional structure is degenerate, the nodal coordinate vectors are linearly dependent and so are the coordinate difference vectors. Therefore, rank of **G** must be less than $(d^2 + d)/2$ and the structure is unstable in d-dimensional space.

Example 4.9 Geometry matrix of the two-dimensional structure as shown in Fig. 4.5.

Using the geometry realization in Example 4.7, the geometry matrix $\mathbf{G} \in \mathbb{R}^{6\times 3}$ of the structure in Fig. 4.5 is

$$\mathbf{G} = \begin{pmatrix} 4 & 4 & -4 \\ 4 & 4 & 4 \\ 4 & 0 & 0 \\ 0 & 4 & 0 \\ 4 & 0 & 0 \\ 0 & 4 & 0 \end{pmatrix}, \tag{4.113}$$

the rank of which is calculated to be

$$\text{rank}(\mathbf{G}) = 3. \tag{4.114}$$

Thus, the structure may be stable, because its geometry matrix has the necessary rank, which is three for a two-dimensional structure.

Example 4.10 Geometry matrix of the three-dimensional structure as shown in Fig. 4.6.

Assigning the angle between planes I and II of the structure in Fig. 4.6 as $\pi/6$, geometry realization of the structure, with the nodal coordinates **x**, **y**, and **z** ($\in\mathbb{R}^6$), is

$$\mathbf{x} = \begin{pmatrix} -1 \\ 1 \\ -1 \\ 1 \\ 1+\sqrt{3} \\ 1+\sqrt{3} \end{pmatrix}, \quad \mathbf{y} = \begin{pmatrix} 1 \\ 1 \\ -1 \\ -1 \\ 1 \\ -1 \end{pmatrix}, \quad \mathbf{z} = \begin{pmatrix} 0 \\ 0 \\ 0 \\ 0 \\ 1 \\ 1 \end{pmatrix}. \tag{4.115}$$

The geometry matrix $\mathbf{G} \in \mathbb{R}^{11\times 6}$ of the structure is

$$\mathbf{G} = \begin{pmatrix} 4 & 4 & 0 & -4 & 0 & 0 \\ 4 & 4 & 0 & 4 & 0 & 0 \\ 4 & 0 & 0 & 0 & 0 & 0 \\ 0 & 4 & 0 & 0 & 0 & 0 \\ 4 & 0 & 0 & 0 & 0 & 0 \\ 0 & 4 & 0 & 0 & 0 & 0 \\ 3 & 0 & 1 & 0 & \sqrt{3} & 0 \\ 0 & 4 & 0 & 0 & 0 & 0 \\ 3 & 0 & 1 & 0 & \sqrt{3} & 0 \\ 3 & 4 & 1 & -2\sqrt{3} & \sqrt{3} & -2 \\ 3 & 4 & 1 & 2\sqrt{3} & \sqrt{3} & 2 \end{pmatrix}, \tag{4.116}$$

the rank of which is calculated to be

$$\text{rank}(\mathbf{G}) = 5. \tag{4.117}$$

Hence, the structure is unstable, because its geometry matrix does not have enough rank, which should be six for a three-dimensional structure from Lemma 4.6.

4.4.2 Sufficient Conditions

This subsection presents the sufficient conditions for super-stability of a tensegrity structure. Note that when we discuss stability of a free-standing structure, the rigid-body motions are not taken into consideration. They are assumed to be properly constrained such that the zero eigenvalues and the corresponding eigenvectors in the stiffness matrices are implicitly ignored.

According to Lemma 4.6, the rank of the geometry matrix must be $(d^2 + d)/2$ for a d-dimensional stable tensegrity structure. The affine motions span the whole null-space of the geometrical stiffness matrix, if the force density matrix has the minimum rank deficiency for non-degeneracy condition. Therefore, the structures are guaranteed to be super-stable, irrespective of selection of materials and level of prestresses, if the force density matrix (the geometrical stiffness matrix as well) is positive semi-definite in addition to the above two conditions [1]. Accordingly, we have the following lemma for the super-stability of a tensegrity structure.

Lemma 4.7 *If the following three conditions are all satisfied, then the d-dimensional tensegrity structure is super-stable:*

1. *Rank of the geometry matrix* $\mathbf{G}$ *is* $(d^2+d)/2$*;*
2. *The force density matrix* $\mathbf{E}$ *has the minimum necessary rank deficiency* $d+1$*;*
3. *The force density matrix* $\mathbf{E}$ *is positive semi-definite.*

Proof The linear stiffness matrix must be positive semi-definite; i.e., its quadratic form Q_E with respect to any displacement is non-negative:

$$Q_\mathrm{E} \geq 0, \tag{4.118}$$

in which the equality holds only when the displacement is a (infinitesimal) mechanism denoted by $\mathbf{d}_\mathrm{m}$.

When the geometry matrix $\mathbf{G}$ has rank of $(d^2+d)/2$ from the first condition, there exists no non-trivial affine motion $\mathbf{d}_\mathrm{a}$ in this space that leads to $Q_\mathrm{E} = 0$, therefore, we have

$$\mathbf{d}_\mathrm{a}^\top \mathbf{K}_\mathrm{E} \mathbf{d}_\mathrm{a} > 0. \tag{4.119}$$

Furthermore, we know that

$$\mathbf{d}_\mathrm{a} \neq \mathbf{d}_\mathrm{m}, \tag{4.120}$$

because $\mathbf{d}_\mathrm{m}^\top \mathbf{K}_\mathrm{E} \mathbf{d}_\mathrm{m} = 0$.

If the second condition is satisfied, then the geometrical stiffness matrix $\mathbf{K}_\mathrm{G}$ has d^2+d zero eigenvalues, corresponding to the affine motions. These affine motions span the whole null-space of $\mathbf{K}_\mathrm{G}$; half of them are rigid-body motions and the other half are non-trivial affine motions.

If the third condition is also satisfied, the quadratic form Q_G of the geometrical stiffness matrix with respect to any displacement is also non-negative, irrespective of level of prestresses; i.e.,

$$Q_\mathrm{G} \geq 0, \tag{4.121}$$

in which the equality holds only when the displacement is a non-trivial affine motion $\mathbf{d}_\mathrm{a}$. Hence, we know that the quadratic form of geometrical stiffness matrix with respect to a mechanism $\mathbf{d}_\mathrm{m}$ that is not a non-trivial affine motion; i.e., $\mathbf{d}_\mathrm{m} \neq \mathbf{d}_\mathrm{a}$, must be positive:

$$\mathbf{d}_\mathrm{m}^\top \mathbf{K}_\mathrm{G} \mathbf{d}_\mathrm{m} > 0. \tag{4.122}$$

If the displacement is neither a mechanism nor a non-trivial affine motion, the quadratic form Q_K of the tangent stiffness matrix with respect to this displacement must be positive, because $Q_K = Q_E + Q_G > 0$ with $Q_E > 0$ and $Q_G > 0$ from our previous discussions.

Consider the other case that the displacement $\mathbf{d}$ is a linear combination of the mechanism $\mathbf{d}_m$ and the non-trivial affine motion $\mathbf{d}_a$. Here, $\mathbf{d}$ can be defined as follows by using the arbitrary parameters α_m and α_a:

$$\mathbf{d} = \alpha_m \mathbf{d}_m + \alpha_a \mathbf{d}_a. \tag{4.123}$$

From Eqs. (4.119) and (4.122), the quadratic form Q_K of the tangent stiffness matrix with respect to $\mathbf{d}$ is positive:

$$\begin{aligned} Q_K &= \mathbf{d}^\top \mathbf{K} \mathbf{d} \\ &= (\alpha_m \mathbf{d}_m + \alpha_a \mathbf{d}_a)^\top (\mathbf{K}_E + \mathbf{K}_G)(\alpha_m \mathbf{d}_m + \alpha_a \mathbf{d}_a) \\ &= \alpha_a^2 \mathbf{d}_a^\top \mathbf{K}_E \mathbf{d}_a + \alpha_m^2 \mathbf{d}_m^\top \mathbf{K}_G \mathbf{d}_m \\ &> 0, \end{aligned} \tag{4.124}$$

because the parameters α_m and α_a cannot be equal to zero at the same time to avoid a zero displacement.

Therefore, the structure is super-stable if all of these three conditions are satisfied at the same time, and the lemma has been proved. □

It is notable that the first condition in Lemma 4.7 is the necessary condition for stability, and the last condition is the necessary condition for super-stability. However, satisfying only the last two conditions does not guarantee a (super-)stable structure. See, for example, the unstable structure in Fig. 4.6 studied in Example 4.10.

4.5 Remarks

Concerning positive definiteness of the stiffness matrices, we have presented and discussed three stability criteria in this chapter:

- *Stability* in the sense of having local minimum total potential energy or strain energy in the state of self-equilibrium;
- *Prestress-stability* in which the members are assumed to have high enough axial stiffness compared to the level of prestresses;
- *Super-stability* which is always stable irrespective of material properties and level of prestresses.

Among these criteria, super-stability is the most robust criterion: super-stability implies stability without regard to level of prestresses, and then implies prestress-stability. However, it is not always true in the reverse direction: a prestress-stable structure might not be stable, especially when the level of prestresses is very high; moreover, a stable structure might not be super-stable.

Super-stable structures have many advantages compared to the structures that are not super-stable: any stretched versions of them by affine motions are also super-stable. Therefore, in the design of tensegrity structures, which will be studied from the next chapter, super-stability is usually preferable, as long as it is available.

A structure cannot be stable if its geometry matrix is not full-rank. Moreover, a tensegrity structure is guaranteed to be super-stable if two additional conditions are satisfied: the force density matrix is positive semi-definite and it has the minimum rank deficiency for non-degeneracy condition. Using these sufficient conditions, we will present a numerical method, making use of the idea of force density method for cable-nets, for the form-finding problem of super-stable tensegrity structures in Chap. 5; furthermore, we will present the analytical super-stability conditions for the structures with high level of symmetry in Chaps. 6–8.

References

1. Connelly, R. (1999). Tensegrity structures: why are they stable? In M. F. Thorpe & P. M. Duxbury (Eds.), *Rigidity theory and applications* (pp. 47–54). New York: Kluwer Academic/Plenum Publishers.
2. Connelly, R., & Whiteley, W. (1996). Second-order rigidity and prestress stability for tensegrity frameworks. *SIAM Journal on Discrete Mathematics*, *9*(3), 453–491.
3. Golub, G. H., & Van Loan, C. F. (1996). *Matrix computations* (3rd ed.). Baltimore: Johns Hopkins University Press.
4. Gray, A. (1997). *Modern differential geometry of curves and surfaces with mathematica* (2nd ed.). Boca Raton: CRC Press.
5. Lanczos, C. (1986). *The variational principles of mechanics* (4th ed.). New York: Dover Publications.

Chapter 5
Force Density Method

Abstract This chapter presents a numerical method for form-finding problem of tensegrity structures, by making use of the idea of force density method originally developed for cable-nets. The basic idea is to find the feasible force densities satisfying the non-degeneracy condition as presented in Chap. 2, and then to determine geometry realization of the structure satisfying the self-equilibrium equations. The method is suitable for finding the self-equilibrated configuration of complex tensegrity structures with a large number of members and nodes, and more importantly, it guarantees super-stable structures in most cases.

Keywords Force density method · Spectral decomposition · Geometrical constraints · Tensegrity tower

5.1 Concept of Force Density Method

To solve the form-finding problem of a prestressed pin-jointed structure, the concept of force density was introduced to simplify the problem by transforming the non-linear self-equilibrium equations into linear equations. This idea was initially developed for form-finding of cable-nets [3], which are attached to fix nodes or supports.

Cable-nets and tensegrity structures share many common mechanical properties: both of cable-nets and tensegrity structures are prestressed pin-jointed structures, carrying only axial forces. However, the force density method for cable-nets cannot be directly extended to form-finding of tensegrity structures. This is because that a tensegrity structure is free-standing without any fixed node.

To make use of the concept of force density for tensegrity structures, different strategy has to be developed. In this section, we will show how the force density method works for form-finding problem of cable-nets as well as tensegrity structures, in quite a different way.

J.Y. Zhang and M. Ohsaki, *Tensegrity Structures*, Mathematics for Industry 6,
DOI 10.1007/978-4-431-54813-3_5

5.1.1 Force Density Method for Cable-nets

In Chap. 2, we presented the self-equilibrium equations of a prestressed pin-jointed structure in two different ways:

- the equation in terms of equilibrium matrix associated with prestress vector; and
- the equations in terms of force density matrices associated with nodal coordinate vectors.

In the force density method for form-finding of prestressed pin-jointed structures, the latter equations with respect to nodal coordinates are extensively utilized.

Consider a prestressed pin-jointed structure, which is composed of n free nodes, n^{f} fixed nodes, and m members. Denote coordinate vectors of free nodes by $\mathbf{x}$, $\mathbf{y}$, $\mathbf{z}$ ($\in\mathbb{R}^{n}$) and those of fixed nodes by $\mathbf{x}^{\mathrm{f}}$, $\mathbf{y}^{\mathrm{f}}$, $\mathbf{z}^{\mathrm{f}}$ ($\in\mathbb{R}^{n^{\mathrm{f}}}$). In Chap. 2, the self-equilibrium equations using force density matrices $\mathbf{E} \in \mathbb{R}^{n\times n}$ for free nodes and $\mathbf{E}^{\mathrm{f}} \in \mathbb{R}^{n\times n^{\mathrm{f}}}$ for fixed nodes, in each direction, are formulated as

$$\begin{aligned}\mathbf{Ex} + \mathbf{E}^{\mathrm{f}}\mathbf{x}^{\mathrm{f}} &= \mathbf{0},\\ \mathbf{Ey} + \mathbf{E}^{\mathrm{f}}\mathbf{y}^{\mathrm{f}} &= \mathbf{0},\\ \mathbf{Ez} + \mathbf{E}^{\mathrm{f}}\mathbf{z}^{\mathrm{f}} &= \mathbf{0}.\end{aligned} \tag{2.110}$$

From the direct definition of the force density matrices $\mathbf{E}$ and $\mathbf{E}^{\mathrm{f}}$ presented in Eq. (2.107), the entries of these matrices are linear functions of the force densities q_k ($k = 1, 2, \dots, m$), or the force density vector $\mathbf{q} \in \mathbb{R}^{m}$. Remember that the force density q_k of the kth member is defined as ratio of its axial force s_k to its member length l_k; i.e., $q_k = s_k/l_k$.

In the form-finding problem of a prestressed pin-jointed structure using Eq. (2.110), only the coordinates $\mathbf{x}^{\mathrm{f}}$, $\mathbf{y}^{\mathrm{f}}$, $\mathbf{z}^{\mathrm{f}}$ of the fixed nodes are known or given a priori, and those $\mathbf{x}$, $\mathbf{y}$, $\mathbf{z}$ of the free nodes as well as the force densities $\mathbf{q}$ are to be determined. Because member lengths are non-linear functions of nodal coordinates, self-equilibrium equations in Eq. (2.110) are also non-linear with respect to nodal coordinates. Consequently, the form-finding problem directly solving these equations is *non-linear*.

However, we can transform these non-linear equations into linear equations. Suppose that the force densities $\mathbf{q}$ are known or given a priori, the force density matrices $\mathbf{E}$ and $\mathbf{E}^{\mathrm{f}}$ are constant, and therefore, the self-equilibrium equations in Eq. (2.110) become linear with respect to coordinates $\mathbf{x}$, $\mathbf{y}$, and $\mathbf{z}$ of free nodes. Furthermore, the unknown coordinates $\mathbf{x}$, $\mathbf{y}$, and $\mathbf{z}$ of the structure can be uniquely determined as follows:

$$\begin{aligned}\mathbf{x} &= -\mathbf{E}^{-1}\mathbf{E}^{\mathrm{f}}\mathbf{x}^{\mathrm{f}},\\ \mathbf{y} &= -\mathbf{E}^{-1}\mathbf{E}^{\mathrm{f}}\mathbf{y}^{\mathrm{f}},\\ \mathbf{z} &= -\mathbf{E}^{-1}\mathbf{E}^{\mathrm{f}}\mathbf{z}^{\mathrm{f}},\end{aligned} \tag{2.111}$$

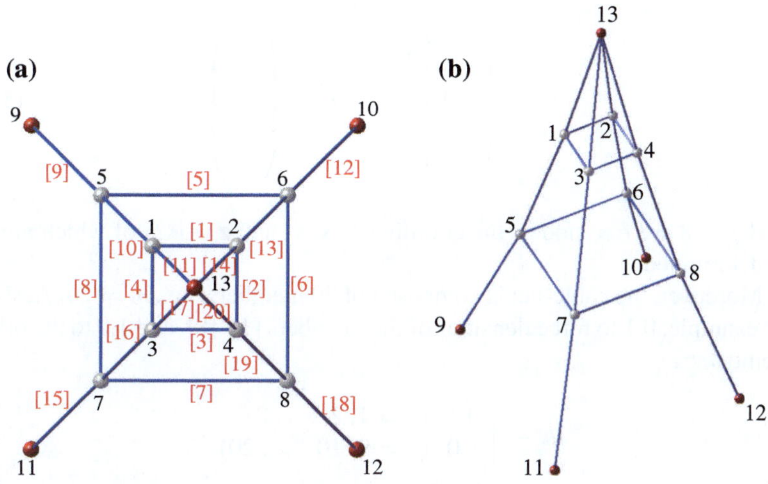

Fig. 5.1 Self-equilibrated configuration of the cable-net obtained by force density method in Example 5.1. Nodes 9–13 are fixed nodes (supports) and the other nodes are free nodes. **a** Top view, **b** diagonal view

as long as the inverse matrix $\mathbf{E}^{-1}$ of $\mathbf{E}$ exists. Using preassigned force densities to derive the linear equations with respect to nodal coordinates is the basic idea of force density method.

In Example 5.1, we demonstrate how the force density method works for form-finding of cable-nets, which carry only tension (positive force densities) in their members. Moreover, in the examples in this chapter, the units will be omitted without doing any harm to discussions on self-equilibrium as well as stability of the structures.

Example 5.1 Form-finding of the cable-net as shown in Fig. 5.1 by using force density method.

The cable-net as shown in Fig. 5.1 is composed of 13 nodes, including five fixed nodes and eight free nodes; i.e., $n^{\mathrm{f}} = 5$ and $n = 8$. Nodes 9–13 are fixed, and their coordinates $\mathbf{x}^{\mathrm{f}}$, $\mathbf{y}^{\mathrm{f}}$, and $\mathbf{z}^{\mathrm{f}}$ $(\in \mathbb{R}^5)$ are given, for example, as follows:

$$\mathbf{x}^{\mathrm{f}} = \begin{pmatrix} x_9 \\ x_{10} \\ x_{11} \\ x_{12} \\ x_{13} \end{pmatrix} = \begin{pmatrix} -10 \\ 10 \\ -10 \\ 10 \\ 0 \end{pmatrix},$$

$$\mathbf{y}^{\mathrm{f}} = \begin{pmatrix} 10 \\ 10 \\ -10 \\ -10 \\ 0 \end{pmatrix}, \quad \mathbf{z}^{\mathrm{f}} = \begin{pmatrix} 0 \\ 0 \\ 0 \\ 0 \\ 50 \end{pmatrix}. \tag{5.1}$$

Nodes 1–8 are free-nodes, the coordinates **x**, **y**, and **z** ($\in \mathbb{R}^8$) of which are to be determined.

Moreover, the cable-net is composed of 20 members; i.e., $m = 20$. Assign, for example, 0.1 to force densities of the members [1]–[8], and 1.0 to the other members:

$$q_k = \begin{cases} 0.1, & (k = 1, 2, \ldots, 8); \\ 1.0, & (k = 9, 10, \ldots, 20). \end{cases} \tag{5.2}$$

Using the direct definition in Eq. (2.107), the force density matrix $\mathbf{E} \in \mathbb{R}^{8\times 8}$ of the free nodes is

$$\mathbf{E} = \begin{pmatrix} 2.2 & -0.1 & -0.1 & 0.0 & -1.0 & 0.0 & 0.0 & 0.0 \\ -0.1 & 2.2 & 0.0 & -0.1 & 0.0 & -1.0 & 0.0 & 0.0 \\ -0.1 & 0.0 & 2.2 & -0.1 & 0.0 & 0.0 & -1.0 & 0.0 \\ 0.0 & -0.1 & -0.1 & 2.2 & 0.0 & 0.0 & 0.0 & -1.0 \\ -1.0 & 0.0 & 0.0 & 0.0 & 2.2 & -0.1 & -0.1 & 0.0 \\ 0.0 & -1.0 & 0.0 & 0.0 & -0.1 & 2.2 & 0.0 & -0.1 \\ 0.0 & 0.0 & -1.0 & 0.0 & -0.1 & 0.0 & 2.2 & -0.1 \\ 0.0 & 0.0 & 0.0 & -1.0 & 0.0 & -0.1 & -0.1 & 2.2 \end{pmatrix}, \tag{5.3}$$

and the force density matrix $\mathbf{E}^{\mathrm{f}} \in \mathbb{R}^{8\times 5}$ of the fixed nodes is

$$\mathbf{E}^{\mathrm{f}} = \begin{pmatrix} 0.0 & 0.0 & 0.0 & 0.0 & -1.0 \\ 0.0 & 0.0 & 0.0 & 0.0 & -1.0 \\ 0.0 & 0.0 & 0.0 & 0.0 & -1.0 \\ 0.0 & 0.0 & 0.0 & 0.0 & -1.0 \\ -1.0 & 0.0 & 0.0 & 0.0 & 0.0 \\ 0.0 & -1.0 & 0.0 & 0.0 & 0.0 \\ 0.0 & 0.0 & -1.0 & 0.0 & 0.0 \\ 0.0 & 0.0 & 0.0 & -1.0 & 0.0 \end{pmatrix}. \tag{5.4}$$

By using Eq. (2.111), coordinates **x**, **y**, and **z** of the free nodes are computed as

$$\mathbf{x} = \begin{pmatrix} -2.6042 \\ 2.6042 \\ -2.6042 \\ 2.6042 \\ -5.7292 \\ 5.7292 \\ -5.7292 \\ 5.7292 \end{pmatrix}, \quad \mathbf{y} = \begin{pmatrix} 2.6042 \\ 2.6042 \\ -2.6042 \\ -2.6042 \\ 5.7292 \\ 5.7292 \\ -5.7292 \\ -5.7292 \end{pmatrix}, \quad \mathbf{z} = \begin{pmatrix} 33.3333 \\ 33.3333 \\ 33.3333 \\ 33.3333 \\ 16.6667 \\ 16.6667 \\ 16.6667 \\ 16.6667 \end{pmatrix}. \tag{5.5}$$

The geometry realization with these nodal coordinates of the structure is shown in Fig. 5.1.

Different values of the force densities will certainly result in different geometry realizations. Furthermore, the linear equations need to be solved only once, with the computation of the inverse matrix $\mathbf{E}^{-1}$ of $\mathbf{E}$ or using Gaussian elimination. Hence, the force density method is very efficient for form-finding of the prestressed pin-jointed structures with fixed nodes.

5.1.2 Force Density Method for Tensegrity Structures

By contrast to cable-nets consisting of fixed nodes, tensegrity structures are free-standing structures, without any fixed node or support. Accordingly, the self-equilibrium equations in Eq. (2.110) for tensegrity structures are reduced to

$$\begin{aligned} \mathbf{Ex} &= \mathbf{0}, \\ \mathbf{Ey} &= \mathbf{0}, \\ \mathbf{Ez} &= \mathbf{0}, \end{aligned} \tag{2.113}$$

in which coordinates $\mathbf{x}^{\mathrm{f}}$, $\mathbf{y}^{\mathrm{f}}$, and $\mathbf{z}^{\mathrm{f}}$ of fixed nodes do not exist.

As has been discussed in Chap. 2, the force density matrix $\mathbf{E}$ of a tensegrity structure is singular, because it is rank deficient with at least one zero eigenvalue due to being self-standing. The formulations in Eq. (2.111) for form-finding of the structures with fixed nodes are not applicable to tensegrity structures, since their force density matrices are not invertible. To make use of the idea of force density, an alternative strategy has to be developed.

One possibility of utilizing the concept of force density has been shown in Chap. 3: the self-equilibrium equations of a symmetric structure are simplified to those of the representative nodes, and the reduced force density matrix is enforced to be singular to guarantee non-trivial solutions for the nodal coordinates. Using this symmetry approach it is possible to derive analytical solutions, however, it is limited to the structures with high symmetry. For the structures with less symmetry, an alternative (numerical) approach becomes necessary. In Chap. 1, we have reviewed a very limited number of existing numerical methods. In the following, we will focus those making use of the concept of force density.

Regarding the force density matrix, we have shown in Chap. 2 that there must be more than one zero eigenvalues for a tensegrity structure. To be specific, its force density matrix should have rank deficiency of at least $d + 1$, so as to guarantee a non-degenerate geometry realization in d-dimensional space. Notice that the force density matrix is related only to the connectivity of the structure and the force densities, as indicated in its definition in Eq. (2.107). Hence, for a non-degenerate tensegrity structure, there must exist some sets of force densities that satisfy the non-degeneracy condition, while its connectivity is assumed to be given. This indicates another approach to form-finding of tensegrity structures by using the concept of force density, which is separated into two consecutive design stages as follows:

Design procedures for form-finding of tensegrity structures by force density method:

First stage: Feasible force densities: Find the force densities satisfying the non-degeneracy condition; i.e., the corresponding force density matrix has the required minimum rank deficiency $d + 1$ for a d-dimensional structure.

Second stage: Self-equilibrated configuration: Determine the geometry realization in terms of nodal coordinates satisfying the self-equilibrium equations in Eq. (2.113). Because the force density matrix has been determined in the first design stage, the nodal coordinates can be derived by solving the linear equations, with an independent set of specified coordinates lying in the null-space of the force density matrix.

It is noticed that the basic idea behind the method for a general tensegrity structure presented here are identical to those for the symmetric structures in Chap. 3: we first find the feasible force densities, and then determine the nodal coordinates. However, practical numerical strategy is necessary for complex structures consisting of a large number of nodes and members. In Example 5.2, we may see that analytical solutions are difficult even for simple structures.

Example 5.2 Form-finding of a two-dimensional tensegrity structure, the connectivity of which is shown in Fig. 5.2.

The structure as shown in Fig. 5.2 is the simplest tensegrity structure in two-dimensional space. It consists of four (free) nodes and six members; i.e., $n = 4$, $m = 6$. Note that members [1] and [2] do not mechanically contact to each other.

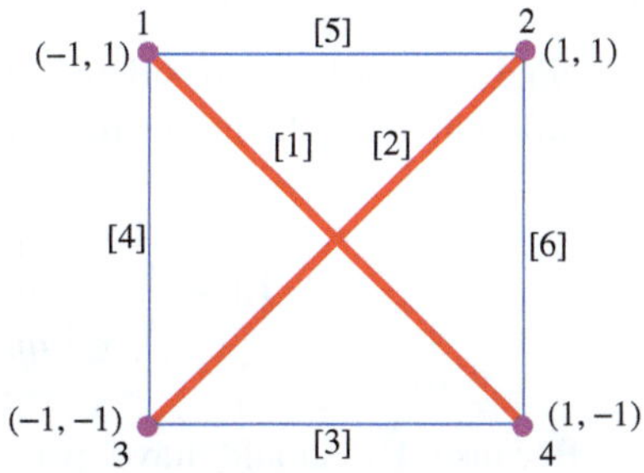

Fig. 5.2 Form-finding of the simplest two-dimensional tensegrity structure by using force density method. This structure is super-stable, when the exterior members are cables in tension and the interior members are struts in compression

Force densities of the members are denoted by q_k $(k = 1, 2, \ldots, 6)$, and the force density matrix $\mathbf{E} \in \mathbb{R}^{4\times 4}$ is written as follows by using its direct definition given in Eq. (2.107):

$$\mathbf{E} = \begin{pmatrix} q_1+q_4+q_5 & -q_5 & -q_4 & -q_1 \\ -q_5 & q_2+q_5+q_6 & -q_2 & -q_6 \\ -q_4 & -q_2 & q_2+q_3+q_4 & -q_3 \\ -q_1 & -q_6 & -q_3 & q_1+q_3+q_6 \end{pmatrix}, \tag{5.6}$$

which can be transformed to $\mathbf{E}_1$, having the same eigenvalues as $\mathbf{E}$, by row and column transformations

$$\mathbf{E}_1 = \begin{pmatrix} q_1+q_4+q_5 & -q_5 & -q_4 & 0 \\ -q_5 & q_2+q_5+q_6 & -q_2 & 0 \\ -q_4 & -q_2 & q_2+q_3+q_4 & 0 \\ 0 & 0 & 0 & 0 \end{pmatrix} = \begin{pmatrix} \hat{\mathbf{E}}_1 & \mathbf{0}_3 \\ \mathbf{0}_3^\top & 0 \end{pmatrix}, \tag{5.7}$$

where $\mathbf{0}_3 \in \mathbb{R}^3$ is a zero vector. To satisfy the non-degeneracy condition for a two-dimensional structure, $\mathbf{E}$ or $\mathbf{E}_1$ should possess three zero eigenvalues among its four eigenvalues. Consequently, the non-zero part $\hat{\mathbf{E}}_1 \in \mathbb{R}^{3\times 3}$ of $\mathbf{E}_1$ should have two zero eigenvalues.

However, it is still difficult to analytically determine the six force densities, although this is the simplest tensegrity structure. To simplify the problem, we assume a symmetric assignment of force densities as follows:

$$\begin{aligned} q_1 &= q_2 = q_{\mathrm{a}}, \\ q_3 &= q_4 = q_5 = q_6 = q_{\mathrm{b}}. \end{aligned} \tag{5.8}$$

Note that we have considered the same (symmetric) case in Chap. 3, but in a different way. The non-zero part $\hat{\mathbf{E}}_1$ of $\mathbf{E}_1$ becomes

$$\hat{\mathbf{E}}_1 = \begin{pmatrix} q_a + 2q_b & -q_b & -q_b \\ -q_b & q_a + 2q_b & -q_a \\ -q_b & -q_a & q_a + 2q_b \end{pmatrix}. \tag{5.9}$$

Because $\hat{\mathbf{E}}_1$ should have two zero eigenvalues, it is singular with zero determinant:

$$\det(\hat{\mathbf{E}}_1) = 4q_b\,(q_a + q_b)^2 = 0. \tag{5.10}$$

Since there exist prestresses in the members, the force density q_b should not be zero, and the following equation holds:

$$q_a + q_b = 0 \quad \Longrightarrow \quad q_a = -q_b. \tag{5.11}$$

Thus, the force density matrix $\mathbf{E}$ of the symmetric structure is

$$\mathbf{E} = q_a \begin{pmatrix} 1.0 & -1.0 & -1.0 & 1.0 \\ -1.0 & 1.0 & 1.0 & -1.0 \\ -1.0 & 1.0 & 1.0 & -1.0 \\ 1.0 & -1.0 & -1.0 & 1.0 \end{pmatrix}, \tag{5.12}$$

the eigenvalues λ_i^{E} of which are

$$\begin{aligned} \lambda_1^{\mathrm{E}} &= 4q_a, \\ \lambda_2^{\mathrm{E}} &= \lambda_3^{\mathrm{E}} = \lambda_4^{\mathrm{E}} = 0, \end{aligned} \tag{5.13}$$

satisfying the non-degeneracy condition for two-dimensional tensegrity structures. The linearly independent solutions, lying in the null-space of $\mathbf{E}$, of the self-equilibrium equations can be written as

$$\mathbf{x}, \mathbf{y} = a \begin{pmatrix} -1 \\ 1 \\ -1 \\ 1 \end{pmatrix} + b \begin{pmatrix} 1 \\ 1 \\ -1 \\ -1 \end{pmatrix}, \tag{5.14}$$

where a and b are arbitrary values that do not vanish simultaneously, and the trivial solution $(1, 1, 1, 1)^{\top}$ has been ruled out. Figure 5.2 shows one of the (infinite) possible geometry realizations with the nodal coordinates

$$\mathbf{x} = \begin{pmatrix} -1 \\ 1 \\ -1 \\ 1 \end{pmatrix}, \quad \mathbf{y} = \begin{pmatrix} 1 \\ 1 \\ -1 \\ -1 \end{pmatrix}. \tag{5.15}$$

Choosing a positive value for the force density of the exterior members [4]–[6]; i.e., $q_{\rm a} > 0$, configuration of the structure is shown in Fig. 5.2. In this case, the exterior members are cables, which are in tension and indicated by thin lines, and the interior members are struts, which are in compression with negative force density $q_{\rm b} = -q_{\rm a} < 0$.

In Example 5.3, we will see another self-equilibrated configuration of the structure with opposite signs of force densities; and furthermore, we will see that selection of signs of the force densities is critical to (super-)stability of the structure.

Note that symmetry, in terms of force densities, has been used in Example 5.2 to simplify the form-finding problem. More systematic usage of symmetry properties of complex tensegrity structures will be given in the next two chapters. For the structures with relatively large number of members, it becomes difficult to manage the force densities analytically, although this can be partly alleviated by using semi-analytical methods [4].

5.1.3 Super-Stability Condition

A super-stable structure is always preferable in the design process, because imperfection involved in the self-equilibrated configuration is unlikely to alter its super-stability property. This comes from the fact that any stretched version (geometry realization), by affine motions discussed in Appendix B.1, of a super-stable structure remains super-stable, because the force density matrix is not affected by changes of geometry realization. This advantage is of great importance since uncertainties in manufacture and construction are unavoidable in practice.

In Chap. 4, we have presented the sufficient conditions for super-stability of tensegrity structures—a d-dimensional tensegritry structure is super-stable if all of the following conditions are satisfied:

1. The force density matrix, or geometrical stiffness matrix, is positive semi-definite;
2. The force density matrix has rank deficiency of $d + 1$;
3. The geometry matrix is full-rank.

The first two conditions are related to force density matrix, or more specifically, they are related to force densities, while the third condition has nothing to do with force densities. Moreover, the first condition is also the necessary condition for super-stability, and the non-degeneracy condition is satisfied if the second condition holds. Therefore, in the first design stage of force density method, we search the force densities that make the force density matrix satisfy the first two conditions. This way, its geometry realization is non-degenerate, and more importantly, the structure is super-stable, if the third condition is also satisfied in the second design stage.

The conditions related to positive semi-definiteness of the force density matrix can be confirmed by looking at the signs of its eigenvalues: except for the $d+1$ zero eigenvalues, the eigenvalues of the force density matrix are positive.

Example 5.3 Super-stability of the two-dimensional X-cross tensegrity structures as in Figs. 5.2 and 5.3.

As presented in Example 5.2 using symmetry conditions, the only non-zero eigenvalue λ_1^{E} of its force density matrix $\mathbf{E}$ is

$$\lambda_1^{\mathrm{E}} = 4q_{\mathrm{a}}. \tag{5.16}$$

Thus, the symmetric X-cross tensegrity structure is super-stable only if $q_{\mathrm{a}} > 0$. This indicates that the four exterior members are cables carrying tension, while the two interior members are struts carrying compression, as indicated in Fig. 5.2.

For the case with opposite signs for the force densities, i.e., $q_{\mathrm{a}} < 0$ and $q_{\mathrm{b}} > 0$, the exterior members become struts and the interior members are cables as shown in Fig. 5.3. This structure is not super-stable,[a] because the non-zero eigenvalue λ_1^{E} becomes negative.

[a] The structure in this case is still prestress-stable, because it is kinematically determinate.

As discussed previously, the analytical method used for Example 5.2 is not efficient enough for relatively complex tensegrity structures, which are composed of a large number of nodes and members. Moreover, super-stability is not easy to guarantee in the form-finding process, and therefore, it is strongly desired by the designers to have a more effective and systematic (numerical) method, which

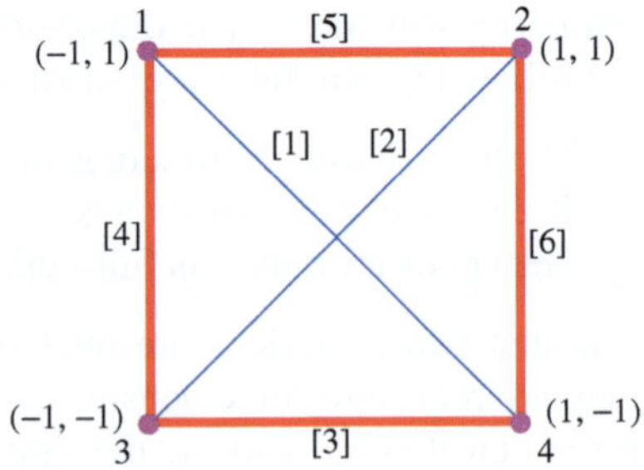

Fig. 5.3 Form-finding of the (two-dimensional) X-cross tensegrity structure. This structure is not super-stable, when the cables in tension are located inside of the struts in compression, although it has the same geometry realization as the structure in Fig. 5.2

guarantees a (super-)stable structure and has good control over its self-equilibrated configuration at the same time. In the next section, we will show that the so-called *adaptive force density method* (AFDM), which makes use of the concept of force density, is a promising method for this purpose.

5.2 Adaptive Force Density Method

In the form-finding process of a tensegrity structure in this chapter, we assume that connectivity of the structure is determined a priori. In the consecutive design stages of force density method for form-finding of a tensegrity structure, the variables are

1. The force densities, which are determined in the first design stage;
2. The geometry realization, which is described by nodal coordinates and determined in the second design stage.

5.2.1 First Design Stage: Feasible Force Densities

In this subsection, we show how to find the feasible force densities satisfying the non-degeneracy condition by applying eigenvalue analysis to the force density matrix. The necessary condition for super-stability, expressed by positive semi-definiteness of the force density matrix, can also be guaranteed in this design stage.

5.2.1.1 Linear Relation Between Force Densities and Force Density Matrix

Suppose that a free-standing d-dimensional tensegrity structure consists of n (free) nodes and m members. Let K_i denote the set of members connected to node i. From the direct definition of force density matrix $\mathbf{E}$ in Eq. (2.107), the ith row $\mathbf{E}_i$ of $\mathbf{E}$ can be written in terms of the force density vector $\mathbf{q}$ via a matrix ${}_i\mathbf{B} \in \mathbb{R}^{n\times m}$ as

$$ {}_i\mathbf{B}\mathbf{q} = (\mathbf{E}_i)^\top, \tag{5.17} $$

where the (j,k)th component ${}_iB_{(j,k)}$ of ${}_i\mathbf{B}$ is defined as

$$ {}_iB_{(j,k)} = \begin{cases} 1 & \text{if } i = j \text{ and } k \in K_i, \\ -1 & \text{if nodes } i \text{ and } j \text{ are connected by member k,} \\ 0 & \text{for other cases.} \end{cases} \tag{5.18} $$

By defining $\mathbf{B} \in \mathbb{R}^{n^2\times m}$ and $\mathbf{g} \in \mathbb{R}^{n^2}$ as

$$\mathbf{B}^\top = ({}_1\mathbf{B}^\top, \ldots, {}_i\mathbf{B}^\top, \ldots, {}_n\mathbf{B}^\top),$$
$$\mathbf{g} = (\mathbf{E}_1, \ldots, \mathbf{E}_i, \ldots, \mathbf{E}_n)^\top, \tag{5.19}$$

the relation between the force density matrix and force density vector $\mathbf{q}$ can be written in a linear form as follows:

$$\mathbf{Bq} = \mathbf{g}. \tag{5.20}$$

From the definition of $\mathbf{B}$, we may notice that there exist at least m rows in $\mathbf{B}$ consisting only one zero entries: -1; and moreover, these m rows are mutually independent. Hence, the rank of $\mathbf{B}$ is m, and furthermore, $\mathbf{B}$ is full-rank because n^2 is obviously larger than m.

Equation (5.20) can be regarded as a linear constraint on the force densities in terms of the force density matrix.

Example 5.4 The matrix ${}_1\mathbf{B}$ corresponding to node 1 of the two-dimensional X-cross tensegrity structure as shown in Fig. 5.2.

Node 1 of the X-cross tensegrity structure as shown in Fig. 5.2 is connected to nodes 2, 3, and 4 by members [5], [4], and [1], respectively. According to the direct definition of force density matrix $\mathbf{E}$ in Eq. (2.107), the first row $\mathbf{E}_1$ of $\mathbf{E}$ is

$$\mathbf{E}_1 = \left(q_1 + q_4 + q_5, \, -q_5, \, -q_4, \, -q_1\right). \tag{5.21}$$

According to the definition of ${}_i\mathbf{B}$ ($i = 1$) given in Eq. (5.18), the fifth, fourth, and first entries in the first row of ${}_1\mathbf{B}$ are all equal to 1, corresponding to members connected to node 1, and all other entries in this row are zero; moreover, the entries ${}_1B_{(2,5)}$, ${}_1B_{(3,4)}$, and ${}_1B_{(4,1)}$ are -1, while all other entries in these rows are zero.

In summary, ${}_1\mathbf{B}$ corresponding to node 1 of the X-cross structure is

$${}_1\mathbf{B} = \begin{pmatrix} 1 & 0 & 0 & 1 & 1 & 0 \\ 0 & 0 & 0 & 0 & -1 & 0 \\ 0 & 0 & 0 & -1 & 0 & 0 \\ -1 & 0 & 0 & 0 & 0 & 0 \end{pmatrix}. \tag{5.22}$$

Furthermore, Eq. (5.17) can be verified by

$${}_1\mathbf{Bq} = \begin{pmatrix} q_1 + q_4 + q_5 \\ -q_5 \\ -q_4 \\ -q_1 \end{pmatrix} = (\mathbf{E}_1)^\top. \tag{5.23}$$

5.2.1.2 Spectral Decomposition

To let the force density matrix have necessary number of zero eigenvalues, its eigenvalues have to be derived in an explicit way, which can be done by spectral decomposition or eigenvalue analysis of the matrix.

Because the force density matrix is symmetric, it can be written as follows by applying spectral decomposition [2]:

$$\mathbf{E} = \boldsymbol{\Phi}\boldsymbol{\Lambda}\boldsymbol{\Phi}^\top, \tag{5.24}$$

where the diagonal entries λ_i^{E} $(i = 1, 2, \ldots, n)$ are the *eigenvalues* of $\mathbf{E}$. Moreover, the eigenvalues are arranged in a non-decreasing order as

$$\lambda_1^{\mathrm{E}} \leq \lambda_2^{\mathrm{E}} \leq \cdots \leq \lambda_n^{\mathrm{E}}. \tag{5.25}$$

The ith column $\boldsymbol{\Phi}_i$ of $\boldsymbol{\Phi}$ is the eigenvector of $\mathbf{E}$ corresponding to its ith eigenvalue λ_i^{E}; i.e.,

$$\mathbf{E}\boldsymbol{\Phi}_i = \lambda_i^{\mathrm{E}}\boldsymbol{\Phi}_i, \tag{5.26}$$

and $\boldsymbol{\Phi}_i$ is orthor-nomalized as

$$\boldsymbol{\Phi}_i^\top\boldsymbol{\Phi}_j = \delta_{ij}, \tag{5.27}$$

where δ_{ij} is the Kronecker delta:

$$\delta_{ij} = \begin{cases} 1, & i = j; \\ 0, & i \neq j. \end{cases} \tag{5.28}$$

It is clear that the number of non-zero eigenvalues of $\mathbf{E}$ is equal to its rank, denoted by r^{E}. Moreover, rank deficiency $\bar{r}^{\mathrm{E}}$ of $\mathbf{E}$ is defined as

$$\bar{r}^{\mathrm{E}} = n - r^{\mathrm{E}}. \tag{5.29}$$

From the non-degeneracy condition for a d-dimensional tensegrity structure in Eq. (2.121), $\bar{r}^{\mathrm{E}}$ should be equal to or larger than $d + 1$:

$$\bar{r}^{\mathrm{E}} \geq h^\star = d + 1. \tag{5.30}$$

For arbitrarily assigned (initial) force densities, the above condition is usually not satisfied. Moreover, as has been discussed in Chap. 4, having the minimum rank deficiency $d + 1$ in the force density matrix is one of the sufficient conditions for super-stability of tensegrity structures. Hence, our task is to find the force densities satisfying

$$\bar{r}^{\mathrm{E}} = h^\star = d + 1, \tag{5.31}$$

with all the non-zero eigenvalues positive. For such purpose, we can directly assign desired values to the eigenvalues of the matrix and then recalculate the matrix by using Eq. (5.24) with the newly assigned eigenvalues.

Let $\hat{r}^{\mathrm{E}}$ denote the number of non-positive eigenvalues of $\mathbf{E}$. It is obvious that $\bar{r}^{\mathrm{E}} \leq \hat{r}^{\mathrm{E}}$. Moreover, we have the following two cases for $\hat{r}^{\mathrm{E}}$ compared to $h^\star$:

Case 1: $\hat{r}^{\mathrm{E}} \leq h^\star$;
Case 2: $\hat{r}^{\mathrm{E}} > h^\star$.

For Case 1, we can simply assign 0 to the first $h^\star$ smallest eigenvalues of $\mathbf{E}$ as

$$\lambda_i^{\mathrm{E}} = 0, \quad (i = 1, 2, \ldots, h^\star), \tag{5.32}$$

to obtain the new $\bar{\boldsymbol{\Lambda}}$ with the modified eigenvalues. The force density matrix is updated as follows by using $\bar{\boldsymbol{\Lambda}}$:

$$\bar{\mathbf{E}} = \boldsymbol{\Phi}\bar{\boldsymbol{\Lambda}}\boldsymbol{\Phi}^\top. \tag{5.33}$$

This way, $\mathbf{E}$ will have the required rank deficiency $h^\star$; moreover, it is positive semi-definite without negative eigenvalues.

However, for Case 2 where $\hat{r}^{\mathrm{E}} > h^\star$, the rank deficiency will be larger than the required number, if the same operation as Case 1 is adopted. For this case, we may have several strategies, for example

1. Assign positive values to some of the non-positive eigenvalues;
2. Specify more than $h^\star$ independent coordinates in the form-finding process.

Since arbitrarily specified initial force densities usually result in $\hat{r}^{\mathrm{E}} \leq h^\star$, we will consider only Case 1 in the following discussions.

Example 5.5 Spectral decomposition of the force density matrix of the two-dimensional X-cross tensegrity structure as shown in Fig. 5.2.

Assign, for example, 0.5 to force densities of the cables and −1.0 to those of the struts:

$$\begin{aligned} q_1 = q_2 &= -1.0, \\ q_3 = q_4 = q_5 = q_6 &= 0.5. \end{aligned} \tag{5.34}$$

The corresponding force density matrix $\mathbf{E}$ is

$$\mathbf{E} = \begin{pmatrix} 0.0 & -0.5 & -0.5 & 1.0 \\ -0.5 & 0.0 & 1.0 & -0.5 \\ -0.5 & 1.0 & 0.0 & -0.5 \\ 1.0 & -0.5 & -0.5 & 0.0 \end{pmatrix}. \tag{5.35}$$

The four eigenvalues of $\mathbf{E}$ are

$$\begin{aligned}\lambda_1^{\mathrm{E}} &= \lambda_2^{\mathrm{E}} = -1.0,\\ \lambda_3^{\mathrm{E}} &= 0.0,\\ \lambda_4^{\mathrm{E}} &= 2.0,\end{aligned} \tag{5.36}$$

and the corresponding eigenvectors in $\boldsymbol{\Phi}$ are

$$\boldsymbol{\Phi} = \begin{pmatrix} 0.0926 & 0.7010 & -0.5000 & -0.5000 \\ 0.7010 & -0.0926 & -0.5000 & 0.5000 \\ -0.7010 & 0.0926 & -0.5000 & 0.5000 \\ -0.0926 & -0.7010 & -0.5000 & -0.5000 \end{pmatrix}. \tag{5.37}$$

The non-degeneracy condition for a two-dimensional tensegrity structure is not satisfied, because there is only one zero eigenvalue in $\mathbf{E}$, which comes from the fact that the structure is free-standing. Therefore, additional two zero eigenvalues in $\mathbf{E}$ are needed.

Since $\hat{r}^{\mathrm{E}} = h^\star = 3$ for a two-dimensional structure, this belongs to Case 1, and we can enforce the two negative eigenvalues λ_1^{E} and λ_2^{E} to zero:

$$\bar{\lambda}_1^{\mathrm{E}} = \bar{\lambda}_2^{\mathrm{E}} = 0.0. \tag{5.38}$$

By applying Eq. (5.33), the force density matrix $\bar{\mathbf{E}}$ is updated as

$$\bar{\mathbf{E}} = \frac{1}{2}\begin{pmatrix} 1 & -1 & -1 & 1 \\ -1 & 1 & 1 & -1 \\ -1 & 1 & 1 & -1 \\ 1 & -1 & -1 & 1 \end{pmatrix}. \tag{5.39}$$

Note that the updated force density matrix $\bar{\mathbf{E}}$ is usually not related to any set of force densities of the members, for which the least square strategy presented in the next subsection is necessary.

5.2.1.3 Updating Force Densities

Suppose that we have obtained the updated force density matrix $\bar{\mathbf{E}}$ by enforcing necessary number of its smallest eigenvalues to zero. Our next task is to find a set of (approximate) force densities $\bar{\mathbf{q}}$ corresponding to $\bar{\mathbf{E}}$.

Because $\mathbf{B}$ is full-rank but not square, $\mathbf{B}^\top\mathbf{B}$ is square and full-rank, and therefore, it is invertible. Assembling $\bar{\mathbf{g}}$ by the components of $\bar{\mathbf{E}}$ using Eq. (5.18), the approximate force densities $\bar{\mathbf{q}}$ can be derived as follows by applying the least square method as presented in Appendix A.4:

$$\bar{\mathbf{q}} = \left(\mathbf{B}^\top \mathbf{B}\right)^{-1} \mathbf{B}^\top \bar{\mathbf{g}}. \tag{5.40}$$

Example 5.6 The approximate force densities corresponding to the updated force density matrix presented in Example 5.5.

Assemble $\bar{\mathbf{g}}$ from the modified force density matrix $\bar{\mathbf{E}}$ given in Eq. (5.39) in Example 5.5. The least square solution of the force densities is then calculated as

$$\begin{aligned} \bar{q}_1 &= \bar{q}_2 = -0.5, \\ \bar{q}_3 &= \bar{q}_4 = \bar{q}_5 = \bar{q}_6 = 0.5. \end{aligned} \tag{5.41}$$

The force density matrix $\mathbf{E}$ corresponding to these updated force densities is

$$\mathbf{E} = \frac{1}{2}\begin{pmatrix} 1.0 & -1.0 & -1.0 & 1.0 \\ -1.0 & 1.0 & 1.0 & -1.0 \\ -1.0 & 1.0 & 1.0 & -1.0 \\ 1.0 & -1.0 & -1.0 & 1.0 \end{pmatrix}, \tag{5.42}$$

which consists of three zero eigenvalues, satisfying the non-degeneracy condition for a two-dimensional structure, and furthermore, the only non-zero eigenvalue is 2.0, which is positive satisfying the necessary and one of the sufficient conditions for super-stability of a tensegrity structure.

In Example 5.6, the force densities satisfying the non-degeneracy and super-stability condition can be derived after only one updating process. However, it should be noted that this is a special case because the structure is simple. For complex structures with large numbers of nodes and members, the feasible force densities have to be derived by application of the least square updating process many times in an iterative manner.

5.2.1.4 Algorithm for Finding Feasible Force Densities

The process of iteratively updating the force densities until they satisfy the non-degeneracy as well as the super-stability conditions is summarized in Algorithm 5.1, where the superscript $[i]$ refers to the ith iteration in the iterative process.

Algorithm 5.1: Finding feasible force densities

Step 0: Specify an initial force density vector $\mathbf{q}^{[0]}$, and assemble its corresponding force density matrix $\mathbf{E}^{[0]}$ by Eq. (2.101). Set the counter of iterations as $i := 0$.

Step 1: Assign zero to the $h^{\star}$ smallest eigenvalues of $\mathbf{E}^{[i]}$, and compute its counterpart $\bar{\mathbf{E}}^{[i]}$ using the modified eigenvalues by Eq. (5.33).

Step 2: Assemble $\bar{\mathbf{g}}^{[i]}$ from the components of $\bar{\mathbf{E}}^{[i]}$, update the force density vector $\bar{\mathbf{q}}^{[i]}$ by Eq. (5.40), and compute its corresponding force density matrix $\mathbf{E}^{[i+1]}$ by Eq. (2.101).

Step 3: If the updated force density matrix $\mathbf{E}^{[i+1]}$ has the required rank deficiency $h^{\star}(=d+1)$, then let $\mathbf{q} = \bar{\mathbf{q}}^{[i]}$ and $\mathbf{E} = \mathbf{E}^{[i+1]}$, and terminate the algorithm; otherwise, let $\mathbf{q}^{[i+1]} := \bar{\mathbf{q}}^{[i]}$, update the counter as $i := i + 1$, and return to Step 1.

By application of Algorithm 5.1, we can adaptively derive the force densities that satisfy the non-degeneracy as well as super-stability conditions for a d-dimensional tensegrity structure. Hence, the form-finding method using this algorithm is called the adaptive force density method.

Example 5.7 Find the feasible force densities for the simplest three-dimensional tensegrity structure as shown in Fig. 5.4.

The structure as shown in Fig. 5.4 is the simplest tensegrity structure in three-dimensional space. The self-equilibrium of a class of this structure having dihedral symmetry has been studied in Chap. 3.

The structure is composed of six nodes and twelve members; i.e., $n = 6$, $m = 12$. There are two types of cables: the horizontal cables [1]–[6] connected to the nodes on the same parallel plane, and the vertical cables [7]–[9] connected to the nodes on different planes. The initial force densities of the different types of cables as well as the struts are given as follows:

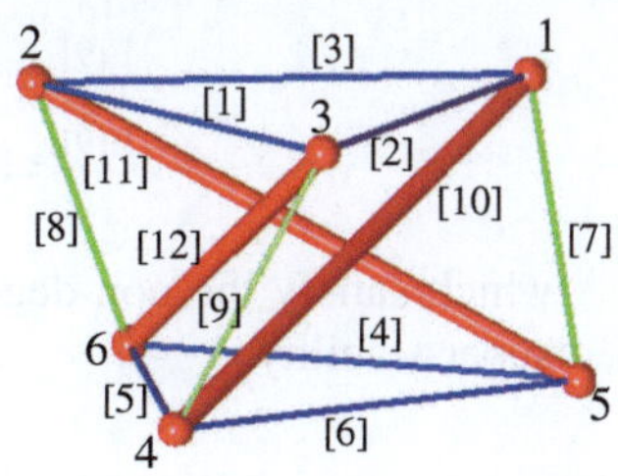

Fig. 5.4 The simplest three-dimensional (prismatic) tensegrity structure

$$
\begin{aligned}
q_k^{[0]} &= 1.0, \quad (k = 1, 2, \ldots, 6); \\
q_k^{[0]} &= 2.0, \quad (k = 7, 8, 9); \\
q_k^{[0]} &= -1.0, \ (k = 10, 11, 12).
\end{aligned} \tag{5.43}
$$

The eigenvalues of the corresponding force density matrix $\mathbf{E}^{[0]}$ are

$$
\begin{aligned}
\lambda_1^{E[0]} &= 0.0000, \quad \lambda_2^{E[0]} = 1.3542, \quad \lambda_3^{E[0]} = 1.3542, \\
\lambda_4^{E[0]} &= 2.0000, \quad \lambda_5^{E[0]} = 6.6458, \quad \lambda_6^{E[0]} = 6.6458.
\end{aligned} \tag{5.44}
$$

As can be observed from Eq. (5.44), the non-degeneracy condition for a three-dimensional tensegrity structure is not satisfied, because there is only one zero eigenvalue coming from having no fixed node, and three additional zero eigenvalues are necessary. By setting the smallest four eigenvalues of $\mathbf{E}$ to zero and updating the force densities as

$$
\begin{aligned}
q_k^{[1]} &= 1.0812, \quad (k = 1, 2, \ldots, 6), \\
q_k^{[1]} &= 1.9258, \quad (k = 7, 8, 9), \\
q_k^{[1]} &= -1.8751, \ (k = 10, 11, 12),
\end{aligned} \tag{5.45}
$$

the eigenvalues of the corresponding force density matrix become

$$
\begin{aligned}
\lambda_1^{E[1]} &= 0.0000, \quad \lambda_2^{E[1]} = 0.0575, \quad \lambda_3^{E[1]} = 0.0575, \\
\lambda_4^{E[1]} &= 0.5024, \quad \lambda_5^{E[1]} = 6.5883, \quad \lambda_6^{E[1]} = 6.5883.
\end{aligned} \tag{5.46}
$$

After nine iterations, Algorithm 5.1 terminates with the force densities

$$
\begin{aligned}
q_k^{[9]} &= 1.0976, \quad (k = 1, 2, \ldots, 6), \\
q_k^{[9]} &= 1.9012, \quad (k = 7, 8, 9), \\
q_k^{[9]} &= -1.9012, \ (k = 10, 11, 12),
\end{aligned} \tag{5.47}
$$

and the eigenvalues of the force density matrix are

$$
\begin{aligned}
\lambda_1^{E[9]} &= \lambda_2^{E[9]} = \lambda_3^{E[9]} = \lambda_4^{E[9]} = 0.0000, \\
\lambda_5^{E[9]} &= \lambda_6^{E[9]} = 6.5859,
\end{aligned} \tag{5.48}
$$

which satisfy the non-degeneracy condition and the necessary condition for super-stability.

As demonstrated in Example 5.7, the force densities as well as the corresponding force density matrix, satisfying the conditions for non-degeneracy and super-stability of a tensegrity structure, can be determined by conducting Algorithm 5.1. However, there are infinite number of possibilities of geometry realization satisfying the self-equilibrium equations associated with the determined force density matrix. The next subsection presents a strategy to uniquely determine the self-equilibrated configuration in terms of nodal coordinates according to requirements by the designers.

5.2.2 Second Design Stage: Self-equilibrated Configuration

When the feasible force densities have been determined by the application of Algorithm 5.1, the next step is to find the nodal coordinates $\mathbf{x}, \mathbf{y}, \mathbf{z}$ satisfying the self-equilibrium equations associated with the determined force density matrix $\mathbf{E}$ in Eq. (2.113).

Furthermore, the nodal coordinates $\mathbf{x}, \mathbf{y}, \mathbf{z}$ should be mutually independent as discussed in Chap. 2; otherwise, geometry realization of the structure will degenerate into the space with lower dimensions. In the following, we consider only the three-dimensional cases for clarity.

Assemble the matrix $\mathbf{H} \in \mathbb{R}^{3n \times 3n}$ by the force density matrix $\mathbf{E}$, and assemble the generalized nodal coordinates $\mathbf{X} \in \mathbb{R}^{3n}$ by the nodal coordinates $\mathbf{x}, \mathbf{y}, \mathbf{z}$ as follows:

$$\mathbf{H} = \begin{pmatrix} \mathbf{E} & \mathbf{O}_n & \mathbf{O}_n \\ \mathbf{O}_n & \mathbf{E} & \mathbf{O}_n \\ \mathbf{O}_n & \mathbf{O}_n & \mathbf{E} \end{pmatrix},$$
$$\mathbf{X} = \begin{pmatrix} \mathbf{x} \\ \mathbf{y} \\ \mathbf{z} \end{pmatrix}, \tag{5.49}$$

where $\mathbf{O}_n \in \mathbb{R}^{n \times n}$ is a zero matrix. Note that $\mathbf{H}$ is identical to the geometrical stiffness matrix $\mathbf{K}_G$. The self-equilibrium equations in each direction can be summarized as

$$\mathbf{HX} = \mathbf{0}. \tag{5.50}$$

Because rank deficiency of the force density matrix $\mathbf{E}$ is $h^\star$ after conducting Algorithm 5.1, rank deficiency of $\mathbf{H}$ is $3h^\star$. Therefore, there are in total $3h^\star$ independent solutions for $\mathbf{X}$ if no constraint is considered.

The solutions of Eq. (5.50) can be summarized in a matrix form as

$$\mathbf{X} = \mathbf{A}\boldsymbol{\beta}, \tag{5.51}$$

where the columns of $\mathbf{A} \in \mathbb{R}^{3n \times 3h^\star}$ are the independent coordinate modes, and $\boldsymbol{\beta} \in \mathbb{R}^{3h^\star}$ is the coefficient vector. Let $\mathbf{A}_i$ denote the ith column of $\mathbf{A}$, we have

$$\mathbf{H}\mathbf{A}_i = \mathbf{0}. \tag{5.52}$$

If we specify an independent set of nodal coordinates $\bar{\mathbf{X}} \in \mathbb{R}^{3h^\star}$ and obtain the corresponding components $\bar{\mathbf{A}} \in \mathbb{R}^{3h^\star \times 3h^\star}$ in $\mathbf{A}$, where $\text{rank}(\bar{\mathbf{A}}) = 3h^\star$, geometry of the structure in terms of nodal coordinates $\mathbf{X}$ can be uniquely determined as

$$\mathbf{X} = \mathbf{A}\bar{\mathbf{A}}^{-1}\bar{\mathbf{X}}. \tag{5.53}$$

Example 5.8 A geometry realization of the prismatic structure in Example 5.7.

For the three-dimensional structure in Example 5.7, rank deficiency of its force density matrix is four, and therefore, there are in total twelve independent components of the coordinates that should be assigned to uniquely determine its self-equilibrated configuration.

From numerical investigations, as given later in Example 5.9, nodal coordinates of the first four nodes 1, 2, 3, and 4 are mutually independent. Assign, for example, the coordinates of these nodes as

$$\bar{\mathbf{x}}_1 = \begin{pmatrix} 10.0 \\ 0.0 \\ -5.0 \end{pmatrix}, \quad \bar{\mathbf{x}}_2 = \begin{pmatrix} -5.0 \\ 8.0 \\ -5.0 \end{pmatrix},$$
$$\bar{\mathbf{x}}_3 = \begin{pmatrix} -5.0 \\ -8.0 \\ -5.0 \end{pmatrix}, \quad \bar{\mathbf{x}}_4 = \begin{pmatrix} 5.0 \\ 0.0 \\ 5.0 \end{pmatrix}. \tag{5.54}$$

By applying Eq. (5.53) together with the assigned components

$$\bar{\mathbf{X}} = \left(\bar{\mathbf{x}}_1^\top, \bar{\mathbf{x}}_2^\top, \bar{\mathbf{x}}_3^\top, \bar{\mathbf{x}}_4^\top\right)^\top, \tag{5.55}$$

the coordinates of nodes 5 and 6 are calculated as

$$\bar{\mathbf{x}}_5 = \begin{pmatrix} 22.3205 \\ 0.0000 \\ 5.0000 \end{pmatrix}, \quad \bar{\mathbf{x}}_6 = \begin{pmatrix} 13.6603 \\ 13.8564 \\ 5.0000 \end{pmatrix}, \tag{5.56}$$

and the geometry realization of the structure is shown in Fig. 5.5.

Independent components of nodal coordinates can be, for instance, identified by using RREF (Reduced Row-Echelon Form) of transpose $\mathbf{A}^\top$ of the matrix $\mathbf{A}$ defined

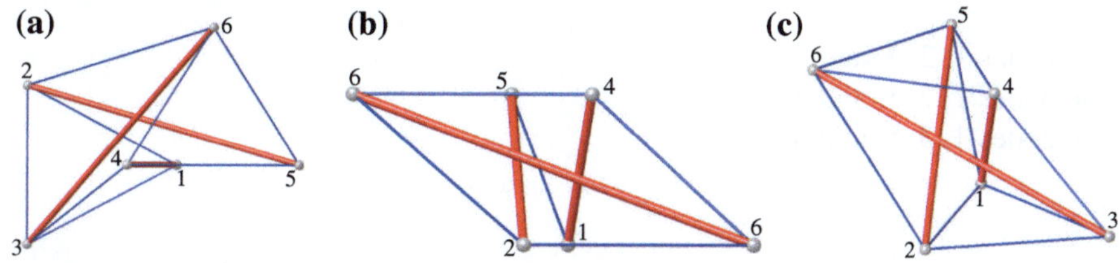

Fig. 5.5 A possible (asymmetric) geometry realization of the prismatic tensegrity structure in Example 5.8. **a** Top view, **b** side view, **c** diagonal view

in Eq. (5.51). More discussions on RREF can be found in Appendix A, and an example is given in Example 5.9.

Example 5.9 Selection of independent components of nodal coordinates for Example 5.8.

Using the force densities obtained in Example 5.7 for the simplest three-dimensional tensegrity structure, Reduced Row-Echelon Form (RREF) of $\mathbf{A}^\top$ is computed as

$$
\mathrm{RREF}(\mathbf{A}^\top)
$$

$$
= \left(\begin{array}{cccccc|cccccc|cccccc}
\multicolumn{6}{c}{x} & \multicolumn{6}{c}{y} & \multicolumn{6}{c}{z} \\
1&0&0&0&1.155&0.577&0&0&0&0&0&0&0&0&0&0&0&0\\
0&1&0&0&-0.577&0.577&0&0&0&0&0&0&0&0&0&0&0&0\\
0&0&1&0&-0.577&-1.155&0&0&0&0&0&0&0&0&0&0&0&0\\
0&0&0&1&1&1&0&0&0&0&0&0&0&0&0&0&0&0\\
0&0&0&0&0&0&1&0&0&0&1.155&0.577&0&0&0&0&0&0\\
0&0&0&0&0&0&0&1&0&0&-0.577&0.577&0&0&0&0&0&0\\
0&0&0&0&0&0&0&0&1&0&-0.577&-1.155&0&0&0&0&0&0\\
0&0&0&0&0&0&0&0&0&1&1&1&0&0&0&0&0&0\\
0&0&0&0&0&0&0&0&0&0&0&0&1&0&0&0&1.155&0.577\\
0&0&0&0&0&0&0&0&0&0&0&0&0&1&0&0&-0.577&0.577\\
0&0&0&0&0&0&0&0&0&0&0&0&0&0&1&0&-0.577&-1.155\\
0&0&0&0&0&0&0&0&0&0&0&0&0&0&0&1&1&1
\end{array}\right). \tag{5.57}
$$

The first six columns correspond to the x-coordinates of the six nodes of the structure, and the remaining twelve columns correspond to y- and z-coordinates. It is observed from Eq. (5.57) that the coordinates in each direction of the first four nodes 1, 2, 3, and 4 are mutually independent. There are of course many other combinations of independent components. Some but not exhaustive examples are the coordinates of

- nodes 1, 2, 3, and 5;
- nodes 1, 2, 3, and 6;
- nodes 1, 3, 5, and 6.

5.2.3 *Remarks*

The two design stages of the force density method for form-finding of tensegrity structures are completely separated. In the first design stage, different initially specified force densities will result in different solutions, which lead to different possible geometry realizations. More importantly, super-stability of the structure can be guaranteed by assigning positive values to the eigenvalues of the corresponding force density matrix, while the necessary number of zero eigenvalues is ensured for non-degeneracy condition.

In the second design stage, the final geometry realization of the structure can be controlled by choosing different sets of independent coordinates, and/or by assigning different values to the selected components.

To have more accurate control over the final geometry realization of a tensegrity structure, geometrical constraints might be assigned in both design stages, which will be discussed in the next section.

5.3 Geometrical Constraints

This section introduces geometrical constraints, including the constraints on (rotational) symmetry properties as well as elevation (z-coordinates) of the structure, to the adaptive force density method. The constraints on force densities are incorporated into the first design stage, and the constraints on nodal coordinates are considered in the second design stage. In particular, we intend to apply the geometrical constraints to the form-finding of tensegrity towers, which are assembled by using the prismatic structures in the vertical direction. More discussions on configuration and connectivity of tensegrity towers can be found in Appendix C.

5.3.1 *Constraints on Rotational Symmetry*

In this subsection, we show that rotational symmetry about z-axis of a tensegrity structure is guaranteed by using linear equations in two different forms:

- equations with respect to force densities, and
- equations with respect to nodal coordinates.

These linear constraints on rotational symmetry of the structure are respectively incorporated into the two design stages.

5.3.1.1 Symmetry of Force Densities

When we say that a structure is of $\bar{n}$-fold (rotational) symmetry, the structure has the same appearance after rotation about a specific axis through any of the $\bar{n}$ angles $2i\pi/\bar{n}$ ($i = 0, 1, 2, \dots, \bar{n}-1$). Furthermore, the members that can be moved to each other by the rotations are said to be in the same *orbit*, having the same lengths and prestresses; therefore, the members in the same orbit have the same force densities. There are usually more than one orbit of members in a symmetric structure.

Example 5.10 Orbits of members in Fig. 5.6.

Figure 5.6 shows the kth layer of a tensegrity tower, where the vertical cables and saddle cables have been removed from the figure for clarity. In each layer, there are five struts. The nodes are located on two parallel planes, the top plane and the bottom plane. The nodes on the top plane are denoted by $\mathrm{P}^{\mathrm{t}}_{k,j}$ ($j = 1, 2, \dots, 5$), and those on the bottom plane are $\mathrm{P}^{\mathrm{b}}_{k,j}$ as shown in the figure.

The five struts in thick lines are connected by pairs of nodes denoted by $[\mathrm{P}^{\mathrm{t}}_{k,j}, \mathrm{P}^{\mathrm{b}}_{k,j}]$. These struts are of rotational symmetry, and they belong to the same orbit. One of the struts can be moved to any other by the rotation about z-axis through a proper angle $2i\pi/5$ ($i = 0, 1, 2, 3, 4$). Note that the (vertical) cables are also of rotational symmetry, and they belong to another orbit.

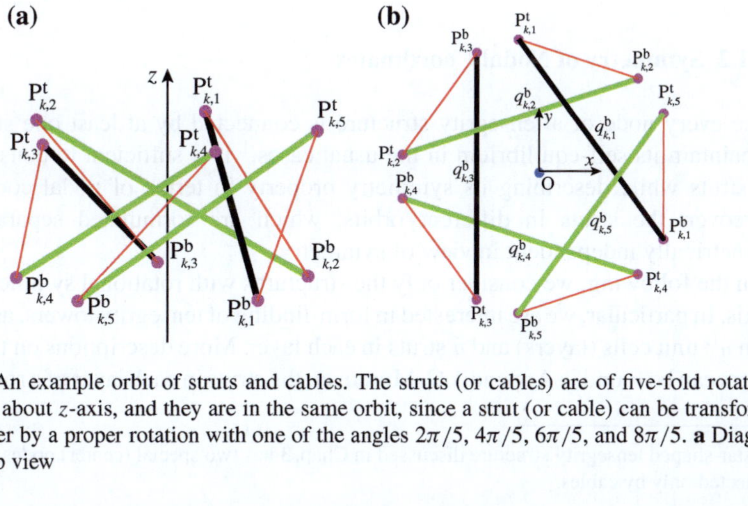

Fig. 5.6 An example orbit of struts and cables. The struts (or cables) are of five-fold rotational symmetry about z-axis, and they are in the same orbit, since a strut (or cable) can be transformed to any other by a proper rotation with one of the angles $2\pi/5$, $4\pi/5$, $6\pi/5$, and $8\pi/5$. **a** Diagonal view, **b** top view

The members of a (rotationally) symmetric structure in the same orbit have the same force densities, which is expressed in a matrix form as

$$\mathbf{Fq} = \mathbf{0}. \tag{5.58}$$

There are only two non-zero entries, 1 and -1, in each row of $\mathbf{F}$: if members i and j ($i < j$) are in the same orbit, there is one row of $\mathbf{F}$, the ith and jth entries of which are respectively 1 and -1 while the other entries in the row are 0. Size of $\mathbf{F}$ depends on number of obits as well as number of members in each orbit.

The linear constraint defined in Eq. (5.58) on the symmetry properties with respect to the force densities will be incorporated in the first design stage for determination of force densities satisfying the non-degeneracy condition.

Example 5.11 Symmetry of the five struts as shown in Fig. 5.6 in terms of force densities.

Consider the linear constraint on force densities of the struts in the kth layer of a tensegrity tower as shown in Fig. 5.6. These struts belong to the same orbit, and their force densities are denoted by $q^{\mathrm{b}}_{k,i}$ ($i = 1, 2, 3, 4, 5$). Since the struts have the same force densities, we have the following linear constraint on their force densities

$$\mathbf{Fq} = \begin{pmatrix} 1 & -1 & 0 & 0 & 0 \\ 0 & 1 & -1 & 0 & 0 \\ 0 & 0 & 1 & -1 & 0 \\ 0 & 0 & 0 & 1 & -1 \end{pmatrix} \begin{pmatrix} q^{\mathrm{b}}_{k,1} \\ q^{\mathrm{b}}_{k,2} \\ q^{\mathrm{b}}_{k,3} \\ q^{\mathrm{b}}_{k,4} \\ q^{\mathrm{b}}_{k,5} \end{pmatrix} = \begin{pmatrix} 0 \\ 0 \\ 0 \\ 0 \end{pmatrix}. \tag{5.59}$$

5.3.1.2 Symmetry of Nodal Coordinates

Since every node of a tensegrity structure is connected by at least one strut so as to maintain its self-equilibrium in the usual cases,[1] it is sufficient to consider only the struts while describing its symmetry property in terms of nodal coordinates. Moreover, the struts in different orbits, which are formulated separately, are geometrically independent in view of symmetry.

In the following, we consider only the structures with rotational symmetry about z-axis. In particular, we are interested in form-finding of tensegrity towers, assembled from n^{L} unit cells (layers) and $\bar{n}$ struts in each layer. More descriptions on tensegrity towers can be found in Appendix C. Moreover, the struts in each layer of a rotationally

[1] A star-shaped tensegrity structure discussed in Chap. 3 has two special (center) nodes, which are connected only by cables.

symmetric tensegrity tower belong to the same orbit, and hence, there are $\bar{n}$ struts in each orbit of struts.

The top (higher) and bottom (lower) nodes of the ith strut $\mathrm{B}_{k,i}$ of layer k are denoted by $\mathrm{P}^{\mathrm{t}}_{k,i}$ and $\mathrm{P}^{\mathrm{b}}_{k,i}$, respectively. Because the members in the same orbit that are rotationally symmetric about z-axis must have the same z-coordinate, we need only to consider the constraints on xy-coordinates. The x- and y-coordinates of nodes $\mathrm{P}^{\mathrm{t}}_i$ and $\mathrm{P}^{\mathrm{b}}_i$ are denoted by the vectors $\mathbf{x}^{\mathrm{t}}_{k,i} = (x^{\mathrm{t}}_{k,i}, y^{\mathrm{t}}_{k,i})^\top$ and $\mathbf{x}^{\mathrm{b}}_{k,i} = (x^{\mathrm{b}}_{k,i}, y^{\mathrm{b}}_{k,i})^\top$, respectively.

The member direction vector $\mathbf{d}_{k,i} \in \mathbb{R}^2$ $(i = 1, 2, \ldots, \bar{n})$ of the struts on xy-plane is defined as

$$\mathbf{d}_{k,i} = \mathbf{x}^{\mathrm{b}}_{k,i} - \mathbf{x}^{\mathrm{t}}_{k,i}, \tag{5.60}$$

which corresponds to an edge of a directed graph [1].

Example 5.12 Directed graph of the five struts in Fig. 5.6.

Figure 5.7 shows the directed graph of the five struts of layer k in Fig. 5.6 on xy-plane. The directed edge corresponding to the ith strut is denoted by $\mathbf{d}_{k,i}$ $(i = 1, 2, \ldots, 5)$, and it is pointing from the top node $\mathrm{P}^{\mathrm{t}}_{k,i}$ to the bottom node $\mathrm{P}^{\mathrm{b}}_{k,i}$.

The member direction vectors $\mathbf{d}_{k,i}$ $(i = 1, 2, \ldots, \bar{n})$ in the kth orbit of struts are combined to $\mathbf{d}_k \in \mathbb{R}^{2\bar{n}}$ as

$$\mathbf{d}_k = \left(\mathbf{d}^\top_{k,1}, \ldots, \mathbf{d}^\top_{k,\bar{n}}\right)^\top, \tag{5.61}$$

and the xy-coordinates $\mathbf{x}^{\mathrm{b}}_i$ and $\mathbf{x}^{\mathrm{t}}_i$ $(i = 1, 2, \ldots, \bar{n})$ of the nodes in the orbit are also combined to $\mathbf{x}_k \in \mathbb{R}^{4\bar{n}}$ as

$$\mathbf{x}_k = \left((\mathbf{x}^{\mathrm{b}}_{k,1})^\top, \ldots, (\mathbf{x}^{b}_{k,\bar{n}})^\top, (\mathbf{x}^{\mathrm{t}}_{k,1})^\top, \ldots, (\mathbf{x}^{\mathrm{t}}_{k,\bar{n}})^\top\right)^\top. \tag{5.62}$$

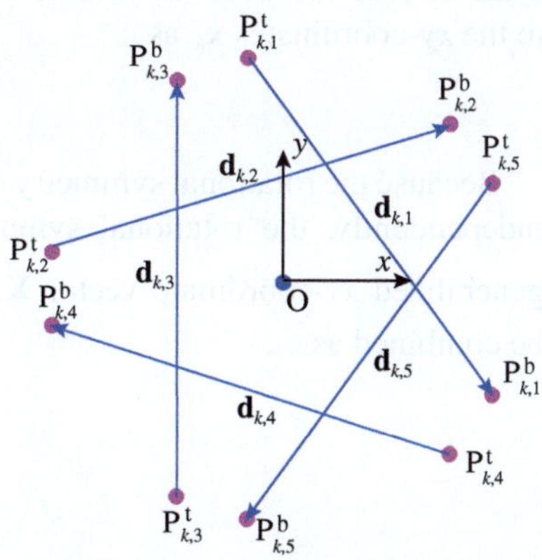

Fig. 5.7 *Top view* of directed graph of the five struts in Fig. 5.6

The relationship between $\mathbf{d}^k$ and $\mathbf{x}^k$ in orbit (layer) k can be written as follows by a matrix $\mathbf{T}_k \in \mathbb{R}^{2\bar{n}\times 4\bar{n}}$:

$$\mathbf{d}_k = \mathbf{T}_k \mathbf{x}_k, \tag{5.63}$$

where the matrix $\mathbf{T}_k$ is constructed by the $2\bar{n}$-by-$2\bar{n}$ identity matrix $\mathbf{I}_{2\bar{n}}$ as

$$\mathbf{T}_k = \left(\mathbf{I}_{2\bar{n}},\ -\mathbf{I}_{2\bar{n}}\right). \tag{5.64}$$

If two member direction vectors $\mathbf{d}_{k,1}$ and $\mathbf{d}_{k,i}$ of orbit k are of rotational symmetry; i.e., $\mathbf{d}_{k,1}$ coincides with $\mathbf{d}_{k,i}$ by the counter-clockwise rotation about z-axis by the angle $2(i-1)\pi/\bar{n}$, which is simply written as

$$\mathbf{d}_{k,i} = \mathbf{R}_{k,i}\mathbf{d}_{k,1}, \tag{5.65}$$

where the rotation matrix $\mathbf{R}_{k,i} \in \mathbb{R}^{2\times 2}$ is defined as

$$\mathbf{R}_{k,i} = \begin{pmatrix} C_i & -S_i \\ S_i & C_i \end{pmatrix}, \tag{5.66}$$

with $C_i = \cos(2(i-1)\pi/\bar{n})$ and $S_i = \sin(2(i-1)\pi/\bar{n})$.

Equation (5.65) can be rewritten with respect to $\mathbf{d}_k$ as follows:

$$\left(\mathbf{R}_{k,i},\ \ldots,\ -\mathbf{I}_2,\ \ldots\right)\mathbf{d}_k = \mathbf{0}, \tag{5.67}$$

where $\mathbf{I}_2$ refers to the 2-by-2 identity matrix.

Combining all the relations of $\mathbf{d}_{k,i}$ $(i \neq 1)$ and $\mathbf{d}_{k,1}$ similar to Eq. (5.67) by using the matrix $\mathbf{S}_k \in \mathbb{R}^{2(\bar{n}-1)\times 2\bar{n}}$, we obtain

$$\mathbf{S}_k\mathbf{d}_k = \mathbf{0}. \tag{5.68}$$

By substituting Eq. (5.63) into Eq. (5.68) and letting $\bar{\mathbf{S}} = \mathbf{S}_k\mathbf{T}_k$, rotational symmetry of the struts in orbit k can be expressed in a linear form with respect to the xy-coordinates $\mathbf{x}_k$ as

$$\bar{\mathbf{S}}\mathbf{x}_k = \mathbf{0}. \tag{5.69}$$

Because the rotational symmetry of the struts in different orbits can be formulated independently, the rotational symmetry of the whole structure in terms of the generalized xy-coordinate vector $\mathbf{X} = \left((\mathbf{x}^1)^\top, \ldots, (\mathbf{x}^{n^{\mathrm{L}}})^\top\right)^\top \in \mathbb{R}^{4n^{\mathrm{L}}\bar{n}}$ can then be combined as

$$\mathbf{S}\mathbf{X} = \mathbf{0}, \tag{5.70}$$

where the elements of $\bar{\mathbf{S}}$ for each orbit have been incorporated into the matrix $\mathbf{S} \in \mathbb{R}^{2n^{\mathrm{L}}(\bar{n}-1)\times 4n^{\mathrm{L}}\bar{n}}$.

If the structure has n^{L} similar orbits of struts, with $\bar{n}$ struts in each orbit, the matrix $\mathbf{S}$ can be calculated by the tensor product of the n^{L}-by-n^{L} identity matrix $\mathbf{I}_{n^{\mathrm{L}}}$ and the matrix $\bar{\mathbf{S}}$; i.e.,

$$\mathbf{S} = \mathbf{I}_{n^{\mathrm{L}}} \otimes \bar{\mathbf{S}}, \tag{5.71}$$

where $\otimes$ denotes tensor product.

This way, the symmetry properties of the whole structure can be formulated as a set of linear equations with respect to the generalized xy-coordinate vector $\mathbf{X}$. The linear constraints on nodal coordinates defined in Eq. (5.70) will be incorporated in the second design stage to ensure rotational symmetry of the structure.

5.3.2 Elevation (z-Coordinates)

From the self-equilibrium equation with respect to the nodal coordinates in z-direction, we can also formulate the linear constraint on elevation of the structure with respect to the force densities.

Suppose that the elevation of the structure is assigned by the designer, which means that z-coordinates of all the nodes are determined a priori. Since the following relation always holds

$$\mathbf{QCz} = \mathrm{diag}(\mathbf{Cz})\mathbf{q} = \mathbf{Wq}, \tag{5.72}$$

where $\mathbf{Q}$ is the diagonal version of $\mathbf{q}$, $\mathbf{C}$ is the connectivity matrix, and $\mathbf{W}$ is the diagonal version of the coordinate difference vector $\mathbf{w}(=\mathbf{Cz})$ in z-direction, the self-equilibrium equation in z-direction can be rewritten with respect to the force density vector $\mathbf{q}$ as

$$\begin{aligned}\mathbf{C}^\top\mathbf{Wq} &= \mathbf{C}^\top\mathbf{QCz} = \mathbf{Ez}\\ &= \mathbf{0}.\end{aligned} \tag{5.73}$$

By letting $\mathbf{N} = \mathbf{C}^\top\mathbf{W}$, the linear constraint on the elevation of a tensegrity structure with respect to the force densities becomes

$$\mathbf{Nq} = \mathbf{0}, \tag{5.74}$$

which is incorporated into the first design stage in the next section for finding the feasible set of force densities. Through Eq. (5.74), we can have exact control over the elevation of the structure.

5.3.3 Summary of Constraints

So far, we have formulated several linear constraints with respect to force densities, including the force density matrix, force densities of specific members, symmetry properties, and elevation. Among these constraints, only the constraint on the force density matrix is mandatory, and the other constraints are optional.

The optional constraints on symmetry in Eq. (5.58) and elevation in Eq. (5.74) with respect to the force density are combined as

$$\begin{pmatrix} \mathbf{F} \\ \mathbf{N} \end{pmatrix} \mathbf{q} = \mathbf{0}. \tag{5.75}$$

Since the matrix in the linear equation (5.75) with respect to $\mathbf{q}$ is usually rank deficient, the solution of Eq. (5.75) can be written as

$$\mathbf{q} = \boldsymbol{\Psi}\boldsymbol{\alpha}, \tag{5.76}$$

where $\boldsymbol{\alpha}$ is the coefficient vector, and the columns of the matrix $\boldsymbol{\Psi}$ span the solution space of Eq. (5.75). Note that $\boldsymbol{\Psi}$ is also a constant matrix when the constraints are given.

Since the force density matrix $\mathbf{E}$ has to satisfy the non-degeneracy condition, the coefficient vector $\boldsymbol{\alpha}$ cannot be selected arbitrarily. Suppose that we have obtained the force density vector $\mathbf{q}^{[i]}$ at the ith step of the iterative Algorithm 5.1. The force density matrix $\mathbf{E}^{[i]}$ corresponding to $\mathbf{q}^{[i]}$ has the necessary rank deficiency. Substituting Eq. (5.76) into Eq. (5.20), we have

$$\mathbf{g}^{[i]} = \mathbf{B}\boldsymbol{\Psi}\boldsymbol{\alpha}^{[i]}. \tag{5.77}$$

$\mathbf{B}\boldsymbol{\Psi}$ in Eq. (5.77) is usually full-rank and not square, and the least square solution of the coefficient vector $\boldsymbol{\alpha}^{[i]}$ is computed as follows:

$$\boldsymbol{\alpha}^{[i]} = (\mathbf{B}\boldsymbol{\Psi})^{-}\mathbf{g}^{[i]}, \tag{5.78}$$

where $(\cdot)^{-}$ denotes the generalized inverse of a matrix. The force density vector $\mathbf{q}^{[i]}$ can be updated to $\mathbf{q}^{[i+1]}$ by Eq. (5.76) as

$$\mathbf{q}^{[i+1]} = \boldsymbol{\Psi}(\mathbf{B}\boldsymbol{\Psi})^{-}\mathbf{g}^{[i]}. \tag{5.79}$$

5.3.4 AFDM with Constraints

By incorporation of the geometrical constraints, form-finding of a tensegrity structure is summarized as follows by making use of Algorithm 5.1:

Algorithm 5.2 – Form-finding process with geometrical constraints

First stage: *Finding feasible force densities*

Step 1: Specify the connectivity.
Step 2: Formulate the geometrical constraints with respect to the force densities.
Step 3: Find the feasible force densities by Algorithm 5.1.

Second stage: *Determination of geometry realization*

Step 4: Formulate the geometrical constraints with respect to the nodal coordinates.
Step 5: Specify an independent set of nodal coordinates, and uniquely determine the self-equilibrated configuration.

As will be demonstrated in the numerical examples, designers are enable to control the self-equilibrated configuration of a tensegrity structure by changing the values of the parameters in Steps 3 and 5. Moreover, symmetry of the structure is guaranteed by the constraints in Steps 2 and 4.

5.4 Numerical Examples

In this section, we consider the form-finding of tesegrity towers by using Algorithm 5.2. A tensegrity tower has one or more layers, and there are at least three struts in each layer. More details on tensegrity towers can be found in Appendix C.

5.4.1 Three-Layer Tensegrity Tower

The first numerical example is the tensegrity tower as shown in Fig. 5.8. The structure consists of three layers, and four struts in each layer belonging to the same orbit; i.e., $n^{\mathrm{L}} = 3$ and $\bar{n} = 4$. Therefore, there are in total 24 nodes; the nodes of each layer lie on two parallel planes—the bottom and the top planes, and the nodes on the same plane have the same z-coordinate. Height of layer k ($k = 1, 2, 3$) is denoted by H_k, overlap between layers k and $k - 1$ is denoted by h_k ($h_1 = 0$), and the total height of the structure is $H(=\sum_{k=1}^{3}(H_k - h_k))$.

The three-layer tensegrity tower in Fig. 5.8 is composed of 64 members, including 12 struts, 8 horizontal cables, 12 vertical cables, 16 saddle cables, and 16 diagonal cables.

To have a symmetric configuration, the saddle cables connecting different layers are classified into two groups, the members of other types in the same obit are classified into different groups. In total, there are 14 groups of members for this

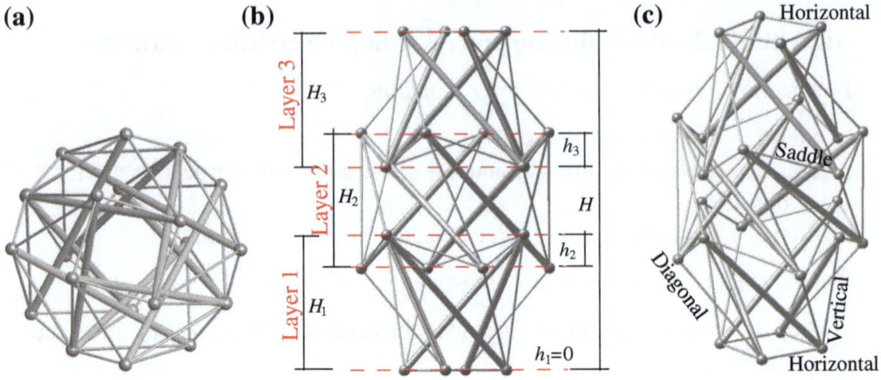

Fig. 5.8 A symmetric three-layer tensegrity tower with four struts in each layer. **a** Top view, **b** side view, **c** diagonal view

structure. Let q_k^{b} and q_k^{v} $(k = 1, 2, 3)$ denote the force densities of the groups of struts and vertical cables, respectively. Denote the force densities of the group of horizontal cables on the bottom and the top planes by q_1^{h} and q_2^{h}, respectively. Let q_{2k-1}^{d} and q_{2k}^{d} denote the force densities of the group of diagonal cables connecting the bottom and the top planes of layer k, respectively.

Example 5.13 Form-finding of the three-layer tensegrity tower in Fig. 5.8 by application of Algorithm 5.2.

Consider the rotationally symmetric three-layer tensegrity tower as shown in Fig. 5.8. The height of each layer is set as 8.0, the overlaps are 2.0, and therefore, the total height of the structure in Fig. 5.8 is 20.0:

$$\begin{aligned} H_1 &= H_2 = H_3 = 8.0, \\ (h_1 &= 0), \quad h_2 = h_3 = 2.0, \\ H &= \sum_{k=1}^{3} (H_k - h_k) = 20.0. \end{aligned} \tag{5.80}$$

To start Algorithm 5.1 for feasible force densities satisfying the non-degeneracy condition, the initial force densities of all struts and all cables are assigned as −1 and 1, respectively. After 382 iterations, Algorithm 5.1 terminates with the feasible force densities as follows:

$$\begin{aligned} \text{Struts:} \quad q_1^{\mathrm{b}} &= q_2^{\mathrm{b}} = -1.1436, \\ q_3^{\mathrm{b}} &= -1.1432, \end{aligned}$$

$$\begin{aligned}
\text{Horizontal: } & q_1^{\mathrm{h}} = q_2^{\mathrm{h}} = 1.2965, \\
\text{Vertical: } & q_1^{\mathrm{v}} = q_2^{\mathrm{v}} = q_3^{\mathrm{v}} = 0.6415, \\
\text{Saddle: } & q_1^{\mathrm{s}} = q_2^{\mathrm{s}} = 1.5058, \\
\text{Diagonal: } & q_1^{\mathrm{d}} = q_2^{\mathrm{d}} = 0.6694, \\
& q_3^{\mathrm{d}} = q_4^{\mathrm{d}} = 0.3345.
\end{aligned} \tag{5.81}$$

The final force density matrix $\hat{\mathbf{E}}$ has four zero eigenvalues satisfying the non-degeneracy condition for a three-dimensional structure, and furthermore, the remaining 20 eigenvalues are all positive such that the structure is super-stable.

In the second design stage, there are up to four independent *xy*-coordinates that can be arbitrarily specified by the designers, while the constraint on symmetry is considered. By specifying the *xy*-coordinates of nodes 5 and 7, which are on the top plane of layer 1, as (10.0, 0) and (2.5, 4.0), respectively, the geometry of the structure is uniquely determined as shown in Fig. 5.8. It is easy to observe from the top view of the structure that the struts in the same layer (orbit) are rotationally symmetric by the angle $\pi/2$.

Note that this is not the only choice for independent set of coordinates. See, for example, Example 5.14 for another possible geometry realization of the tensegrity tower with the same force densities.

Example 5.14 Self-equilibrated configuration of the three-layer tensegrity tower in Fig. 5.9.

In this example, we use the same settings of force densities as in Example 5.13, and specify the same *xy*-coordinates (10.0, 0) and (2.5, 4.0) for different nodes 1 and 3, which are located on the bottom plane of layer 1. The self-equilibrated configuration of the three-layer tensegrity tower is shown in Fig. 5.9.

The structure is super-stable because it has the same force densities as the structure in Example 5.13, such that its force density matrix is positive semi-definite with four zero eigenvalues satisfying the non-degeneracy condition.

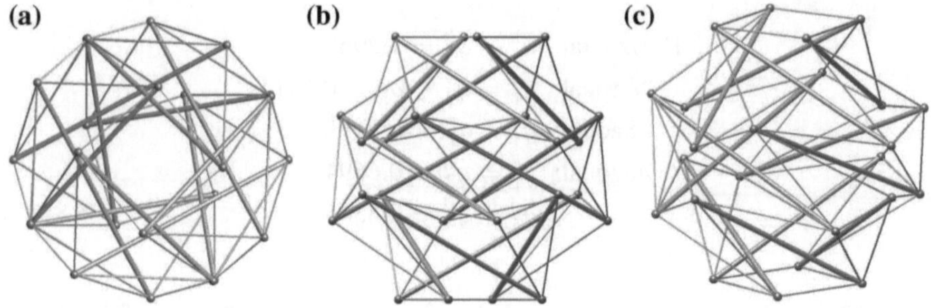

Fig. 5.9 New configuration of the symmetric three-layer tensegrity tower in Example 5.14. The structure has the same force densities and coordinates in z-direction as the structure in Example 5.13, but they have different xy-coordinates. **a** Top view, **b** side view, **c** diagonal view

5.4.2 Ten-Layer Tensegrity Tower

As a more complex example, we consider a ten-layer tensegrity tower as shown in Fig. 5.10 with four struts in each layer; i.e., $n^{\mathrm{L}} = 10$ and $\bar{n} = 4$. The structure is composed of 80 nodes and 232 members.

Example 5.15 Form-finding of the rotationally symmetric ten-layer tensegrity tower as shown in Fig. 5.10 by application of Algorithm 5.2.

For simplicity, the heights and overlaps are uniformly assigned as $H_i = 10.0$ and $h_i = 2.0$ $(i = 1, 2, \ldots, 10)$, respectively, except that $h_1 = 0.0$. Hence, the total height H of the structure is 82.0. Constraint on rotational symmetry of the structure about z-axis is incorporated.

The initial force densities of all struts and cables are given as -1 and 1, respectively. After 511 iterations of applying Algorithm 5.1, we find the feasible force densities satisfying the non-degeneracy condition. Moreover, in order to uniquely determine the self-equilibrated configuration, we need to specify four independent xy-coordinates.

By specifying the xy-coordinates of nodes 5 and 7, which are on the top plane of layer 1, as $(10.0, 0)$ and $(2.5, 4.0)$, respectively, self-equilibrated configuration of the structure is uniquely determined as shown in Fig. 5.10.

The structure is super-stable, because the eigenvalues of its force density matrix are positive, except for the four zero eigenvalues necessary for the non-degeneracy condition.

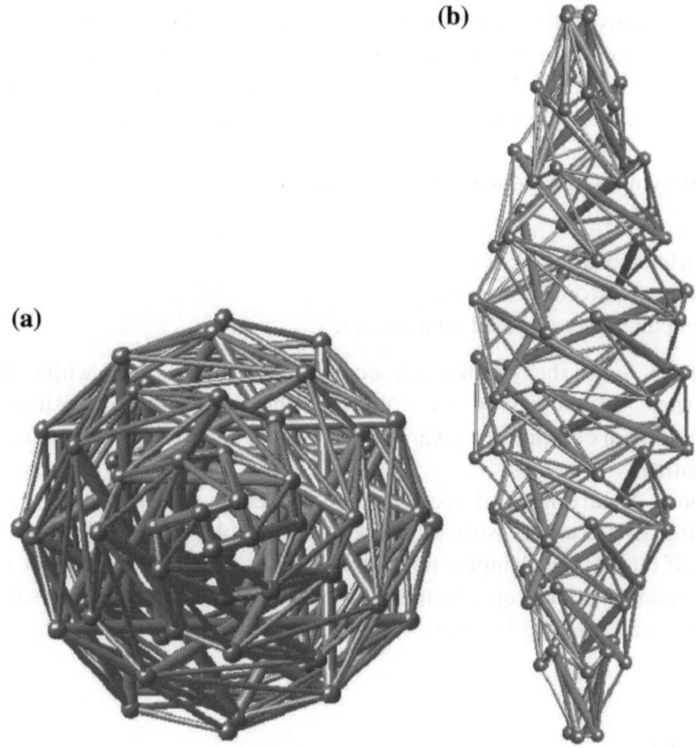

Fig. 5.10 The self-equilibrated configuration of the ten-layer tensegrity tower considered in Example 5.15. The structure is rotationally symmetric about z-axis as specified. **a** Top view, **b** side view

5.5 Remarks

In this chapter, we extended the concept of force density method, which was originally developed for cable-nets, to the form-finding problem of tensegrity structures. Hence, it also has the advantage of the force density method in dealing with non-linear equations in a linear manner in the second design stage. By introducing geometrical constraints into the form-finding process, the method shows a strong capability in controlling geometrical properties of the self-equilibrated configuration. More importantly, the proposed method guarantees a super-stable structure during the process of searching for feasible force densities.

The form-finding process is divided into two interrelated design stages: finding the feasible force densities, and determining the geometry realization. To control self-equilibrated configuration of a tensegrity structure, the constraints are formulated in linear forms with respect to force densities and nodal coordinates, respectively.

The constraints with respect to force densities are incorporated into the first design stage to constrain the direction of finding the feasible force densities from the initially given values. On the other hand, the linear constraints on nodal coordinates are incorporated into the second design stage to uniquely determine its geometry realization.

The following parameters are needed as inputs in the form-finding process:

1. connectivity;
2. geometrical constraints;
3. an initial set of force densities;
4. an independent set of nodal coordinates.

Among these, only the geometrical constraints are optional, while the others are necessary. By modifying the values of the initial force densities and the independent nodal coordinates, a variety of new self-equilibrated configurations can be systematically found.

The force density method presented in this chapter is a general method for form-finding of tensegrity structures: it is applicable to any kind of tensegrity structures, if the necessary inputs listed above are available. However, the method is unlikely to control all aspects of a tensegrity structure. For example, it is not easy to exactly assign all member lengths.

References

1. Harary, F. (1969). *Graph theory*. Reading: Addison-Wesley.
2. Lay, D. C. (2011). *Linear algebra and its applications* (4th ed.). New Jersey: Pearson.
3. Schek, H.-J. (1974). The force density method for form finding and computation of general networks. *Computer Methods in Applied Mechanics and Engineering*, *3*(1), 115–134.
4. Vassart, N., & Motro, R. (1999). Multiparametered formfinding method: application to tensegrity systems. *International Journal of Space Structures*, *14*(2), 147–154.

Chapter 6
Prismatic Structures of Dihedral Symmetry

Abstract This chapter presents the conditions for self-equilibrium and super-stability of prismatic tensegrity structures with dihedral symmetry in an analytical manner. These conditions are derived based on the analytical symmetry-adapted force density matrix given in Appendix D.

Keywords Prismatic structure · Dihedral symmetry · Self-equilibrium condition · Super-stability condition · Stability catalogue

6.1 Configuration and Connectivity

Let m and n respectively denote the number of members as well as the number of nodes. We consider a prismatic structure with dihedral symmetry $\mathbf{D}_N$, which consists of (n=)$2N$ nodes and (m=)$4N$ members. The nodes on the upper circle are numbered from 0 to $N - 1$, while the nodes on the lower circle are numbered from N to $2N - 1$.

There exist in total $2N$ symmetry operations in the dihedral group $\mathbf{D}_N$, as given in Sect. 3.3 and Appendix D:

- N-fold rotations C_N^i ($i = 0, 1, \ldots, N - 1$) of angles $2i\pi/N$ about the principal axis, and
- N two-fold rotations $\mathrm{C}_{2,i}$ ($i = 0, 1, \ldots, N - 1$) of a half circle π about the axes going through the origin and perpendicular to the principal axis.

The nodes of a symmetric prismatic tensegrity structure are symmetrically located on the two (horizontal) circles, and they belong to a *regular orbit*. The nodes belonging to a regular orbit indicates that a node is transformed to the position of any other node in the same orbit by only one proper symmetry operation of the group. Therefore, the nodes in a regular orbit have one-to-one correspondence to the symmetry operations.

Moreover, there are three orbits of members: horizontal cables, vertical cables, and struts. Each node is connected by two horizontal cables, one vertical cable, and one strut. A horizontal cable connects nodes lying on the same horizontal plane, while the vertical cable and strut connect nodes on different planes.

J.Y. Zhang and M. Ohsaki, *Tensegrity Structures*, Mathematics for Industry 6,
DOI 10.1007/978-4-431-54813-3_6

Furthermore, connectivity of the members of a prismatic tensegrity structure is defined as follows by using the parameters h and v:

- *Struts*: A strut connects node i ($i = 0, 1, \dots, N-1$) on the upper circle to node $N+i$ on the lower circle.
- *Horizontal cables*: On the upper plane, a horizontal cable connects node i ($i = 0, 1, \dots, N-1$) to node $h+i$, or $h+i-N$ when $h+i \geq N$; on the lower plane, a horizontal cable connects node i ($i = N, N+1, \dots, 2N-1$) to node $h+i$, or $h+i-N$ when $h+i \geq 2N$.
- *Vertical cables*: A vertical cable connects node i ($i = 0, 1, \dots, N$) on the upper circle to node $N+v+i$, or $v+i$ when $v+i \geq N$, on the lower circle.

Accordingly, there are $2N$ horizontal cables, which form a regular orbit. There are, however, only N vertical cables and N struts; and there is a one-to-two correspondence between the vertical cables (or struts) and the symmetry operations.

Furthermore, a prismatic tensegrity structure with dihedral symmetry $\mathbf{D}_N$ is denoted by $\mathbf{D}_N^{h,v}$, with its connectivity defined by the parameters h and v.

Example 6.1 Node and member orbits of the structure $\mathbf{D}_3^{1,1}$ in Fig. 6.1.

The structure $\mathbf{D}_3^{1,1}$ as shown in Fig. 6.1 is of $\mathbf{D}_3$ symmetry. Any node, e.g., node 0, of the structure can be transformed to the position of any other node, including itself, by one of the six symmetry operations of $\mathbf{D}_3$ given in Example 3.5.

Transformations of node 0, horizontal cable [1], vertical cable [7], and strut [10], by applying the symmetry operations of $\mathbf{D}_3$, are listed below.

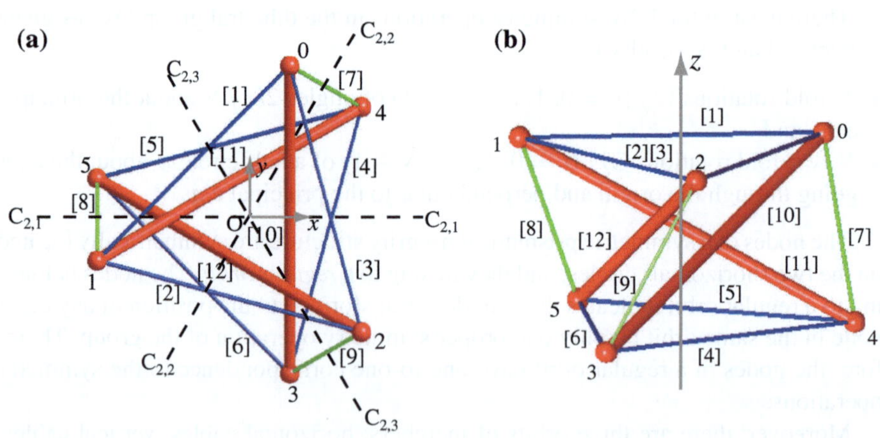

Fig. 6.1 The simplest prismatic tensegrity structure $\mathbf{D}_3^{1,1}$. The structure is of dihedral symmetry $\mathbf{D}_3$, and consists of six nodes, six horizontal cables, three vertical cables, and three struts. **a** Top view, **b** diagonal view

	E (C_3^0)	C_3^1	C_3^2	$C_{2,1}$	$C_{2,2}$	$C_{2,3}$	
Node 0	0	1	2	3	4	5	
Member [1]	[1]	[2]	[3]	[6]	[4]	[5]	Horizontal cables
Member [7]	[7]	[8]	[9]	[9]	[7]	[8]	Vertical cables
Member [10]	[10]	[11]	[12]	[10]	[11]	[12]	Struts

There is only one proper symmetry operation that can transform one horizontal cable to another. However, for the vertical cables and struts, any one of them can be transformed to the position of any other one by *two* of the six symmetry operations, and therefore, they have one-to-two correspondence to the symmetry operations.

6.2 Preliminary Study on Stability

Using the modified Maxwell's rule in Eq. (2.61), the number n^{s} of independent prestress modes and the number n^{m} of infinitesimal mechanisms of a three-dimensional ($d = 3$) prismatic structure have the following relation:

$$\begin{aligned} n^{\mathrm{s}} - n^{\mathrm{m}} &= m - dn + n^{\mathrm{b}} \\ &= 4N - 3 \times 2N + 6 \\ &= -2N + 6. \end{aligned} \tag{6.1}$$

As will be discussed later, a symmetric prismatic structure is statically indeterminate, consisting of only single prestress mode; i.e., $n^{\mathrm{s}} = 1$. Thus, the number n^{m} of (infinitesimal) mechanisms is proportional to the number n (=$2N$) of nodes as

$$n^{\mathrm{m}} = 2N - 5. \tag{6.2}$$

Example 6.2 Kinematic indeterminacy of the prismatic structures with $\mathbf{D}_3$ and $\mathbf{D}_8$ symmetries.

The structure $\mathbf{D}_3^{1,1}$ as shown in Fig. 6.1 is the simplest prismatic tensegrity structure. It consists of six nodes and twelve members, including six horizontal cables, three vertical cables, and three struts; i.e., $n = 6$ and $m = 12$. From Eq. (6.2), there exists only one infinitesimal mechanism in the structure:

$$\begin{aligned} n^{\mathrm{m}} &= 2N - 5 \\ &= 2 \times 3 - 5 \\ &= 1. \end{aligned} \tag{6.3}$$

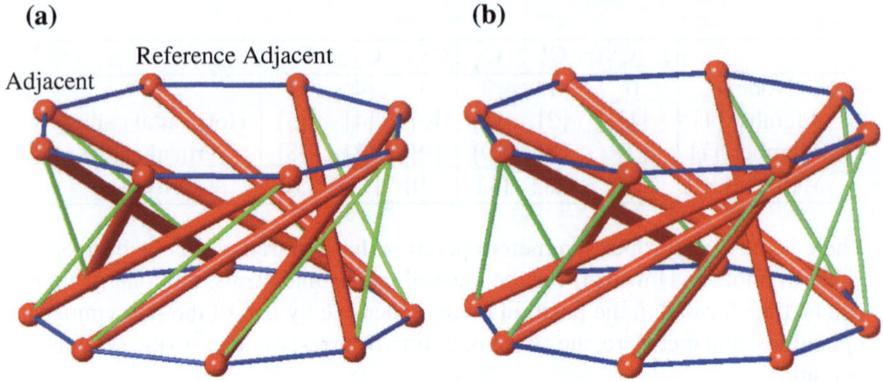

Fig. 6.2 Super-stable structures, horizontal cables of which are connected by adjacent nodes lying on the same circle. **a** Structure $\mathbf{D}_8^{1,1}$, **b** structure $\mathbf{D}_8^{1,2}$

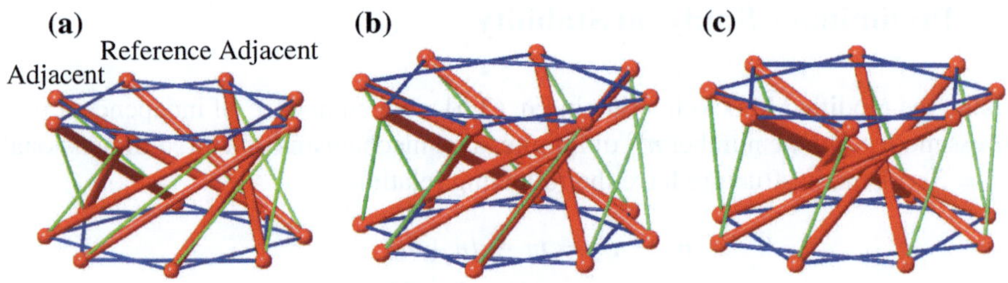

Fig. 6.3 Prismatic tensegrity structures with $\mathbf{D}_8$ symmetry. The structure $\mathbf{D}_8^{2,3}$ is possible to be prestress-stable, depending on height-to-radius ratio, the structure $\mathbf{D}_8^{2,2}$ is unstable because it can be physically divided into two identical substructures $\mathbf{D}_4^{1,1}$, and the structure $\mathbf{D}_8^{2,1}$ is unstable although it is indivisible. **a** Unstable $\mathbf{D}_8^{2,1}$, **b** divisible $\mathbf{D}_8^{2,2}$, **c** prestress-stable $\mathbf{D}_8^{2,3}$

In contrast to the structure $\mathbf{D}_3^{1,1}$, the structures with the same $\mathbf{D}_8$ symmetry in Figs. 6.2 and 6.3 are composed of 16 nodes and 32 members, including 16 horizontal cables, eight vertical cables, and eight struts; i.e., $n = 16$ and $m = 32$. From Eq. (6.2), there exist eleven infinitesimal mechanisms in the structure:

$$\begin{aligned} n^{\mathrm{m}} &= 2N - 5 \\ &= 2 \times 8 - 5 \\ &= 11. \end{aligned} \tag{6.4}$$

Despite of existence of many infinitesimal mechanisms, the symmetric prismatic structures might still be super-stable if *the horizontal cables are connected to adjacent nodes lying on the same circle*; i.e., the parameter h for connectivity of horizontal cables satisfies $h = 1$. This super-stability condition for the prismatic tensegrity structures with dihedral symmetry will be proved later in this chapter, making use of the symmetry-adapted force density matrix.

Example 6.3 Example structures that are super-stable and that are not super-stable.

According to the super-stability conditions for symmetric prismatic structures presented in the coming Sect. 6.6.2, a prismatic structure is super-stable if its horizontal cables are connected to the adjacent nodes on the same plane. Therefore, the structures $\mathbf{D}_8^{1,1}$ and $\mathbf{D}_8^{1,2}$ in Fig. 6.2, which are of $\mathbf{D}_8$ symmetry, are super-stable.

The structures $\mathbf{D}_8^{2,1}$, $\mathbf{D}_8^{2,2}$, and $\mathbf{D}_8^{2,3}$ in Fig. 6.3 have the same $\mathbf{D}_8$ symmetry as the structures in Fig. 6.2; However, they are different in connectivity of horizontal cables: the horizontal cables of the structures in Fig. 6.3 are connected by the nodes next to the adjacent nodes; i.e., $h = 2$. According to the super-stability condition for symmetric prismatic structures, the structures $\mathbf{D}_8^{2,1}$, $\mathbf{D}_8^{2,2}$, and $\mathbf{D}_8^{2,3}$ are not super-stable.

It should be noted that the structures that are not super-stable may still be stable, in the sense of having positive definite tangent stiffness. For example the structure $\mathbf{D}_8^{2,1}$ as shown in Fig. 6.3a is not super-stable, however, it is prestress-stable and is stable if level of prestresses is low enough. Furthermore, some structures that are not super-stable cannot even be (locally) stable: for instance the divisible structure $\mathbf{D}_8^{2,2}$ in Fig. 6.3b can be divided into two independent sub-structures, the divisibility condition for which will be presented in the next section; and the structure $\mathbf{D}_8^{2,3}$ in Fig. 6.3c is not prestress-stable.

6.3 Conventional Symmetry-adapted Approach

As have been discussed in Chap. 4, stability of a structure can be verified by observing positive definiteness (signs of the eigenvalues) of its tangent stiffness matrix. Since the size of the stiffness matrix increases in proportion to the number of nodes, it is difficult to analytically investigate (super-)stability of a complex structure with a relatively large number of nodes. One way to study such complex structures is to make use of their high symmetry. Symmetric structures are aesthetically beautiful, and have been extensively accepted in practical applications.

On the other hand, symmetry properties of a structure are very useful in structural analysis, because the analysis process can be significantly simplified by considering only a small portion of the structure.

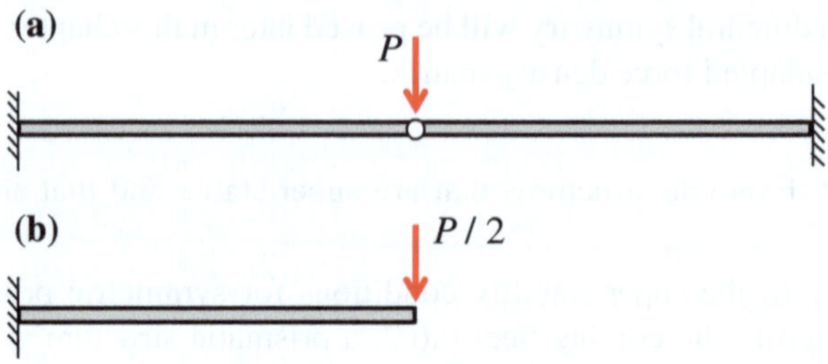

Fig. 6.4 Structural analysis by using symmetry. The analysis problem in **a** can be simplified as that of the cantilever in **b** due to the reflection symmetry, if axial deformations of the members can be ignored. **a** Original problem, **b** simplified problem

Example 6.4 Structural analysis of the structure as shown in Fig. 6.4 with reflection symmetry.

The analysis problem in Fig. 6.4a is symmetric because both of the structure and the external load are of reflection symmetry. Thus, the problem can be reduced to that in Fig. 6.4b, with half of the external load applied at the free end of a cantilever, if the axial deformations in the two members are neglected. The cantilever problem in Fig. 6.4b is a basic problem in structural mechanics, and can be solved easily.

Symmetry properties of tensegrity structures have also been extensively utilized in their self-equilibrium analyses, see for example [6, 7]. In Chap. 3, we have analytically studied self-equilibrium of the structures with high level of symmetry. For the super-stability investigation of tensegrity structures, their symmetry properties can also be of great help.

Since the positive semi-definiteness of the force density matrix, or equivalently the geometrical stiffness matrix, is the necessary condition for super-stability, the basic idea in utilizing symmetry for super-stability investigation is to rewrite the force density matrix into block-diagonal (symmetry-adapted) forms with respect to the symmetry-adapted coordinate system.

In the symmetry-adapted form of a matrix, the independent sub-matrices (or blocks) with much smaller sizes are located in the leading diagonal. Since eigenvalues of the matrix are independent of coordinate systems, its positive definiteness can be verified by observing signs of the eigenvalues of the independent blocks in its symmetry-adapted form. Because the independent blocks are of much smaller sizes than the original matrix, it is possible to analytically derive the eigenvalues even for complex structures.

Many researchers have tried to obtain the symmetry-adapted (or block-diagonal) forms of stiffness matrices. An introduction and review of the conventional (numerical) methods for block-diagonalization of the stiffness matrices were given in Ref. [4]. Transformation matrices that transform the initial coordinate system into the symmetry-adapted coordinate system are usually necessary in the conventional methods. For instance, the symmetry-adapted force density matrix $\tilde{\mathbf{E}} \in \mathbb{R}^{n \times n}$ is (usually numerically) derived by applying transformation matrix $\mathbf{T} \in \mathbb{R}^{n \times n}$ on both sides of the force density matrix $\mathbf{E}$:

$$\tilde{\mathbf{E}} = \mathbf{T}\mathbf{E}\mathbf{T}^{\top}, \tag{6.5}$$

where $\mathbf{T}$ is a unitary matrix:

$$\mathbf{T}^{\top}\mathbf{T} = \mathbf{I}_n, \tag{6.6}$$

and $\mathbf{I}_n \in \mathbb{R}^{n \times n}$ is an identity matrix.

See, for example, the numerical example in Example 6.5. Note that, in the numerical examples in this chapter, the units are usually omitted without any loss of generality on discussions on equilibrium as well as stability.

Example 6.5 Symmetry-adapted force density matrix of the symmetric prismatic structure $\mathbf{D}_3^{1,1}$ as shown in Fig. 6.1.

The structure $\mathbf{D}_3^{1,1}$ in Fig. 6.1 is composed of six nodes. Using the results in Example 3.11, the force densities of vertical cables q_{v}, horizontal cables q_{h}, and struts q_{s} are given as

$$q_{\mathrm{v}} = 1, \quad q_{\mathrm{s}} = -1, \quad q_{\mathrm{h}} = \frac{\sqrt{3}}{3}. \tag{3.50}$$

From the direct definition in Eq. (2.107), the force density matrix $\mathbf{E} \in \mathbb{R}^{6 \times 6}$ is given as

$$\mathbf{E} = \frac{1}{\sqrt{3}} \begin{pmatrix} 2 & -1 & -1 & \sqrt{3} & -\sqrt{3} & 0 \\ -1 & 2 & -1 & 0 & \sqrt{3} & -\sqrt{3} \\ -1 & -1 & 2 & -\sqrt{3} & 0 & \sqrt{3} \\ \sqrt{3} & 0 & -\sqrt{3} & 2 & -1 & -1 \\ -\sqrt{3} & \sqrt{3} & 0 & -1 & 2 & -1 \\ 0 & -\sqrt{3} & \sqrt{3} & -1 & -1 & 2 \end{pmatrix}. \tag{6.7}$$

The six eigenvalues of $\mathbf{E}$ are

$$\begin{aligned} \lambda_1^{\mathrm{E}} &= \lambda_2^{\mathrm{E}} = \lambda_3^{\mathrm{E}} = \lambda_4^{\mathrm{E}} = 0, \\ \lambda_5^{\mathrm{E}} &= \lambda_6^{E} = 3.4641. \end{aligned} \tag{6.8}$$

The non-degeneracy condition is satisfied because there are four zero eigenvalues, and the structure is super-stable because the remaining two eigenvalues are positive.

Using the transformation matrix $\mathbf{T} \in \mathbb{R}^{6\times 6}$ defined as

$$\mathbf{T} = \begin{pmatrix} 0.4082 & 0.4082 & 0.4082 & 0.4082 & 0.4082 & 0.4082 \\ 0.4082 & 0.4082 & 0.4082 & -0.4082 & -0.4082 & -0.4082 \\ 0.5774 & -0.2887 & -0.2887 & 0.5774 & -0.2887 & -0.2887 \\ 0.0000 & -0.5000 & 0.5000 & 0.0000 & 0.5000 & -0.5000 \\ 0.0000 & 0.5000 & -0.5000 & 0.0000 & 0.5000 & -0.5000 \\ 0.5774 & -0.2887 & -0.2887 & -0.5774 & 0.2887 & 0.2887 \end{pmatrix}, \tag{6.9}$$

the symmetry-adapted force density matrix $\tilde{\mathbf{E}}$ is calculated as follows by applying Eq. (6.5):

$$\tilde{\mathbf{E}} = \begin{pmatrix} 0 & 0 & 0 & 0 & 0 & 0 \\ 0 & 0 & 0 & 0 & 0 & 0 \\ 0 & 0 & 3.2321 & -0.8660 & 0 & 0 \\ 0 & 0 & -0.8660 & 0.2321 & 0 & 0 \\ 0 & 0 & 0 & 0 & 3.2321 & -0.8660 \\ 0 & 0 & 0 & 0 & -0.8660 & 0.2321 \end{pmatrix}. \tag{6.10}$$

It is immediately observed from Eq. (6.10) that there exist two zero eigenvalues in $\tilde{\mathbf{E}}$, and the remaining four eigenvalues can be calculated by considering the two copies of the independent block $\hat{\mathbf{E}} \in \mathbb{R}^{2\times 2}$:

$$\hat{\mathbf{E}} = \begin{pmatrix} 3.2321 & -0.8660 \\ -0.8660 & 0.2321 \end{pmatrix}. \tag{6.11}$$

The two eigenvalues of $\hat{\mathbf{E}}$ can be easily calculated as

$$\begin{aligned} \lambda_1^{\hat{\mathrm{E}}} &= 0, \\ \lambda_2^{\hat{\mathrm{E}}} &= 3.4641, \end{aligned} \tag{6.12}$$

which coincide with the eigenvalues of $\mathbf{E}$ given in Eq. (6.8).

It is obvious that the numerical approaches can only deal with every specific structure but not a whole class of structures with the same symmetry. For example, the prismatic structure $\mathbf{D}_{20}^{6,3}$ in Fig. 6.5 is also of dihedral symmetry, with much more nodes and members compared to the structure $\mathbf{D}_3^{1,1}$. In order to derive the symmetry-adapted force density matrix, the corresponding transformation matrix $\mathbf{T} \in \mathbb{R}^{60\times 60}$ has much larger size.

More importantly, numerical methods may lose the opportunity to have more insights into structural properties, and it is impossible to derive stability conditions for a whole class of structures. Hence, the *analytical* methods for symmetry-adapted matrices are always desirable, as long as they are available.

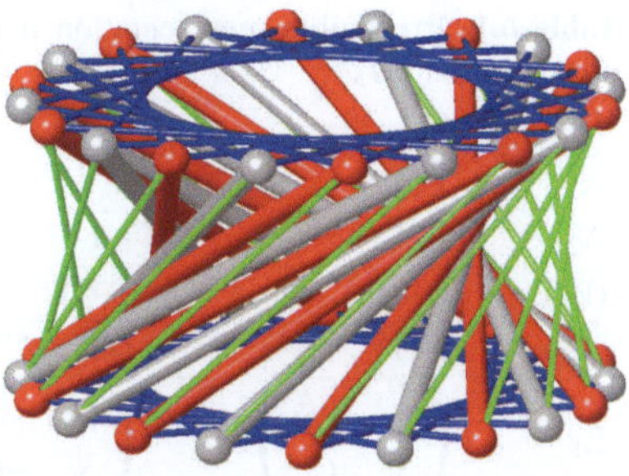

Fig. 6.5 Symmetric prismatic structure $\mathbf{D}_{20}^{6,3}$

The direct strategy to derive the analytical symmetry-adapted force density matrix is presented in Appendix D.5. It is notable in this strategy that transformation matrices, and therefore, the matrix multiplications are unnecessary. Self-equilibrium analysis as well as super-stability investigation of the symmetric tensegrity structures can then be conducted based on the symmetry-adapted force density matrix.

6.4 Symmetry-adapted Force Density Matrix

In this section, we present the direct formulation of the symmetry-adapted force density matrix, structure of which is identified by using the linear combination of representations of nodes [3, 5].

6.4.1 Matrix Representation of Dihedral Group

Group multiplication table describes combinations of two operations (elements) of a group. If a set of matrices obeys the group multiplication table of a group, these matrices are said to form a *matrix representation* of that group. A matrix representation that can be reduced to a linear combination (direct sum) of several matrix representations is called *reducible matrix representation*, otherwise, it is an *irreducible matrix representation*. *Characters* are defined as traces of the irreducible representation matrices. They will be shown to be important in identifying the block structures of the symmetry-adapted matrices.

A dihedral group $\mathbf{D}_N$ consists of at least two one-dimensional representations A_1 and A_2. When N is an even number, there are two more one-dimensional representations B_1 and B_2, in addition to A_1 and A_2. Moreover, there exist p two-dimensional representations E_k ($k = 1, 2, \dots, p$) where

$$p = \begin{cases} (N-1)/2, & N \text{ odd;} \\ (N-2)/2, & N \text{ even.} \end{cases} \tag{6.13}$$

Table 6.1 Irreducible representation matrices $\mathbf{R}_i^{\mu}$ $(i = 1, 2, \ldots, N)$ of representations μ of dihedral group $\mathbf{D}_N$

$\mathbf{D}_N$	C_N^i	$C_{2,i}$		
A_1	1	1		
A_2	1	-1	z, R_z	
(B_1)	$(-1)^i$	$(-1)^i$		N even
(B_2)	$(-1)^i$	$(-1)^{(i+1)}$		N even
E_1	$\begin{pmatrix} C_i & -S_i \\ S_i & C_i \end{pmatrix}$	$\begin{pmatrix} C_i & S_i \\ S_i & -C_i \end{pmatrix}$	(x,y) (R_x, R_y)	
E_k	$\begin{pmatrix} C_{ik} & -S_{ik} \\ S_{ik} & C_{ik} \end{pmatrix}$	$\begin{pmatrix} C_{ik} & S_{ik} \\ S_{ik} & -C_{ik} \end{pmatrix}$		$k = 2, 3, \ldots, p$
	$\mathbf{R}_i^{\mu}$	$\mathbf{R}_{N+i}^{\mu}$		$i = 0, 1, \ldots, N-1$

The first row denotes the symmetry operations of $\mathbf{D}_N$. C_{ik} and S_{ik} respectively denote $\cos(2ik\pi/N)$ and $\sin(2ik\pi/N)$. x, y, z and R_x, R_y, R_z respectively stand for symmetry operations of the corresponding coordinates and rotations about those axes

Note that the standard notation is E for $\mathbf{D}_3$ and $\mathbf{D}_4$, for which there is only one $(p = 1)$ two-dimensional representation, however, we will use the notation E_1 also for these cases.

The irreducible matrix representations of a dihedral group $\mathbf{D}_N$ are listed in Table 6.1. The one-dimensional matrix representations are unique, and their characters are the representation matrices themselves; characters of the two-dimensional representation matrices are also unique—character of the cyclic rotation C_N^i for E_k is $2C_{ik}(=2\cos(2ik\pi/N))$, and that of the two-fold rotation $C_{2,i}$ for any E_k is zero, but we may have some limited choices for their representation matrices.

In Table 6.1, we chose the positive z-direction as the positive direction of rotations to present the two-dimensional representation matrices. The symbols x, y, and z in the fourth column of the table respectively stand for x-, y-, and z-coordinates, and R_x, R_y, and R_z stand for rotations about these axes [1]. We will show later in this chapter that the A_2 and E_1 blocks of the symmetry-adapted force density matrix should be singular to ensure a non-degenerate geometry realization.

6.4.2 Structure of Symmetry-adapted Force Density Matrix

Linear combination of representations for transformation of nodes is helpful in identifying structure of the symmetry-adapted force density matrix. For this purpose, every node is considered to be physically distinct, unlike the case where all nodes of the same type are regarded to be physically indistinguishable when we described symmetry of the structure previously.

Example 6.6 Transformation of nodes of the prismatic structure $\mathbf{D}_3^{1,1}$ in Fig. 6.1 subjected to the three-fold symmetry operation C_3^1 of dihedral group $\mathbf{D}_3$.

To consider transformation of nodes under symmetry operations, we use the structure $\mathbf{D}_3^{1,1}$ with $\mathbf{D}_3$ symmetry in Fig. 6.1 as an example structure. The three-fold rotation C_3^1 exchanges positions of nodes as

$$\begin{aligned} &0 \rightarrow 1 \rightarrow 2 \rightarrow 0, \\ &3 \rightarrow 4 \rightarrow 5 \rightarrow 3, \end{aligned} \tag{6.14}$$

or by using the permutation notations as

$$(0, 1, 2)\ (3, 4, 5). \tag{6.15}$$

Transformations of the nodes under each symmetry operation can also be written in a matrix form, see, for instance, Example 6.7.

Example 6.7 Transformation (reducible representation) matrices corresponding to dihedral group $\mathbf{D}_3$.

Transformations of the nodes of the prismatic structure $\mathbf{D}_3^{1,1}$ in Fig. 6.1 subjected to the six symmetry operations of dihedral group $\mathbf{D}_3$ can be written in matrix form as follows by using the matrices $\mathbf{R}_i$ $(i = 0, 1, \ldots, 5)$:

Operation	C_3^0	$C_{2,0}$
Transformation	$\mathbf{R}_0 = \begin{pmatrix} 1 & & & & & \\ & 1 & & & & \\ & & 1 & & & \\ & & & 1 & & \\ & & & & 1 & \\ & & & & & 1 \end{pmatrix}$	$\mathbf{R}_3 = \begin{pmatrix} 0 & & & 1 & & \\ & 0 & & & & 1 \\ & & 0 & & 1 & \\ 1 & & & 0 & & \\ & & 1 & & 0 & \\ & 1 & & & & 0 \end{pmatrix}$
Trace	$\mathrm{trace}(\mathbf{R}_0) = 6$	$\mathrm{trace}(\mathbf{R}_3) = 0$

Operation	C_3^1	$C_{2,1}$
Transformation	$\mathbf{R}_1 = \begin{pmatrix} 0 & 1 & & & & \\ & 0 & 1 & & & \\ & & 0 & 1 & & \\ & & & 0 & 1 & \\ & & & & 0 & 1 \\ 1 & & & & & 0 \end{pmatrix}$	$\mathbf{R}_4 = \begin{pmatrix} 0 & & & & 1 & \\ & 0 & & 1 & & \\ & & 0 & & & 1 \\ & 1 & & 0 & & \\ 1 & & & & 0 & \\ & & 1 & & & 0 \end{pmatrix}$
Trace	$\mathrm{trace}(\mathbf{R}_1) = 0$	$\mathrm{trance}(\mathbf{R}_4) = 0$

Operation	C_3^2	$C_{2,2}$
Transformation	$\mathbf{R}_2 = \begin{pmatrix} 0 & & 1 & & & \\ & 0 & & 1 & & \\ & & 0 & & 1 & \\ & & & 0 & & 1 \\ 1 & & & & 0 & \\ & 1 & & & & 0 \end{pmatrix}$	$\mathbf{R}_5 = \begin{pmatrix} 0 & & & & & 1 \\ & 0 & & & 1 & \\ & & 0 & 1 & & \\ & & 1 & 0 & & \\ & 1 & & & 0 & \\ 1 & & & & & 0 \end{pmatrix}$
Trace	$\text{trace}(\mathbf{R}_2) = 0$	$\text{trace}(\mathbf{R}_5) = 0$

It should be noted that trace of the transformation matrix $\mathbf{R}_i$ is equal to the number of nodes that remain unchanged under the corresponding transformation (symmetry operation). The set of matrices $\mathbf{R}_i$ indeed forms a *reducible* matrix representation of the group $\mathbf{D}_3$, since they satisfy its multiplication table given in Appendix D.

This reducible matrix representation can be rewritten as a linear combination (direct sum) $\Gamma(\text{N})$ of its irreducible matrix representations making use of the important property that a change in coordinate system will not change the trace of a representation matrix.

Example 6.8 Linear combination of representations of the dihedral group $\mathbf{D}_3$ for the structure with $\mathbf{D}_3$ symmetry.

For the structure with $\mathbf{D}_3$ symmetry, traces of the representation matrices are obtained as

Operations	C_3^0	C_3^1	C_3^2	$C_{2,0}$	$C_{2,1}$	$C_{2,2}$
$\Gamma(\text{N})$	{ 6,	0,	0;	0,	0,	0 }

Based on the fact that the trace of the transformation (reducible representation) matrix under a symmetry operation is equal to the sum of those of irreducible matrices under the same operation, the number of copies of each representation present in its linear combination $\Gamma(\text{N})$ is identified as

$\Gamma(\text{N})$	{6,	0,	0;	0,	0,	0}
$= A_1$	$= \{1,$	1,	1;	1,	1,	1}
$+A_2$	$+\{1,$	1,	1;	$-1,$	$-1,$	$-1\}$
$+2E_1$	$+2\{2,$	$2 \times (-\frac{1}{2}),$	$2 \times (-\frac{1}{2});$	0,	0,	0}

From which, we learn that the reducible matrix representation of the nodes is the direct sum of one copy of each one-dimensional irreducible matrix representation and two copies of each two-dimensional representation; i.e., $\Gamma(\text{N}) = A_1 + A_2 + 2E_1$, for the structure with $\mathbf{D}_3$ symmetry.

In general, any node of a prismatic tensegrity structure with $\mathbf{D}_N$ symmetry is transformed to a different node by any symmetry operation except for the identity operation: all nodes, in total $2N$, remain unchanged under the identity operation such that the trace of $\mathbf{R}_i$ corresponding to it is $2N$, and $\mathbf{R}_i$ have zero traces under all other symmetry operations of the group. Hence, we have

$$\Gamma(\mathrm{N}) = \{2N, 0, \ldots, 0;\ 0, \ldots, 0\}. \tag{6.16}$$

From characters of the irreducible matrices of dihedral group listed in Table 6.2, the reducible matrix representation of the nodes can be written as a linear combination $\Gamma(\mathrm{N})$ of the irreducible representations in a general form as follows:

$$\begin{aligned}
\Gamma(\mathrm{N}) =\ & \mathrm{A}_1 + \mathrm{A}_2 + (\mathrm{B}_1 + \mathrm{B}_2) + 2\sum_{k=1}^{p} \mathrm{E}_k \\
=\ & \{1, \ldots, 1;\ 1, \ldots, 1\} && \mathrm{A}_1 \\
& + \{1, \ldots, 1;\ -1, \ldots -1\} && \mathrm{A}_2 \\
& + (\{1, \ldots, (-1)^i, \ldots, (-1)^N;\ 1, \ldots, (-1)^i, \ldots, (-1)^N\}) && (\mathrm{B}_1) \\
& + (\{1, \ldots, (-1)^i, \ldots, (-1)^N;\ -1, \ldots, (-1)^{i+1}, \ldots, (-1)^{N+1}\}) && (\mathrm{B}_2) \\
& + 2\sum_{k=1}^{p} \{2C_{0k}, \ldots, 2C_{ik}, \ldots, 2C_{(N-1)k};\ 0, \ldots, 0\} && 2\mathrm{E}_k \\
=\ & \{2N, 0, \ldots, 0;\ 0, \ldots, 0\}.
\end{aligned} \tag{6.17}$$

$\Gamma(\mathrm{N})$ characterizes structure of the symmetry-adapted force density matrix $\tilde{\mathbf{E}}$ as follows:

1. The number of representations μ in $\Gamma(\mathrm{N})$ indicates dimensions of $\tilde{\mathbf{E}}^{\mu}$, which is the block in the symmetry-adapted force density matrix corresponding the representation μ. Hence, we learn from Eq. (6.17) that the blocks corresponding to the one-dimensional representations are 1-by-1 matrices, and those of two-dimensional representations are 2-by-2 matrices.
2. Dimensions of a representation indicate times of its corresponding block appearing in the symmetry-adapted form; thus, each one-dimensional representation has only one copy, and each two-dimensional representation has two copies of blocks lying in the leading diagonal of $\tilde{\mathbf{E}}$.

Table 6.2 Traces of the reducible representation matrices $\mathbf{R}_i$, and characters of the irreducible representation matrices of dihedral group $\mathbf{D}_N$

i	0	1	j	$N-1$	N	$N+1$	$N+j$	$2N-1$
trace($\mathbf{R}_i$)	$2N$	0	0	0	0	0	0	0
$\mu\backslash$operation	E	C_N^1	C_N^j	C_N^{N-1}	$C_{2,0}$	$C_{2,1}$	$C_{2,j}$	$C_{2,N-1}$
A_1	1	1	1	1	1	1	1	1
A_2	1	1	1	1	-1	-1	-1	-1
(B_1)	1	-1	$(-1)^j$	$(-1)^N$	1	-1	$(-1)^j$	$(-1)^N$
(B_2)	1	-1	$(-1)^j$	$(-1)^N$	1	-1	$(-1)^{j+1}$	$(-1)^{N+1}$
E_k	2	$2C_k$	$2C_{jk}$	$2C_{(N-1)k}$	0	0	0	0

The block structure of $\tilde{\mathbf{E}}$ can be written in a general form as follows according to the linear combination $\Gamma(\mathrm{N})$ of representations for transformation of nodes in Eq. (6.17):

$$\underset{2N\times 2N}{\tilde{\mathbf{E}}} = \begin{pmatrix} \underset{1\times1}{\tilde{\mathbf{E}}^{\mathrm{A}_1}} & & & & & & & & \\ & \underset{1\times1}{\tilde{\mathbf{E}}^{\mathrm{A}_2}} & & & & \mathbf{O} & & & \\ & & \underset{1\times1}{(\tilde{\mathbf{E}}^{\mathrm{B}_1})} & & & & & & \\ & & & \underset{1\times1}{(\tilde{\mathbf{E}}^{\mathrm{B}_2})} & & & & & \\ & & & & \underset{2\times2}{\tilde{\mathbf{E}}^{\mathrm{E}_1}} & & & & \\ & & & & & \underset{2\times2}{\tilde{\mathbf{E}}^{\mathrm{E}_1}} & & & \\ & & \mathbf{O} & & & & \ddots & & \\ & & & & & & & \underset{2\times2}{\tilde{\mathbf{E}}^{\mathrm{E}_p}} & \\ & & & & & & & & \underset{2\times2}{\tilde{\mathbf{E}}^{\mathrm{E}_p}} \end{pmatrix}, \tag{6.18}$$

where the blocks $\tilde{\mathbf{E}}^{\mathrm{B}_1}$ and $\tilde{\mathbf{E}}^{\mathrm{B}_2}$ corresponding to representations B_1 and B_2 exist only if N is even.

Equation (6.18) can also be written as a direct sum of the independent blocks $\tilde{\mathbf{E}}^{\mu}$.

$$\underset{2N\times 2N}{\tilde{\mathbf{E}}} = \underset{1\times1}{\tilde{\mathbf{E}}^{\mathrm{A}_1}} \oplus \underset{1\times1}{\tilde{\mathbf{E}}^{\mathrm{A}_2}} \oplus \underset{1\times1}{(\tilde{\mathbf{E}}^{\mathrm{B}_1})} \oplus \underset{1\times1}{(\tilde{\mathbf{E}}^{\mathrm{B}_2})} \oplus 2\underset{2\times2}{\tilde{\mathbf{E}}^{\mathrm{E}_1}} \oplus \cdots \oplus 2\underset{2\times2}{\tilde{\mathbf{E}}^{\mathrm{E}_p}}. \tag{6.19}$$

Example 6.9 Structure of the symmetry-adapted force density matrix of the structures with $\mathbf{D}_3$ symmetry.

There are two one-dimensional representations A_1 and A_2, and only one two-dimensional representation E_1 in the dihedral group $\mathbf{D}_3$. Thus, for a prismatic tensegrity structure with $\mathbf{D}_3$ symmetry, its force density matrix can be formulated in the symmetry-adapted form $\tilde{\mathbf{E}} \in \mathbb{R}^{6\times6}$ as follows by using Eq. (6.18):

$$\underset{6\times6}{\tilde{\mathbf{E}}} = \underset{1\times1}{\tilde{\mathbf{E}}^{\mathrm{A}_1}} \oplus \underset{1\times1}{\tilde{\mathbf{E}}^{\mathrm{A}_2}} \oplus 2\underset{2\times2}{\tilde{\mathbf{E}}^{\mathrm{E}_1}}. \tag{6.20}$$

According to the assignment of force densities in Example 6.5, we have the blocks corresponding to each representation as follows:

$$\begin{aligned} \tilde{\mathbf{E}}^{\mathrm{A}_1} &= 0, \\ \tilde{\mathbf{E}}^{\mathrm{A}_2} &= 0, \\ \tilde{\mathbf{E}}^{\mathrm{E}_1} &= \begin{pmatrix} 3.2321 & -0.8660 \\ -0.8660 & 0.2321 \end{pmatrix}. \end{aligned} \tag{6.21}$$

Table 6.3 Selected irreducible representation matrices corresponding to the nodes connecting to the reference node 0

	Sum	Horizontal		Strut	Vertical
μ	$\mathbf{R}_0^\mu$	$\mathbf{R}_h^\mu$	$\mathbf{R}_{N-h}^\mu$	$\mathbf{R}_N^\mu$	$\mathbf{R}_{N+v}^\mu$
A_1	1	1	1	1	1
A_2	1	1	1	-1	-1
B_1	1	$(-1)^h$	$(-1)^{N-h}$	1	$(-1)^v$
B_2	1	$(-1)^h$	$(-1)^{N-h}$	-1	$(-1)^{v+1}$
E_k	$\mathbf{I}_2$	$\begin{pmatrix} C_{hk} & -S_{hk} \\ S_{hk} & C_{hk} \end{pmatrix}$	$\begin{pmatrix} C_{hk} & S_{hk} \\ -S_{hk} & C_{hk} \end{pmatrix}$	$\begin{pmatrix} 1 & 0 \\ 0 & -1 \end{pmatrix}$	$\begin{pmatrix} C_{vk} & S_{vk} \\ S_{vk} & -C_{vk} \end{pmatrix}$

6.4.3 Blocks of Symmetry-adapted Force Density Matrix

Denote the force densities of horizontal cable, vertical cables, and struts as q_h, q_v, and q_s, respectively. Moreover, denote q as the sum of force densities of all members connected to the reference node; i.e.,

$$q = 2q_\mathrm{h} + q_\mathrm{s} + q_\mathrm{v}, \tag{6.22}$$

since any node of a symmetric prismatic tensegrity structures is connected by two horizontal cables, one vertical cable, and one strut.

The block $\tilde{\mathbf{E}}^\mu$ corresponding to representation μ of the symmetry-adapted force density matrix $\tilde{\mathbf{E}}$ can be written in a general form as follows by using the irreducible representation matrices as listed in Table 6.3:

$$\tilde{\mathbf{E}}^\mu = q\mathbf{R}_0^\mu - q_\mathrm{h}\mathbf{R}_h^\mu - q_\mathrm{h}\mathbf{R}_{N-h}^\mu - q_\mathrm{s}\mathbf{R}_N^\mu - q_\mathrm{v}\mathbf{R}_{N+v}^\mu, \tag{6.23}$$

which is presented in Eq. (D.20), and the detailed proof of this direct formulation for dihedral group can be found in Lemma D.1 in Appendix D.

From Eq. (6.23), the block $\tilde{\mathbf{E}}^{A_1}$ is always equal to zero, since all representation matrices $\mathbf{R}_i^{A_1}$ of A_1 are equal to 1:

$$\begin{aligned} \tilde{\mathbf{E}}^{A_1} &= q\mathbf{R}_0^{A_1} - q_\mathrm{h}\mathbf{R}_h^{A_1} - q_\mathrm{h}\mathbf{R}_{N-h}^{A_1} - q_\mathrm{s}\mathbf{R}_N^{A_1} - q_\mathrm{v}\mathbf{R}_{N+v}^{A_1} \\ &= q - 2q_\mathrm{h} - q_\mathrm{s} - q_\mathrm{v} \\ &= 0. \end{aligned} \tag{6.24}$$

For the $\mathbf{A}_2$ block $\tilde{\mathbf{E}}^{\mathrm{A}_2}$ corresponding to the representation $\mathbf{A}_2$, we have

$$\begin{aligned}\tilde{\mathbf{E}}^{\mathrm{A}_2} &= q - q_{\mathrm{h}} - q_{\mathrm{h}} - q_{\mathrm{s}}(-1) - q_{\mathrm{v}}(-1)\\ &= 2(q_{\mathrm{s}} + q_{\mathrm{v}}).\end{aligned} \tag{6.25}$$

The blocks $\tilde{\mathbf{E}}^{\mathrm{B}_1}$ and $\tilde{\mathbf{E}}^{\mathrm{B}_2}$ corresponding to the representations B_1 and B_2, when they exist for N even, are

$$\begin{aligned}\tilde{\mathbf{E}}^{\mathrm{B}_1} &= q - (-1)^h q_{\mathrm{h}} - (-1)^{N-h} q_{\mathrm{h}} - q_{\mathrm{s}} - (-1)^{\nu} q_{\mathrm{v}}\\ &= [2 - 2(-1)^h] q_{\mathrm{h}} - q_{\mathrm{s}} - (-1)^{\nu} q_{\mathrm{v}},\\ \tilde{\mathbf{E}}^{\mathrm{B}_2} &= q - (-1)^h q_{\mathrm{h}} - (-1)^{N-h} q_{\mathrm{h}} - (-1) q_{\mathrm{s}} - (-1)^{\nu+1} q_{\mathrm{v}}\\ &= [2 - 2(-1)^h] q_{\mathrm{h}} + q_{\mathrm{s}} + (-1)^{\nu} q_{\mathrm{v}}.\end{aligned} \tag{6.26}$$

Moreover, the two-dimensional blocks $\tilde{\mathbf{E}}^{\mathrm{E}_k}$ $(k = 1, 2, \dots, p)$ are

$$\begin{aligned}\tilde{\mathbf{E}}^{\mathrm{E}_k} &= q\begin{pmatrix}1 & 0\\ 0 & 1\end{pmatrix} - q_{\mathrm{h}}\begin{pmatrix}C_{hk} & -S_{hk}\\ S_{hk} & C_{hk}\end{pmatrix} - q_{\mathrm{h}}\begin{pmatrix}C_{hk} & S_{hk}\\ -S_{hk} & C_{hk}\end{pmatrix}\\ &\quad - q_{\mathrm{s}}\begin{pmatrix}1 & 0\\ 0 & -1\end{pmatrix} - q_{\mathrm{v}}\begin{pmatrix}C_{\nu k} & S_{\nu k}\\ S_{\nu k} & -C_{\nu k}\end{pmatrix}\\ &= \begin{pmatrix}2(1 - C_{hk})q_{\mathrm{h}} + (1 - C_{\nu k})q_{\mathrm{v}} & -S_{\nu k} q_{\mathrm{v}}\\ -S_{\nu k} q_{\mathrm{v}} & 2(1 - C_{hk})q_{\mathrm{h}} + 2q_{\mathrm{s}} + (1 + C_{\nu k})q_{\mathrm{v}}\end{pmatrix}.\end{aligned} \tag{6.27}$$

6.5 Self-equilibrium Conditions

As an alternative approach to self-equilibrium analysis of the symmetric prismatic structures in Chap. 3, this section makes use of the non-degeneracy condition in terms of rank deficiency of the force density matrix in Chap. 2, as well as the A_2 and E_1 blocks of the symmetry-adapted force density matrix.

To ensure a non-degenerate tensegrity structure in three-dimensional space, the force density matrix $\mathbf{E}$, or equivalently its symmetry-adapted form $\tilde{\mathbf{E}}$, should have rank deficiency of at least four (see Chap. 2 for the non-degeneracy condition for free-standing structures). Rank deficiency of a symmetric matrix is the number of its zero eigenvalues. Thus, there should be no less than four zero eigenvalues in the force density matrix.

From Eq. (6.24), we know that the A_1 block $\tilde{\mathbf{E}}^{\mathrm{A}_1}$ is always zero, with one zero eigenvalue, coming from the fact that tensegrity structures are free-standing without any support or fixed node. The other three zero eigenvalues should come from $\tilde{\mathbf{E}}^{\mathrm{A}_2}$ and the two copies of $\tilde{\mathbf{E}}^{\mathrm{E}_1}$, because representations A_2 and E_1 respectively stand for transformation of z- and xy-coordinates as indicated in Table 6.1. This means that their determinants are zero, such that we have

$$|\tilde{\mathbf{E}}^{\mathrm{A}_2}| = |\tilde{\mathbf{E}}^{\mathrm{E}_1}| = 0, \tag{6.28}$$

where $|\cdot|$ denotes determinant of a matrix.

It is easy to have the following relation between the force density q_{v} of vertical cables and the force density q_{s} of struts from Eqs. (6.25) and (6.28):

$$q_{\mathrm{v}} = -q_{\mathrm{s}} \ (> 0). \tag{6.29}$$

Using Eq. (6.29), the E_1 block $\tilde{\mathbf{E}}^{\mathrm{E}_1}$ defined in Eq. (6.28) with $k = 1$ becomes

$$\tilde{\mathbf{E}}^{\mathrm{E}_1} = \begin{pmatrix} 2(1 - C_h)q_{\mathrm{h}} + (1 - C_v)q_{\mathrm{v}} & -S_v q_{\mathrm{v}} \\ -S_v q_{\mathrm{v}} & 2(1 - C_h)q_{\mathrm{h}} - (1 - C_v)q_{\mathrm{v}} \end{pmatrix}, \tag{6.30}$$

the determinant of which is computed as

$$\begin{aligned} |\tilde{\mathbf{E}}^{\mathrm{E}_1}| &= [2(1 - C_h)]^2 q_{\mathrm{h}}^2 - 2(1 - C_v)q_{\mathrm{v}}^2 \\ &= 0. \end{aligned} \tag{6.31}$$

Therefore, we have

$$t = \frac{q_{\mathrm{h}}}{q_{\mathrm{v}}} = \frac{\sqrt{2 - 2C_v}}{2(1 - C_h)}, \tag{6.32}$$

where the negative solution has been neglected since q_{v} and q_{h}, and therefore, t, are positive to guarantee that cables carry tension (positive prestress). This way, the force densities in a self-equilibrium state are derived by making the relevant blocks of the symmetry-adapted force density matrix to be singular, so as to have enough rank deficiency for satisfying the non-degeneracy condition. The results in Eqs. (6.29) and (6.32) coincide with those obtained from the self-equilibrium equations of the representative node as presented in Chap. 3.

Furthermore, the prismatic structure $\mathbf{D}_N^{h,v}$ has only one prestress mode, because all force densities are uniquely determined if any of them is assigned.

Example 6.10 Verify non-degeneracy condition of the prismatic structure $\mathbf{D}_3^{1,1}$ with the force densities determined in Eqs. (6.29) and (6.32).

From Eqs. (6.29) and (6.32), we have the force densities for the symmetric prismatic structure $\mathbf{D}_3^{1,1}$ as

$$\begin{aligned} q_{\mathrm{s}} &= -q_{\mathrm{v}}, \\ q_{\mathrm{h}} &= \frac{\sqrt{3}}{3} q_{\mathrm{v}} \quad \text{or} \quad t = \frac{q_{\mathrm{h}}}{q_{\mathrm{v}}} = \frac{\sqrt{3}}{3}. \end{aligned} \tag{6.33}$$

Using the force densities in Eq. (6.33), $\tilde{\mathbf{E}}^{\mathrm{A}_2}$ in Eq. (6.25) and $\tilde{\mathbf{E}}^{\mathrm{E}_1}$ in Eq. (6.27) are

$$\tilde{\mathbf{E}}^{\mathrm{A}_2} = 2(q_{\mathrm{s}} + q_{\mathrm{v}}) = 2(-q_{\mathrm{v}} + q_{\mathrm{v}}) = 0, \tag{6.34}$$

$$\tilde{\mathbf{E}}^{\mathrm{E}_1} = \frac{1}{2}\begin{pmatrix} 2\sqrt{3}+3 & -\sqrt{3} \\ -\sqrt{3} & 2\sqrt{3}-3 \end{pmatrix} q_{\mathrm{v}}, \tag{6.35}$$

in which $C_h = \cos(2\pi/3) = -1/2$ has been used for $h = 1$.

It is easy to verify that $\tilde{\mathbf{E}}^{\mathrm{E}_1}$ is singular, providing one zero eigenvalue. Therefore, including the zero eigenvalues in $\tilde{\mathbf{E}}^{\mathrm{A}_1}$ and $\tilde{\mathbf{E}}^{\mathrm{A}_2}$, there are in total four zero eigenvalues in the (symmetry-adapted) force density matrix together with the two zero eigenvalues in the two copies of $\tilde{\mathbf{E}}^{\mathrm{E}_1}$. This satisfies the non-degeneracy condition for a three-dimensional tensegrity structure as discussed in Chap. 2.

6.6 Stability Conditions

Depending on the connectivity of members, a prismatic tensegrity structure may be completely separated into several identical substructures that have no mechanical relation with each other. The substructures have lower symmetry compared to the original structure. Such divisible structures should be excluded from stability investigation, because there is nothing to prevent the substructures moving relative to one another; the stability of the substructures themselves should be considered anyway in the lower symmetry case.

6.6.1 Divisibility Conditions

This subsection presents the necessary and sufficient divisibility conditions for prismatic tensegrity structures. It is demonstrated that divisibility of these structures depends on connectivity of the horizontal cables as well as vertical cables, while connectivity of struts is assumed to be fixed.

6.6.1.1 Divisibility of Horizontal Cables

Suppose that we randomly select one node as the starting node, and travel to the next along the horizontal cables on the same horizontal plane. If we repeat this trip in a consistent direction, eventually, we must come back to the starting node. The nodes and horizontal cables that have been visited in the trip are said to belong to the same *circuit*. If there is more than one circuit on the same plane, the horizontal cables are said to be *divisible*; otherwise, they are *indivisible*.

Denote by n^{c} the number of circuits of the horizontal cables lying on one plane, and the number of nodes in a circuit by N^{n}. Each time we travel from one node to the next node in the same circuit, we pass by h nodes, and hence by the time we return to the starting node, we have passed hN^{n} nodes.

Example 6.11 Divisibility of horizontal cables of the prismatic structure $\mathbf{D}_6^{2,1}$.

Consider the structure $\mathbf{D}_6^{2,1}$ as shown in Fig. 6.6a, with $N = 6$. It can be observed from the figure that node 0 is connected to nodes 2 and 4 by the horizontal cables on the upper plane. These three nodes form a circuit; i.e., $N^{\mathrm{n}} = 3$. Each time we travel around the circuit, we pass by two nodes, which comes from its notation $h = 2$ for connectivity of horizontal cables. Therefore, when we travel around the circuit and come back to the starting node, we have passed by $hN^{\mathrm{n}} = 6$ nodes.

Moreover, there are $n^{\mathrm{c}} = 2$ circuits on the upper plane, the other circuit is composed of nodes 1, 3, and 5. These two circuits do not have any mechanical relation with each other. The same situation occurs for the horizontal cables on the bottom plane. Therefore, the structure has in total four circuits of nodes connected by horizontal cables, two on each of the two parallel planes:

Circuit	Nodes
1	0, 2, 4
2	1, 3, 5
3	6, 8, 10
4	7, 9, 11

(6.36)

Suppose that, in one of the circuits, we have travelled around the center point of the circle h^{n} times, and hence, have passed by Nh^{n} nodes, which should be equal to the number $(N^{\mathrm{n}}h)$ of nodes passed by in the trip along the horizontal cables.

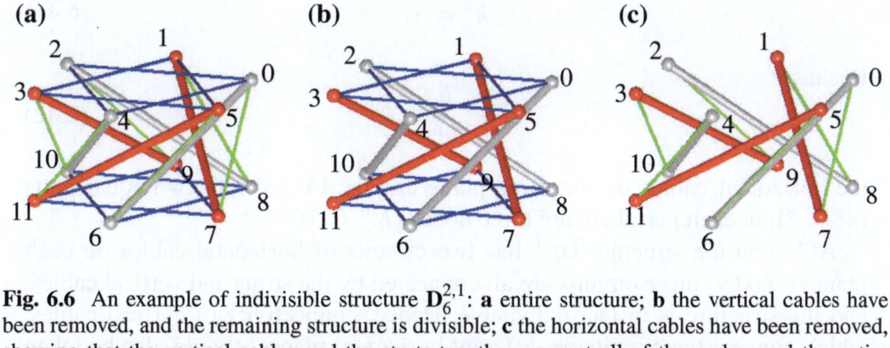

Fig. 6.6 An example of indivisible structure $\mathbf{D}_6^{2,1}$: **a** entire structure; **b** the vertical cables have been removed, and the remaining structure is divisible; **c** the horizontal cables have been removed, showing that the vertical cables and the struts together connect all of the nodes, and the entire structure is therefore indivisible. **a** Indivisible, **b** divisible, **c** indivisible

Accordingly, we have

$$N^{\mathrm{n}} h = N h^{\mathrm{n}}. \tag{6.37}$$

Note that N^{n} and h^{n} are the smallest possible positive integers satisfying Eq. (6.37). Moreover, the number of circuits n^{c} lying on each horizontal plane is given by

$$n^{\mathrm{c}} = \frac{N}{N^{\mathrm{n}}} = \frac{h}{h^{\mathrm{n}}}. \tag{6.38}$$

Necessary and sufficient condition for divisibility of horizontal cables on the same plane:

For the horizontal cables on the same plane, they are divisible if there is more than one circuit of nodes; i.e.,

$$n^{\mathrm{c}} > 1 \quad \text{or equivalently} \quad h \neq h^{\mathrm{n}}. \tag{6.39}$$

If the structure is divisible, the above parameters give useful information about the substructures: there are n^{c} substructures, and they will have N^{n} nodes on each plane, with a connectivity of the horizontal cables of h^{n}.

Example 6.12 Divisibility of horizontal cables of the prismatic structure $\mathbf{D}_6^{2,1}$.

Consider the prismatic structure $\mathbf{D}_6^{2,1}$ as shown in Fig. 6.6a, where $N = 6$ and $h = 2$. From Eq. (6.37), we have

$$2N^{\mathrm{n}} = 6h^{\mathrm{n}}, \tag{6.40}$$

for which the smallest positive integers N^{n} and h^{n} satisfying Eq. (6.40) are

$$N^{\mathrm{n}} = 3, \qquad h^{\mathrm{n}} = 1. \tag{6.41}$$

Because

$$n^{\mathrm{c}} = \frac{h}{h^{\mathrm{n}}} = 2, \tag{6.42}$$

the horizontal cables on the same plane can be divided into two equal parts ($n^{\mathrm{c}} = 2$); in each part there are three nodes ($N^{\mathrm{n}} = 3$).

Although the structure $\mathbf{D}_6^{2,1}$ has two circuits of horizontal cables on each plane of nodes, those circuits are all connected by the struts and vertical cables, and the structure is in fact indivisible. Hence, connectivity of vertical cables, which connect the circuits on different horizontal planes, should also be taken into consideration as discussed later.

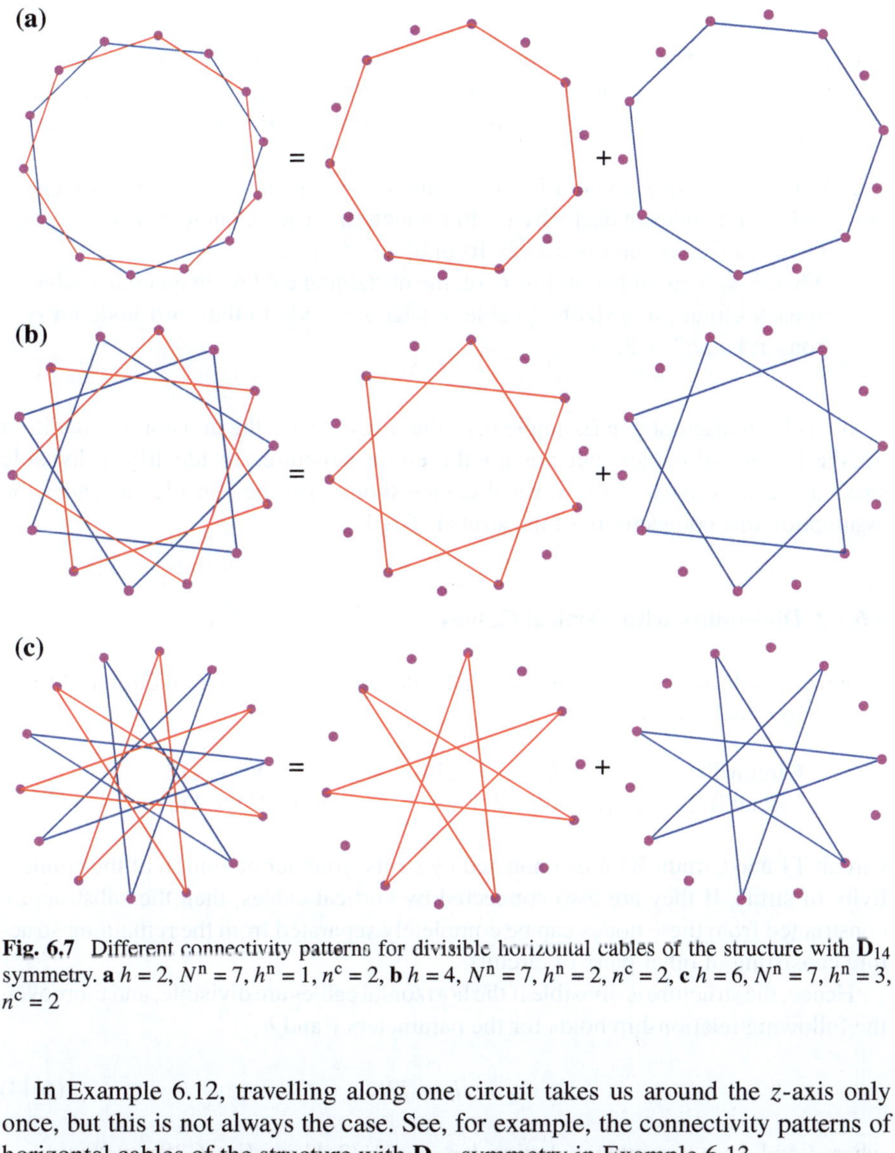

Fig. 6.7 Different connectivity patterns for divisible horizontal cables of the structure with $\mathbf{D}_{14}$ symmetry. **a** $h = 2$, $N^{\mathrm{n}} = 7$, $h^{\mathrm{n}} = 1$, $n^{\mathrm{c}} = 2$, **b** $h = 4$, $N^{\mathrm{n}} = 7$, $h^{\mathrm{n}} = 2$, $n^{\mathrm{c}} = 2$, **c** $h = 6$, $N^{\mathrm{n}} = 7$, $h^{\mathrm{n}} = 3$, $n^{\mathrm{c}} = 2$

In Example 6.12, travelling along one circuit takes us around the z-axis only once, but this is not always the case. See, for example, the connectivity patterns of horizontal cables of the structure with $\mathbf{D}_{14}$ symmetry in Example 6.13.

Example 6.13 Divisibility of horizontal cables of the structure with $\mathbf{D}_{14}$ symmetry.

Consider one of the planes of the structure with $\mathbf{D}_{14}$ symmetry as shown in Fig. 6.7. We have the following cases, where the horizontal cables are divisible with two circuits of nodes on the same plane:

1. In the case of $h = 2$, as shown in Fig. 6.7a, the horizontal cables lying on the plane can be divided into two circuits ($n^{\mathrm{c}} = 2$), seven nodes in each ($N^{\mathrm{n}} = 7$). In each circuit, a horizontal cable connects a node to its adjacent node; $h^{\mathrm{n}} = 1$.
2. When $h = 4$, as shown in Fig. 6.7b, the horizontal cables are also divisible, with seven nodes in each circuit. In each circuit, a horizontal cable connects a node to the second node away from it; i.e., $h^{\mathrm{n}} = 2$.
3. When $h = 6$, as shown in Fig. 6.7c, the horizontal cables are again divisible. In each circuit, a horizontal cable connects a node to the third node away from it; i.e., $h^{\mathrm{n}} = 3$.

It has been discussed in Example 6.12 that Eq. (6.39) is the divisibility condition for the horizontal cables, but not for the entire structure. To identify a divisible structure, connectivity of the vertical cables should also be considered, under the assumption that connectivity of the struts is fixed.

6.6.1.2 Divisibility with Vertical Cables

Suppose that the horizontal cables are divisible: the nodes in a pair of circuits T1 and B1 of horizontal cables are

$$\begin{array}{llllll} \text{Circuit T1:} & 0, & h, & 2h, & \ldots, & (N^{\mathrm{n}}-1)h \\ \text{Circuit B1:} & N, & N+h, & N+2h, & \ldots, & N+(N^{\mathrm{n}}-1)h \end{array} \tag{6.43}$$

Circuit T1 and Circuit B1 are connected by struts from our definition of the connectivity of struts. If they are also connected by vertical cables, then the substructure constructed from these nodes can be completely separated from the remaining structure consisting of other pairs of circuits.

Hence, the structure is divisible if the horizontal cables are divisible, and moreover, the following relationship holds for the parameters v and h

$$v = ih - jN, \tag{6.44}$$

where i and j are two integers letting $1 \leq v \leq N/2$. Using Eq. (6.38), we have

$$v = (ih^{\mathrm{n}} - jN^{\mathrm{n}})n^{\mathrm{c}}. \tag{6.45}$$

$(ih^{\mathrm{n}} - jN^{\mathrm{n}})$ can be any integer in between 1 and $N/2$, since i and j are arbitrary integers. Therefore, Eq. (6.45) holds if v can be divisible by n^{c}.

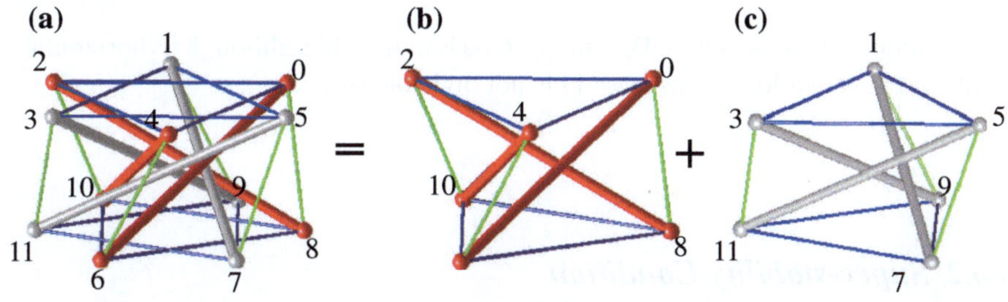

Fig. 6.8 Divisible structure $\mathbf{D}_6^{2,2}$ and its substructures $\mathbf{D}_3^{1,1}$. The structure can be completely divided into two substructures, which have their own self-equilibrium force mode and there is no physical relation between them such that they can have relative (finite) motions. **a** $\mathbf{D}_6^{2,2}$, **b** $\mathbf{D}_3^{1,1}$, **c** $\mathbf{D}_3^{1,1}$

Necessary and sufficient conditions for a divisible prismatic tensegrity structure $\mathbf{D}_N^{h,v}$:

A prismatic structure $\mathbf{D}_N^{h,v}$ is divisible if the following two conditions are satisfied:

- The horizontal cables are divisible; i.e.,

$$h \neq h^{\mathrm{n}}. \tag{6.39}$$

- The connectivity parameter v is divisible by n^{c}.

Moreover, the original structure $\mathbf{D}_N^{h,v}$ can be divided into n^{c} identical substructures $\mathbf{D}_{N^{\mathrm{n}}}^{h^{\mathrm{n}},v^{\mathrm{n}}}$, where v^{n} is the connectivity parameter of each substructure computed by

$$v^{\mathrm{n}} = \frac{v}{n^{\mathrm{c}}}. \tag{6.46}$$

Example 6.14 Divisible prismatic structure $\mathbf{D}_6^{2,2}$.

For the prismatic structure $\mathbf{D}_6^{2,2}$ in Fig. 6.8a, which has the same connectivity of horizontal cables $h = 2$ as the structure $\mathbf{D}_6^{2,1}$ in Fig. 6.6a, we have learned from Example 6.12 that its horizontal cables are divisible, with $n^{\mathrm{c}} = 2$ and $h^{\mathrm{n}} = 1$.

It is obvious that $v(=2)$ is divisible by $n^{\mathrm{c}}(=2)$, because they have the same value. Hence, the structure $\mathbf{D}_6^{2,2}$ is divisible. Moreover, it can be divided into two identical substructures $\mathbf{D}_3^{1,1}$ as shown in Fig. 6.8b, c, since

$$v^{\mathrm{n}} = \frac{v}{n^{\mathrm{c}}} = \frac{2}{2} = 1. \tag{6.47}$$

By contrast, the structure $\mathbf{D}_6^{2,1}$ in Fig. 6.6a is indivisible, although its horizontal cables are divisible, because $v(=1)$ is not divisible by n^{c}.

6.6.2 Super-stability Condition

In this subsection, we present the analytical conditions for super-stability of prismatic structures with dihedral symmetry. In Chap. 4, we have presented the following three sufficient conditions for super-stability of a three-dimensional tensegrity structure:

1. The geometry matrix of the structure has (full-)rank of six;
2. The force density matrix $\mathbf{E}$ is positive semi-definite;
3. Rank deficiency of $\mathbf{E}$ is four.

The first condition is usually satisfied, if a prismatic tensegrity structure is indivisible, and hence, only the last two conditions need to be considered for verification of its super-stability.

Moreover, since the A_1 and A_2 blocks are always zero as discussed in Sect. 6.5 for self-equilibrium analysis, we need only to investigate positive (semi-)definiteness of the B_1, B_2, and E_k blocks to present the super-stability conditions.

From the analytical formulation of each block in Eq. (6.23), the blocks $\tilde{\mathbf{E}}^{\mathrm{B}_1}$ and $\tilde{\mathbf{E}}^{\mathrm{B}_2}$ corresponding to the representations B_1 and B_2, when they exist for N even, are

$$\begin{aligned}\frac{1}{q_{\mathrm{v}}}\tilde{\mathbf{E}}^{\mathrm{B}_1} &= [2-2(-1)^h]t + 1 - (-1)^v,\\ \frac{1}{q_{\mathrm{v}}}\tilde{\mathbf{E}}^{\mathrm{B}_2} &= [2-2(-1)^h]t + (-1)^{v+1}.\end{aligned} \tag{6.48}$$

The two-dimensional blocks $\tilde{\mathbf{E}}^{\mathrm{E}_k}$ $(k = 1, 2, \dots, p)$ are

$$\frac{1}{q_{\mathrm{v}}}\tilde{\mathbf{E}}^{\mathrm{E}_k} = \begin{pmatrix} 2t(1-C_{hk}) + 1 - C_{vk} & -S_{vk} \\ -S_{vk} & 2t(1-C_{hk}) - (1-C_{vk}) \end{pmatrix}, \tag{6.49}$$

the two eigenvalues of which are easily computed as

$$\begin{aligned}\frac{\lambda_1^{\mathrm{E}_k}}{q_{\mathrm{v}}} &= 2t(1-C_{hk}) + \sqrt{2(1-C_{vk})} > 0,\\ \frac{\lambda_2^{\mathrm{E}_k}}{q_{\mathrm{v}}} &= 2t(1-C_{hk}) - \sqrt{2(1-C_{vk})}.\end{aligned} \tag{6.50}$$

$\lambda_1^{\mathrm{E}_k} > 0$ always holds since $t > 0$, $1 - C_{hk} > 0$, and $1 - C_{vk} \geq 0$.

For representation E_1, we know from Eq. (6.32) that $\lambda_2^{\mathrm{E}_1} = 0$ holds, satisfying the non-degeneracy condition. Therefore, the second sufficient condition, having the minimum necessary rank deficiency, for super-stability of tensegrity structures is satisfied.

Because $\lambda_1^{\mathrm{E}_k}$ for any k is positive as indicated in Eq. (6.50), we need that $\lambda_2^{\mathrm{E}_k} > 0$ for $k > 1$ holds so as to have a positive semi-definite force density matrix.

Connelly obtained the same two-dimensional blocks making use of the special properties of the force density matrix as a circulant matrix [2]. He further proved that all other two-dimensional blocks (for $k > 1$) are positive definite if $h = 1$.

From the divisibility condition in Eq. (6.39), the structure is indivisible when $h = 1$. To further verify whether $h = 1$ is actually the super-stability condition for prismatic structures, we need to investigate the one-dimensional blocks: $\tilde{\mathbf{E}}^{\mathrm{A}_1} = \tilde{\mathbf{E}}^{\mathrm{A}_2} = 0$ always holds as discussed previously; and $\tilde{\mathbf{E}}^{\mathrm{B}_1}$ and $\tilde{\mathbf{E}}^{\mathrm{B}_2}$ exist only when N is even, for which we have the following relation from Eq. (6.48) for $h = 1$:

$$\begin{aligned}\frac{1}{q_{\mathrm{v}}}\tilde{\mathbf{E}}^{\mathrm{B}_1} &= 4t + 1 - (-1)^{v} \\ &\geq 4t \\ &> 0,\end{aligned} \tag{6.51}$$

and moreover,

$$\begin{aligned}\frac{1}{q_{\mathrm{v}}}\tilde{\mathbf{E}}^{\mathrm{B}_2} &= 4t + (-1)^{v+1} = 2\frac{\sqrt{2-2C_v}}{1-C_1} - 1 - (-1)^{v} \\ &\geq 2\frac{\sqrt{2-2C_1}}{1-C_1} - 1 - (-1)^{1} = \frac{2\sqrt{2}}{\sqrt{1-C_1}} - 2 \\ &> 0.\end{aligned} \tag{6.52}$$

In summary, $h = 1$ guarantees two of the sufficient conditions for super-stability of a prismatic tensegrity structure: its force density matrix has rank deficiency of four (one in $\tilde{\mathbf{E}}^{\mathrm{A}_1}$, one in $\tilde{\mathbf{E}}^{\mathrm{A}_2}$, and two in the two copies of $\tilde{\mathbf{E}}^{\mathrm{E}_1}$); and moreover, the force density matrix is positive semi-definite.

Super-stability condition for a symmetric prismatic tensegrity structure $\mathbf{D}_N^{h,v}$:

A prismatic tensegrity structure $\mathbf{D}_N^{h,v}$ is super-stable with the force densities in Eqs. (6.29) and (6.32), if the following condition about connectivity of horizontal cables is satisfied

$$h = 1, \tag{6.53}$$

i.e., the horizontal cables are connected to adjacent nodes.

Example 6.15 Verification of super-stability of the prismatic structure $\mathbf{D}_3^{1,1}$.

For the prismatic structure $\mathbf{D}_3^{1,1}$, the sub-matrices $\tilde{\mathbf{E}}^{A_1}$ and $\tilde{\mathbf{E}}^{A_2}$ of its force density matrix corresponding to the one-dimensional representations A_1 and A_2, respectively, are zero.

The two eigenvalues of the sub-matrix $\tilde{\mathbf{E}}^{E_1}$ corresponding to the two-dimensional representation E_1 are

$$\frac{\lambda_1^{E_1}}{q_v} = 2t(1 - C_1) + \sqrt{2(1 - C_1)} = 2\sqrt{3},$$
$$\frac{\lambda_2^{E_1}}{q_v} = 2t(1 - C_1) - \sqrt{2(1 - C_1)} = 0, \tag{6.54}$$

by using Eq. (6.50), where $C_1 = -1/2$, $S_1 = \sqrt{3}/2$, and $t = \sqrt{3}/3$ as presented in Example 6.10.

In summary, there exist four zero eigenvalues in the (symmetry-adapted) force density matrix of the structure $\mathbf{D}_3^{1,1}$, and two positive eigenvalues, both of which are equal to $2\sqrt{3}$. Thus, the structure is super-stable, because the last two of the three sufficient conditions for a super-stable tensegrity structure are also satisfied.

6.7 Prestress-stability and Stability

Let $\mathbf{Q}_G$ denote the quadratic form of the geometrical stiffness matrix $\mathbf{K}_G$ with respect to the mechanisms. Definition of $\mathbf{Q}_G$ can be found in Eq. (4.78). In Chap. 4, we have shown that $\mathbf{Q}_G$ has to be positive definite for a prestress-stable structure. Remember that a (super-)stable structure is prestress-stable, however, a prestress-stable structure might not be (super-)stable.

When a prismatic structure is divisible, it is unstable, because $\mathbf{Q}_G$ must have at least one zero eigenvalue corresponding to the relative motion of the substructures. However, the structure may, or may not, be prestress-stable, even it is indivisible, depending upon the interplay between the geometrical stiffness matrix and the mechanisms.

In this section, we will demonstrate that (prestress-)stability of the prismatic structures with dihedral symmetry that are not super-stable may be influenced by the height/radius ratio, connectivity, as well as material and level of prestress.

In the following examples, units are omitted for clarity without any loss of generality on discussions on stability; moreover, the *minimum relative eigenvalue* is defined as the ratio of the minimum eigenvalue of $\tilde{\mathbf{Q}}_G^{\mu}$ to the force density q_v of the vertical cables, where $\tilde{\mathbf{Q}}_G^{\mu}$ is the symmetry-adapted quadratic form of the geometrical stiffness matrix with respect to infinitesimal mechanisms for representation μ. Formulations of

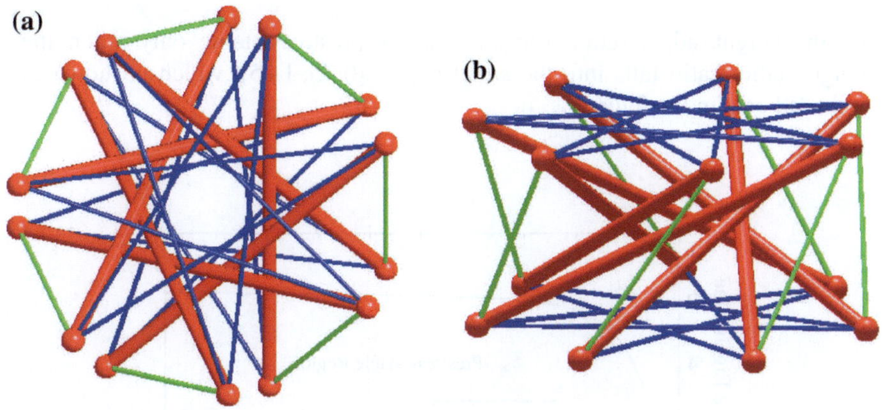

Fig. 6.9 Indivisible prismatic structure $\mathbf{D}_7^{3,2}$. The structure might be prestress-stable with certain conditions satisfied, although it is not super-stable. **a** Top view, **b** diagonal view

$\tilde{\mathbf{Q}}_{\mathrm{G}}^{\mu}$ is not given here for simplicity; interested readers may find more details in our previous study [8].

There always exist zero eigenvalues in $\tilde{\mathbf{Q}}_{\mathrm{G}}^{\mathrm{A}_2}$ and $\tilde{\mathbf{Q}}_{\mathrm{G}}^{\mathrm{E}_1}$ corresponding to affine motions of the structure [8]. Hence, their eigenvalues will not be considered in the discussions on prestress-stability of the structure in the following examples.

6.7.1 Height/Radius Ratio

Height of a prismatic structure $\mathbf{D}_N^{h,v}$ refers to the distance between the two parallel circles, on which the nodes are lying; and radius of the structure refers to radius of the circles. To demonstrate the influence of height/radius ratio on prestress-stability of the structure, we consider the indivisible structure $\mathbf{D}_7^{3,2}$ in Fig. 6.9 as an example.

Example 6.16 Influence of the height/radius ratio on prestress-stability of the prismatic structure $\mathbf{D}_7^{3,2}$.

The prismatic structure $\mathbf{D}_7^{3,2}$ is composed of 14 nodes and 28 members. It is not super-stable associated with the force densities given in Eqs. (6.29) and (6.32), because the connectivity of its horizontal cables ($h = 2$) does not satisfy the super-stability condition ($h = 1$) presented in Sect. 6.6.2. However, it is prestress-stable.

The relationship between the minimum eigenvalue of each block $\tilde{\mathbf{Q}}_{\mathrm{G}}^{\mu}$ ($\mu = \mathrm{A}_1, \mathrm{E}_2, \mathrm{E}_3$) and the height/radius ratio is plotted in Fig. 6.10.

The symmetry-adapted quadratic form $\tilde{\mathbf{Q}}_{\mathrm{G}}^{\mathrm{A}_1}$ of representation A_1 is always positive definite, while positive definiteness of $\tilde{\mathbf{Q}}_{\mathrm{G}}^{\mathrm{E}_2}$ and $\tilde{\mathbf{Q}}_{\mathrm{G}}^{\mathrm{E}_3}$ varies depending on the height/radius ratio. The structure is prestress-stable only when the height/radius ratio falls into the small region [0.75, 1.05], which is shown as a shaded area in the figure.

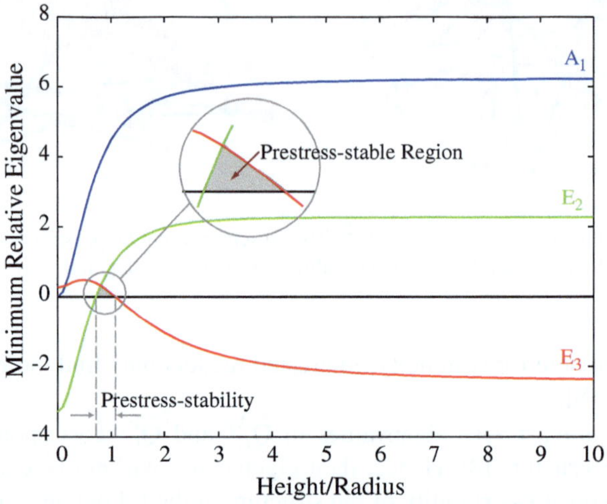

Fig. 6.10 Influence of the height/radius ratio on prestress-stability of the prismatic structure $\mathbf{D}_7^{3,2}$. The structure is prestress-stable when the ratio is in the range [0.75, 1.05]. In order to non-dimensionalize the results, the eigenvalues of $\tilde{\mathbf{Q}}_G^{\mu}$ ($\mu = \mathrm{A}_1, \mathrm{E}_2, \mathrm{E}_3$) are plotted relative to force density of the vertical cables

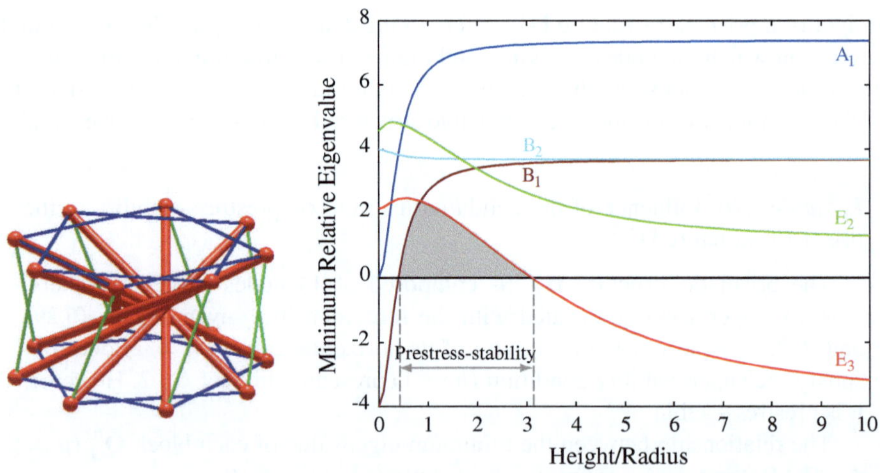

Fig. 6.11 Influence of the height/radius ratio on prestress-stability of the structure $\mathbf{D}_8^{2,3}$. The structure is prestress-stable when the ratio is in the range [0.40, 3.10]. The eigenvalues of $\tilde{\mathbf{Q}}_G^{\mu}$ ($\mu = \mathrm{A}_1, \mathrm{B}_1, \mathrm{B}_2, \mathrm{E}_2, \mathrm{E}_3$) are plotted relative to force density of the vertical cables

Example 6.17 Influence of height/radius ratio on prestress-stability of the prismatic structure $\mathbf{D}_8^{2,3}$.

Consider another indivisible structure $\mathbf{D}_8^{2,3}$ with 16 nodes and 32 members as shown in Fig. 6.11. The dihedral group $\mathbf{D}_8$ has four one-dimensional (A_1, A_2, B_1, B_2) and three two-dimensional representations (E_1, E_2, E_3).

The minimum eigenvalues of $\tilde{\mathbf{Q}}_G^{\mu}$ ($\mu = A_1, B_1, B_2, E_2, E_3$) are plotted with respect to the height/radius ratio in Fig. 6.11. The region of a prestress-stable structure ranges from 0.4 to 3.1, which is much wider than that of the structure $\mathbf{D}_7^{3,2}$.

In Examples 6.16 and 6.17, we have illustrated that the height/radius ratio of the structure can be crucial to prestress-stability of the prismatic tensegrity structures with dihedral symmetry, when they are not super-stable.

6.7.2 Connectivity

Since a prismatic tensegrity structure is super-stable associated with the force densities given in Eqs. (6.29) and (6.32) if $h = 1$, it is clear that stability of this class of structures is directly related to connectivity pattern of the horizontal cables. It has also been illustrated in Sect. 6.7.1 that in some special cases with the right height/radius ratio, the structure might still be prestress-stable although it is not super-stable. However, this is dependent upon the connectivity of both the horizontal and the vertical cables. As an example, consider the structures $\mathbf{D}_8^{2,1}$ and $\mathbf{D}_8^{2,3}$ in Fig. 6.12.

Example 6.18 Influence of connectivity of vertical cables on prestress-stability of the structures $\mathbf{D}_8^{2,1}$ and $\mathbf{D}_8^{2,3}$.

Due to the connectivity of horizontal cables, both of the structures $\mathbf{D}_8^{2,1}$ and $\mathbf{D}_8^{2,3}$ are not super-stable associated with the force densities given in Eqs. (6.29) and (6.32). These two structures are different in connectivity of vertical cables.

As we have seen in Fig. 6.11, the structure $\mathbf{D}_8^{2,3}$ is prestress-stable for a limited range of height/radius ratio. By contrast, the structure $\mathbf{D}_8^{2,1}$ is never prestress-stable; i.e., it is always unstable, because there always exists a negative eigenvalue in its force density matrix as indicated in Fig. 6.12.

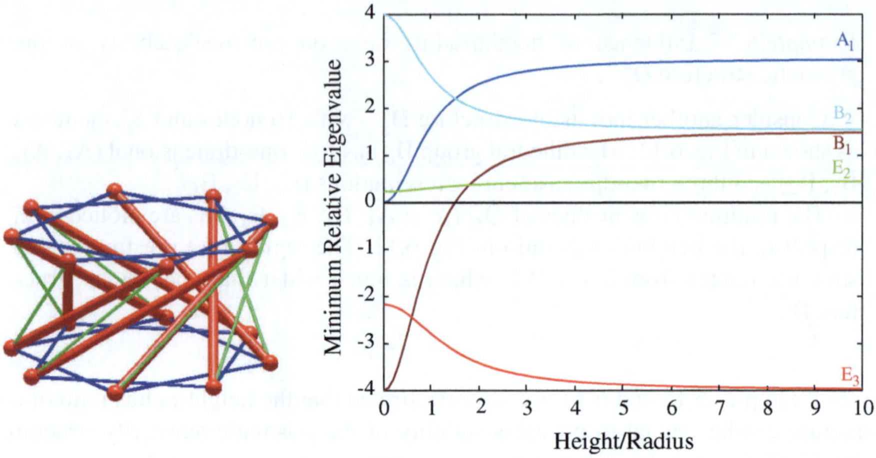

Fig. 6.12 Influence of the height/radius ratio on the prestress-stability of the structure $\mathbf{D}_8^{2,1}$. The eigenvalues of $\tilde{\mathbf{Q}}_G^{\mu}$ ($\mu = \mathrm{A}_1, \mathrm{B}_1, \mathrm{B}_2, \mathrm{E}_2, \mathrm{E}_3$) are plotted relative to the force density in the vertical cables. The structure is never stable, because there always exists one negative eigenvalue in $\tilde{\mathbf{Q}}_G^{\mathrm{E}_3}$

6.7.3 Materials and Level of Prestresses

So far, the prestress-stability is investigated based on the positive definiteness of the quadratic form $\tilde{\mathbf{Q}}_G$ of the geometrical stiffness matrix with respect to the mechanisms, where the members are assumed to have high enough stiffness. Here we show that selection of materials and level of prestresses also have effects on the stability of the structures when they are not super-stable.

Fig. 6.13 Influence of the stiffness/prestress ratio k on the stability of the structure $\mathbf{D}_7^{3,2}$. The minimum eigenvalues of the tangent stiffness matrix are plotted with respect to height/radius ratio of the structure

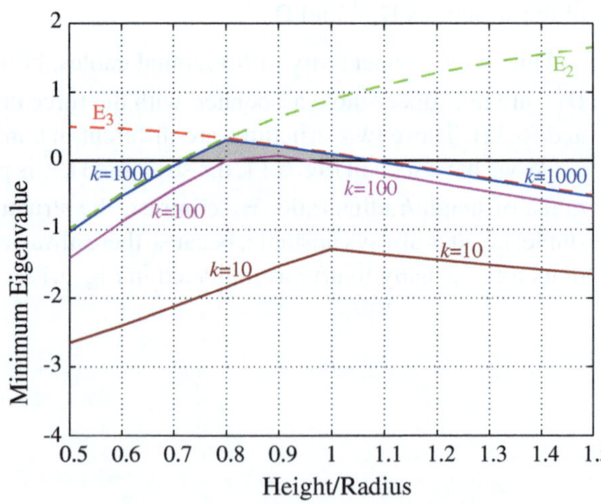

Example 6.19 Influence of material properties and level of prestresses on stability of the structure $\mathbf{D}_7^{3,2}$.

We make a simplification such that all of the struts and cables have the same axial stiffness AE/l, where A is cross-section area, E is Young's modulus, and l is member length. The key parameter is then the ratio of the axial stiffness to the prestress in the structure.

To illustrate influence of selection of material and/or level of prestresses on its stability, we consider the stiffness for several different values of $k = AE/(lq_{\text{v}})$, where k is dimensionless. If material of the structure is linearly elastic, the strain due to a particular prestress will be $1/k$, and thus even values corresponding to $k = 100$ are too small to be realistic for conventional materials.

Figure 6.13 shows the smallest relative eigenvalues of the tangent stiffness matrix for the structure $\mathbf{D}_7^{3,2}$, which is prestress-stable with the height/radius ratio of 1.0. Results are plotted for $k = 10$, $k = 100$, $k = 1{,}000$, and infinite value. As k is reduced, the structure becomes more flexible with smaller eigenvalue, and eventually loses stability. Therefore, the selection of materials and level of prestresses are also critical factors to stability of tensegrity structures that are not super-stable.

6.8 Catalog of Stability of Symmetric Prismatic Structures

After the stability investigation, we are now in the position to present a catalog describing stability of the prismatic tensegrity structures $\mathbf{D}_N^{h,v}$ with dihedral symmetry for small N:

- $h = 1$: The structures are super-stable, and therefore, they are also prestress-stable.
- $h \neq 1$: There are two cases:
  - *divisible*:
    The structures are divisible, and hence, they are unstable, if both of the divisibility conditions are satisfied.
  - *indivisible*:
    Prestress-stability can be verified based on the quadratic form $\mathbf{Q}_{\text{G}}$, or its symmetry-adapted form $\tilde{\mathbf{Q}}_{\text{G}}$, of the geometrical stiffness matrix with respect to the mechanisms.

We present in Table 6.4 a complete catalog of prismatic tensegrity structures with symmetry $\mathbf{D}_N$ for $N \leq 10$.

Table 6.4 Catalog of stability of prismatic tensegrity structures $\mathbf{D}_N^{h,v}$

N	h	v = 1	2	3	4	5
3	1	s	-	-	-	-
4	1	s	s	-	-	-
	2	u	$2\mathbf{D}_2^{1,1}$	-	-	-
5	1	s	s	-	-	-
	2	u	u	-	-	-
6	1	s	s	s	-	-
	2	u	$2\mathbf{D}_3^{1,1}$	p	-	-
	3	u	u	$3\mathbf{D}_2^{1,1}$	-	-
7	1	s	s	s	-	-
	2	u	u	u	-	-
	3	u	[0.75, 1.05]	u	-	-
8	1	s	s	s	s	-
	2	u	$2\mathbf{D}_4^{1,1}$	[0.40, 3.10]	$2\mathbf{D}_4^{1,2}$	-
	3	u	u	u	[0.35, 2.35]	-
	4	u	$2\mathbf{D}_4^{2,1}$	u	$4\mathbf{D}_2^{1,1}$	-
9	1	s	s	s	s	-
	2	u	u	u	u	-
	3	u	u	$3\mathbf{D}_3^{1,1}$	[0.20, 1.60]	-
	4	u	u	u	u	-
10	1	s	s	s	s	s
	2	u	$2\mathbf{D}_5^{1,1}$	[0.70, 1.35]	$2\mathbf{D}_5^{1,2}$	p
	3	u	u	u	u	p
	4	u	$2\mathbf{D}_5^{2,1}$	[0.75, 1.25]	$2\mathbf{D}_5^{2,2}$	p
	5	u	u	u	u	$5\mathbf{D}_2^{1,1}$

's' denotes super-stability, 'u' denotes instability, and 'p' indicates that the structure is not super-stable but is always prestress-stable with arbitrary height/radius ratio. If the structure is prestress-stable only in a specific region of height/radius ratio, then this region is given; and if the structure can be divided, its substructures are given

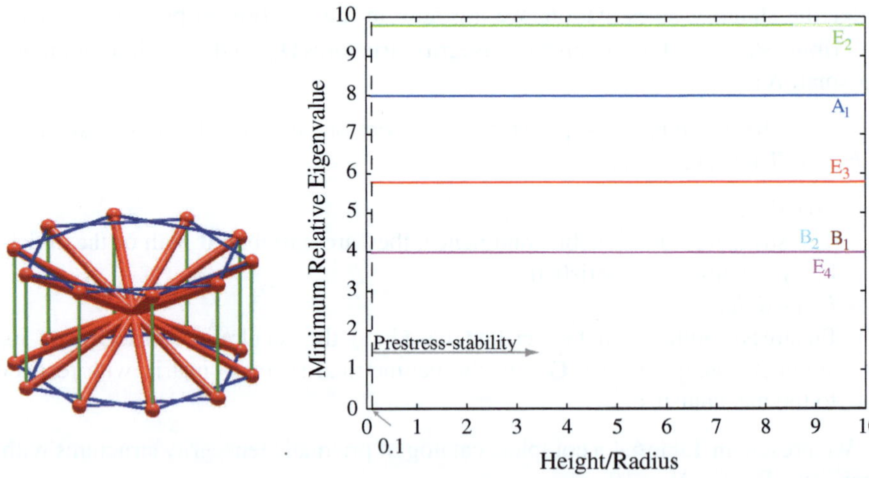

Fig. 6.14 Structure $\mathbf{D}_{10}^{2,5}$ that is not super-stable but is always prestress-stable

Example 6.20 Stability of the prismatic structures with $\mathbf{D}_{10}$ symmetry.

Consider the structures with dihedral $\mathbf{D}_{10}$ symmetry, horizontal cables of which are connected to the nodes next to their adjacent nodes; i.e., $h = 2$. From Table 6.4, we known that the structure $\mathbf{D}_{10}^{2,3}$ is prestress-stable in the region [0.70, 1.35] of height/radius ratio, and the structure $\mathbf{D}_{10}^{2,5}$ in Fig. 6.14 is always prestress-stable. Note that all struts of the structure $\mathbf{D}_{10}^{2,5}$ run across the central point (origin), but they are not connected with each other.

6.9 Remarks

The necessary and sufficient conditions for divisibility of the prismatic tensegrity structures with dihedral symmetry have been presented. It has been shown that divisibility of the structures is influenced by connectivity of the horizontal cables as well as that of the vertical cables, while connectivity of the struts is assumed to be fixed. Divisible structures have their own states of prestresses and rigid-body motions so that they can be physically separated into several identical substructures. Stability of the substructures might be investigated in the cases with lower level of symmetry.

It has been shown that super-stability of a prismatic structure with dihedral symmetry is dependent on its connectivity: the structure is super-stable, if the horizontal cables are connected to adjacent nodes.

Moreover, stability of the structures that are not super-stable is influenced by geometry realization (height/radius ratio), connectivity of vertical cables as well as materials and level of prestresses.

References

1. Atkins, P. W., Child, M. S., & Phillips, C. S. G. (1970). *Tables for group theory*. Oxford: Oxford University Press.
2. Connelly, R. (1995). Globally rigid symmetric tensegrities. *Structural Topology*, *21*, 59–78.
3. Fowler, P. W., & Guest, S. D. (2000). A symmetry extension of Maxwell's rule for rigidity of frames. *International Journal of Solids and Structures*, *37*(12), 1793–1804.
4. Kangwai, R. D., & Guest, S. D. (1999). Detection of finite mechanisms in symmetric structures. *International Journal of Solids and Structures*, *36*(36), 5507–5527.
5. Kettle, S. F. A. (2007). *Symmetry and structure: Readable group theory for chemists*. New York: Wiley.
6. Raj, P. R. (2008). *Novel symmetric tensegrity structures*. PhD thesis, University of Cambridge.
7. Sultan, C., Corless, M., & Skelton, R. E. (2002). Symmetrical reconfiguration of tensegrity structures. *International Journal of Solids and Structures*, *39*(8), 2215–2234.
8. Zhang, J. Y., Guest, S. D., & Ohsaki, M. (2009). Symmetric prismatic tensegrity structures. Part II: Symmetry-adapted formulations. *International Journal of Solids and Structures*, *46*(1), 15–30.

Chapter 7
Star-Shaped Structures of Dihedral Symmetry

Abstract This chapter presents self-equilibrium as well as super-stability conditions for star-shaped structures that have dihedral symmetry. Star-shaped tensegrity structures have the same dihedral symmetry and similar configurations to the prismatic structures studied in Chap. 6. The star-shaped structures have more infinitesimal mechanisms than prismatic structures due to the two additional (center) nodes, nevertheless, they are super-stable if certain connectivity conditions are satisfied. Moreover, some star-shaped structures that are not super-stable might be multi-stable; i.e., they may have more than one stable configuration.

Keywords Star-shaped structure · Dihedral symmetry · Self-equilibrium condition · Super-stability condition · Multi-stable

7.1 Introduction

The star-shaped tensegrity structures considered in this chapter and prismatic tensegrity structures studied in Chap. 6 have the same type of symmetry—both of these two classes of structures are of dihedral symmetry, but they are slightly different in connectivity. There are $2N$ symmetry operations in a dihedral group $\mathbf{D}_N$:

- N cyclic rotations C_N^i $(i = 0, 1, \ldots, N-1)$ about the principal z-axis through an angle $2i\pi/N$;
- N two-fold rotations $\mathrm{C}_{2,i}$ $(i = 0, 1, \ldots, N-1)$ about the axis perpendicular to the principal z-axis.

The cyclic rotation C_N^0 is also called identity operation, because nothing is done by this operation.

A prismatic structure has only one type (orbit) of nodes, such that any node of the structure can be moved to any other by a proper symmetry operation of the dihedral group. However, the center nodes and the boundary nodes of a star-shaped structure are two different types of nodes, since there exists no such a symmetry operation in dihedral group that can move a center node to the position of a boundary node or vice versa.

J.Y. Zhang and M. Ohsaki, *Tensegrity Structures*, Mathematics for Industry 6,
DOI 10.1007/978-4-431-54813-3_7

Accordingly, a star-shaped structure that is of dihedral symmetry $\mathbf{D}_N$ has (n=) $2N + 2$ nodes: $2N$ boundary nodes and two center nodes. All nodes are located on the two parallel planes; moreover, the boundary nodes on each plane actually lie on a circle, and the two center nodes are respectively located on centers of the two circles.

In the following, we repeat the definitions, given in Sect. 3.4, of nodes (and members) for a general star-shaped structure $\mathbf{D}_N^{\nu}$, which has dihedral symmetry $\mathbf{D}_N$. The nodes of a star-shaped structure are numbered as follows:

- the boundary nodes on the upper circles are $\{0, 1, \ldots, N-1\}$;
- the boundary nodes on the lower circles are $\{N, N+1, \ldots, 2N-1\}$;
- the center node on the upper plane is $2N$, and the center node on the lower plane is $2N+1$.

Moreover, a star-shaped structure consists of three types of members—radial cables, vertical cables, and struts. The members that connect the boundary nodes and the center nodes are called *radial cables*; the *struts* and *vertical cables* connect the boundary nodes on different circles. Hence, there are in total (m=)$4N$ members, including $2N$ horizontal cables, N vertical cables, and N struts.

Connectivity of radial cables is unique, and we may have some choices for connectivity of vertical cables defined by a parameter ν (=$1, 2, \ldots, (N-1)/2$), while connectivity of the struts is fixed. A member is defined as follows by a pair $[i, j]$ of nodes i and j:

$$\begin{aligned} \text{Radial cable}: &\ [2N, i] \text{ and } [2N+1, N+i], \\ \text{Vertical cable}: &\ [i, N+i+\nu], \qquad\qquad (i = 0, 1, \ldots, N-1). \\ \text{Strut}: &\ [i, N+i], \end{aligned} \tag{7.1}$$

Note that for the connectivity of vertical cables, we have

$$N + i + \nu := i + \nu, \qquad \text{if} \quad N + i + \nu \geq 2N. \tag{7.2}$$

Example 7.1 Difference between the star-shaped tensegrity structure $\mathbf{D}_3^1$ and the prismatic tensegrity structures $\mathbf{D}_3^{1,1}$.

Consider the prismatic tensegrity structure $\mathbf{D}_3^{1,1}$ as shown in Fig. 7.1a and the star-shaped tensegrity structure $\mathbf{D}_3^1$ as shown in Fig. 7.1b. Both of these two structures are of dihedral symmetry $\mathbf{D}_3$.

The prismatic structure $\mathbf{D}_3^{1,1}$ has six nodes, while the star-shaped structure $\mathbf{D}_3^1$ has eight nodes including six boundary nodes and two center nodes. The boundary nodes lie on two parallel circles, and the two center nodes lie on the centers of the circles.

Any node of the prismatic structure $\mathbf{D}_3^{1,1}$ can be transformed to another node by applying a proper symmetry operation of the dihedral group $\mathbf{D}_3$, and so are the boundary nodes of the star-shaped structure $\mathbf{D}_3^1$. However, it is clear that the center nodes cannot be transformed to the position of any boundary node by any of the six symmetry operations in $\mathbf{D}_3$; hence, the boundary nodes and center nodes belong to different orbits. Furthermore, a center node is transformed to itself by all of the three three-fold (cyclic) rotations, and it is transformed to the other center node by all of the three two-fold rotations.

Both of the structures $\mathbf{D}_3^{1,1}$ and $\mathbf{D}_3^1$ are composed of 12 members. Boundary nodes of the star-shaped structure are connected to center nodes by the radial cables, unlike the prismatic structure where the (boundary) nodes are connected to each other by horizontal cables.

Due to existence of the additional two center nodes in star-shaped structures, more (infinitesimal) mechanisms exist in these structures compared to the prismatic structures with the same symmetry. Because every boundary node is connected by three members, including one radial cable, one vertical cable, and one strut, these three members have to lie on the same plane to achieve self-equilibrium. It is easy to observe that every boundary node has one mechanism in three-dimensional space, perpendicular to the plane on which the three members connected to it are lying. Thus, there are at least $2N$ infinitesimal mechanisms, and in fact there is one more infinitesimal mechanism corresponding to existence of the prestress mode.

As will be discussed later, there exists only one prestress mode in the star-shaped structures; i.e., degree $n^{\rm s}$ of static indeterminacy is 1. According to the modified Maxwell's rule in Sect. 2.3, the number $n^{\rm m}$ of infinitesimal mechanisms in the star-shaped structure $\mathbf{D}_N^{\nu}$ is

$$\begin{aligned} n^{\rm m} &= n^{\rm s} - m + dn - n^{\rm b} \\ &= 1 - 4N + 3(2N + 2) - 6 \\ &= 2N + 1, \end{aligned} \tag{7.3}$$

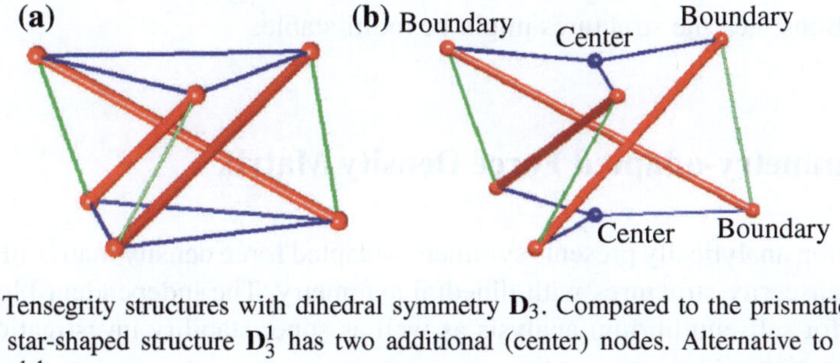

Fig. 7.1 Tensegrity structures with dihedral symmetry $\mathbf{D}_3$. Compared to the prismatic structure $\mathbf{D}_3^{1,1}$, the star-shaped structure $\mathbf{D}_3^1$ has two additional (center) nodes. Alternative to horizontal cables in $\mathbf{D}_3^{1,1}$, all cables lying on the horizontal planes of $\mathbf{D}_3^1$ are connected to the center nodes by radial cables. Both of these two structures are super-stable. **a** Prismatic structure $\mathbf{D}_3^{1,1}$, **b** star-shaped structure $\mathbf{D}_3^1$

which is much larger than the number $2N - 5$ of infinitesimal mechanisms for the prismatic structure $\mathbf{D}_N^{h,v}$ with the same dihedral symmetry $\mathbf{D}_N$ as presented in Eq. (6.2).

Example 7.2 Numbers of infinitesimal mechanisms of the prismatic structure $\mathbf{D}_3^{1,1}$ and the star-shaped structure $\mathbf{D}_3^1$ as shown in Fig. 7.1.

The prismatic structure $\mathbf{D}_3^{1,1}$ as shown in Fig. 7.1a consists of six nodes; i.e., $n = 2N = 6$. According to Eq. (6.2), the number of its infinitesimal mechanisms is 1:

$$\begin{aligned} n^{\mathrm{m}} &= 2N - 5 \\ &= 2 \times 3 - 5 \\ &= 1. \end{aligned} \tag{7.4}$$

The star-shaped structure $\mathbf{D}_3^1$ as shown in Fig. 7.1b consists of eight nodes; i.e., $n = 2N + 2 = 8$. According to Eq. (7.3), the number of its infinitesimal mechanisms is 7:

$$\begin{aligned} n^{\mathrm{m}} &= 2N + 1 \\ &= 2 \times 3 + 1 \\ &= 7. \end{aligned} \tag{7.5}$$

However, it is notable that both of the structures $\mathbf{D}_3^{1,1}$ and $\mathbf{D}_3^1$ are super-stable.

Despite of the large number of infinitesimal mechanisms, the star-shaped structures are super-stable if certain (connectivity) conditions are satisfied. These conditions will be presented later in this chapter using the symmetry-adapted force density matrix given in the next section. Furthermore, we will demonstrate that the structures that are not super-stable may have multiple stable configurations subjected to large deformations; i.e., the structures might be multi-stable.

7.2 Symmetry-adapted Force Density Matrix

This section analytically presents symmetry-adapted force density matrix of the star-shaped tensegrity structures with dihedral symmetry. The independent blocks will be used for self-equilibrium analysis as well as super-stability investigation in the coming sections.

7.2.1 Force Density Matrix

Denote force densities of the strut, vertical cable, and radial cable by q_s, q_v, and q_r, respectively. The force density matrix $\mathbf{E} \in \mathbb{R}^{(2N+2)\times(2N+2)}$ of a star-shaped tensegrity structure $\mathbf{D}_N^v$ can be obtained following the connectivity matrix and force densities in Eq. (2.101) or the direct definition in Eq. (2.107).

Example 7.3 Force density matrix of the star-shaped tensegrity structure $\mathbf{D}_3^1$ as shown in Fig. 7.1b.

The structure $\mathbf{D}_3^1$ as shown in Fig. 7.1b is composed of eight nodes. Using the direct definition in Eq. (2.107), its force density matrix $\mathbf{E} \in \mathbb{R}^{8\times8}$ is

$$\mathbf{E} = \left(\begin{array}{ccc|ccc|cc} q & 0 & 0 & -q_s & -q_v & 0 & -q_r & 0 \\ 0 & q & 0 & 0 & -q_s & -q_v & -q_r & 0 \\ 0 & 0 & q & -q_v & 0 & -q_s & -q_r & 0 \\ \hline -q_s & 0 & -q_v & q & 0 & 0 & 0 & -q_r \\ -q_v & -q_s & 0 & 0 & q & 0 & 0 & -q_r \\ 0 & -q_v & -q_s & 0 & 0 & q & 0 & -q_r \\ \hline -q_r & -q_r & -q_r & 0 & 0 & 0 & 3q_r & 0 \\ 0 & 0 & 0 & -q_r & -q_r & -q_r & 0 & 3q_r \end{array}\right), \tag{7.6}$$

where q is the sum of the force densities of the members connected to a boundary node; i.e.,

$$q = q_s + q_v + q_r, \tag{7.7}$$

because every boundary node is connected by one strut, one vertical cable, and one radial cable. From the numbering of the nodes defined previously in Sect. 7.1, the last two columns and the last two rows of the force density matrix correspond to the center nodes.

7.2.2 Structure of Symmetry-adapted Force Density Matrix

The structure of the symmetry-adapted force density matrix can be determined by considering the permutation representation of the nodes, written in terms of irreducible representations [1, 3].

For a dihedral group $\mathbf{D}_N$, the irreducible representations are denoted as A_1, A_2, B_1, B_2, E_k ($k = 1, \dots, p$), where

$$p = \begin{cases} (N-1)/2, & \text{for } N \text{ odd}, \\ (N-2)/2, & \text{for } N \text{ even}. \end{cases} \tag{7.8}$$

The representations A_1, A_2, B_1, B_2 are one-dimensional representations and E_k ($k = 1, 2, \ldots, p$) is a two-dimensional representation; moreover, B_1 and B_2 exist only when N is even.

In general, any boundary node of a star-shaped tensegrity structure with $\mathbf{D}_N$ symmetry is transformed to a different node by any symmetry operation *except for* the identity operation (C_N^0 or E). Under the identity operation, all boundary nodes, in total $2N$, remain at the same locations. Hence, we have the linear combination $\Gamma(N^b) = \{2N, 0, \ldots, 0;\ 0, \ldots, 0\}$ of the irreducible representations for the boundary nodes.

From characters of the irreducible matrices of dihedral group listed in Table 7.1, the reducible representation of the boundary nodes can be written as a linear combination $\Gamma(N^b)$ of the irreducible representations in a general form as follows:

$$\begin{aligned}
\Gamma(N^b) = &\quad A_1 + A_2 + (B_1 + B_2) + 2\sum_{k=1}^{p} E_k \\
= &\quad \{1, \ldots, 1;\ 1, \ldots, 1\} && A_1 \\
&+ \{1, \ldots, 1;\ -1, \ldots -1\} && A_2 \\
&+ (\{1, \ldots, (-1)^i, \ldots, (-1)^N;\ 1, \ldots, (-1)^i, \ldots, (-1)^N\}) && (B_1) \\
&+ (\{1, \ldots, (-1)^i, \ldots, (-1)^N;\ -1, \ldots, (-1)^{i+1}, \ldots, (-1)^{N+1}\}) && (B_2) \\
&+ 2\sum_{k=1}^{p}\{2C_{0k}, \ldots, 2C_{ik}, \ldots, 2C_{(N-1)k};\ 0, \ldots, 0\} && 2E_k \\
= &\quad \{2N, 0, \ldots, 0;\ 0, \ldots, 0\},
\end{aligned} \tag{7.9}$$

because $C_{0k} = C_0 = 1$ and $\sum_{k=1}^{p} C_{ik} = 0$.

On the other hand, the two center nodes remain at the same places while subjected to the cyclic rotations C_N^i ($i = 0, 1, \ldots, N-1$), and exchange their positions while subjected to the two-fold rotations $C_{2,i}$ ($i = 0, 1, \ldots, N-1$). Thus, we have $\Gamma(N^c) = \{2, 2, \ldots, 2;\ 0, \ldots, 0\}$ for the center nodes and

$$\begin{aligned}
\Gamma(N^c) = &\quad A_1 + A_2 \\
= &\quad \{1, \ldots, 1;\ 1, \ldots, 1\} && A_1 \\
&+ \{1, \ldots, 1;\ -1, \ldots -1\} && A_2 \\
= &\quad \{2, 0, \ldots, 0;\ 0, \ldots, 0\}.
\end{aligned} \tag{7.10}$$

Table 7.1 The linear combinations of the irreducible representations for the boundary nodes $\Gamma(N^b)$ and center nodes $\Gamma(N^c)$ of the star-shaped structure $\mathbf{D}_N^v$ and characters of the irreducible representation matrices of dihedral group $\mathbf{D}_N$

Operation	C_N^0 (E)	C_N^1	C_N^i	C_N^{N-1}	$C_{2,0}$	$C_{2,1}$	$C_{2,i}$	$C_{2,N-1}$
$\Gamma(N^b)$	$2N$	0	0	0	0	0	0	0
$\Gamma(N^c)$	2	2	2	2	0	0	0	0
A_1	1	1	1	1	1	1	1	1
A_2	1	1	1	1	−1	−1	−1	−1
(B_1)	1	−1	$(-1)^i$	$(-1)^N$	1	−1	$(-1)^i$	$(-1)^N$
(B_2)	1	−1	$(-1)^i$	$(-1)^N$	1	−1	$(-1)^{i+1}$	$(-1)^{N+1}$
E_k	2	$2C_k$	$2C_{ik}$	$2C_{(N-1)k}$	0	0	0	0

C_i denotes $\cos(2i\pi/N)$

From Eqs. (7.9) and (7.10), representation of all nodes $\varGamma(\mathrm{N})$ can then be summarized as follows:

$$\varGamma(\mathrm{N}) = \varGamma(\mathrm{N}^{\mathrm{b}}) + \varGamma(\mathrm{N}^{\mathrm{c}}) = 2\mathrm{A}_1 + 2\mathrm{A}_2 + (\mathrm{B}_1 + \mathrm{B}_2) + \sum_{k=1}^{p} 2\mathrm{E}_k, \tag{7.11}$$

which characterizes the structure of the symmetry-adapted force density matrix $\tilde{\mathbf{E}}$:

1. The number of representations μ in $\varGamma(\mathrm{N})$ indicates dimensions of $\tilde{\mathbf{E}}^{\mu}$. Hence, we learn from Eq. (7.11) that the blocks $\tilde{\mathbf{E}}^{\mathrm{A}_1}$ and $\tilde{\mathbf{E}}^{\mathrm{A}_2}$ corresponding to the one-dimensional representations A_1 and A_2 are 2-by-2 matrices, the blocks $\tilde{\mathbf{E}}^{\mathrm{B}_1}$ and $\tilde{\mathbf{E}}^{\mathrm{B}_2}$ corresponding to the one-dimensional representations B_1 and B_2 are 1-by-1 matrices if they exist when N is even, and the blocks $\tilde{\mathbf{E}}^{\mathrm{E}_k}$ $(k = 1, 2, \dots, p)$ of two-dimensional representations E_k are 2-by-2 matrices.
2. Dimension of a representation μ indicates number of copies of its corresponding block $\tilde{\mathbf{E}}^{\mu}$ appearing in the block-diagonal form; thus, each one-dimensional representation has only one copy, and each two-dimensional representation has two copies of blocks lying in the leading diagonal of $\tilde{\mathbf{E}}$.

It is notable that the structure of symmetry-adapted force density matrix of the star-shaped structures is the same as that of the prismatic structures studied in Chap. 6, except for the one-dimensional representations A_1 and A_2. The blocks $\tilde{\mathbf{E}}^{\mathrm{A}_1}$ and $\tilde{\mathbf{E}}^{\mathrm{A}_2}$ respectively corresponding to the one-dimensional representations A_1 and A_2 are 2-by-2 matrices, because both of the A_1 and A_2 representations are involved in boundary nodes and center nodes. By contrast, the A_1 and A_2 blocks are 1-by-1 matrices for the prismatic structures.

According to Eq. (7.11), structure of the symmetry-adapted force density matrix $\tilde{\mathbf{E}}$ of the star-shaped structure $\mathbf{D}_N^v$ can be written as

$$\underset{(2N+2)\times(2N+2)}{\tilde{\mathbf{E}}} = \begin{pmatrix} \underset{2\times2}{\tilde{\mathbf{E}}^{\mathrm{A}_1}} & & & & & & & & \\ & \underset{2\times2}{\tilde{\mathbf{E}}^{\mathrm{A}_2}} & & & & & \mathbf{O} & & \\ & & \underset{1\times1}{(\tilde{\mathbf{E}}^{\mathrm{B}_1})} & & & & & & \\ & & & \underset{1\times1}{(\tilde{\mathbf{E}}^{\mathrm{B}_2})} & & & & & \\ & & & & \underset{2\times2}{\tilde{\mathbf{E}}^{\mathrm{E}_1}} & & & & \\ & & & & & \underset{2\times2}{\tilde{\mathbf{E}}^{\mathrm{E}_1}} & & & \\ & & \mathbf{O} & & & & \ddots & & \\ & & & & & & & \underset{2\times2}{\tilde{\mathbf{E}}^{\mathrm{E}_p}} & \\ & & & & & & & & \underset{2\times2}{\tilde{\mathbf{E}}^{\mathrm{E}_p}} \end{pmatrix}, \tag{7.12}$$

which is simply written as follows by using the direct sum notation $\oplus$:

$$\tilde{\mathbf{E}} = \tilde{\mathbf{E}}^{\mathrm{A}_1} \oplus \tilde{\mathbf{E}}^{\mathrm{A}_2} \oplus (\tilde{\mathbf{E}}^{\mathrm{B}_1} \oplus \tilde{\mathbf{E}}^{\mathrm{B}_2}) \oplus 2\tilde{\mathbf{E}}^{\mathrm{E}_1} \oplus \cdots \oplus 2\tilde{\mathbf{E}}^{\mathrm{E}_p}. \tag{7.13}$$

7.2.3 Blocks of Symmetry-adapted Force Density Matrix

Because the boundary nodes and the center nodes belong to different orbits, the symmetry-adapted blocks have to be formulated separately corresponding to different irreducible representations.

For the irreducible representations μ (=B_1, B_2, or E_k) that exist only for the boundary nodes, the following direct formulation presented in Eq. (D.20) can be applied to directly write down the symmetry-adapted block $\tilde{\mathbf{E}}^{\mu}$ of the force density matrix:

$$\tilde{\mathbf{E}}^{\mu} = (q_{\mathrm{v}} + q_{\mathrm{r}} + q_{\mathrm{s}})\mathbf{R}_0^{\mu} - q_{\mathrm{s}}\mathbf{R}_N^{\mu} - q_{\mathrm{v}}\mathbf{R}_{N+v}^{\mu}, \tag{7.14}$$

because the reference (boundary) node 0 is connected to node N by a strut, to node $N+v$ by a vertical cable according to the definitions of nodes and members. In Eq. (7.14), the matrices $\mathbf{R}_0^{\mu}$, $\mathbf{R}_N^{\mu}$, and $\mathbf{R}_{N+v}^{\mu}$ are the irreducible representation matrices corresponding to the identity, struts, and vertical cables for the representation μ, respectively, as listed in Table 7.2.

However, Eq. (7.14) is not applicable to the blocks $\tilde{\mathbf{E}}^{\mathrm{A}_1}$ and $\tilde{\mathbf{E}}^{\mathrm{A}_2}$, because both the center nodes and boundary nodes of the star-shaped structures contribute to the A_1 and A_2 representations as indicated in Eqs. (7.9) and (7.10). The formulation of these two blocks relies on conventional approach:

$$\tilde{\mathbf{E}}^{\mathrm{A}_1} \oplus \tilde{\mathbf{E}}^{\mathrm{A}_2} = \mathbf{TET}^{\top}, \tag{7.15}$$

Table 7.2 Selected irreducible representation matrices corresponding to the nodes connected to the reference node 0

	Identity	Strut	Vertical		
μ	$\mathbf{R}_0^{\mu}$	$\mathbf{R}_N^{\mu}$	$\mathbf{R}_{N+v}^{\mu}$		
A_1	1	1	1		
A_2	1	-1	-1	z, R_z	
(B_1)	1	1	$(-1)^v$		N even
(B_2)	1	-1	$(-1)^{v+1}$		N even
E_1	$\begin{pmatrix} 1 & 0 \\ 0 & 1 \end{pmatrix}$	$\begin{pmatrix} 1 & 0 \\ 0 & -1 \end{pmatrix}$	$\begin{pmatrix} C_k & S_k \\ S_k & -C_k \end{pmatrix}$	(x,y), $(\mathrm{R}_x,\mathrm{R}_y)$	
E_k	$\begin{pmatrix} 1 & 0 \\ 0 & 1 \end{pmatrix}$	$\begin{pmatrix} 1 & 0 \\ 0 & -1 \end{pmatrix}$	$\begin{pmatrix} C_{vk} & S_{vk} \\ S_{vk} & -C_{vk} \end{pmatrix}$		$k = 2, \ldots, p$

C_i denotes $\cos(2i\pi/N)$ and S_i denotes $\sin(2i\pi/N)$

where $\mathbf{T} \in \mathbb{R}^{4\times 2N}$ is the normalized transformation matrix, and its non-normalized version $\bar{\mathbf{T}} \in \mathbb{R}^{4\times 2N}$ is formulated as

$$\bar{\mathbf{T}} = \begin{pmatrix} \bar{\mathbf{T}}_{\mathrm{b}}^{\mathrm{A}_1} \\ \bar{\mathbf{T}}_{\mathrm{c}}^{\mathrm{A}_1} \\ \bar{\mathbf{T}}_{\mathrm{b}}^{\mathrm{A}_2} \\ \bar{\mathbf{T}}_{\mathrm{c}}^{\mathrm{A}_2} \end{pmatrix}. \tag{7.16}$$

On the other hand, all other blocks of the symmetry-adapted force density matrix $\tilde{\mathbf{E}}$ can be directly found according to Eq. (7.14), because only the boundary nodes contribute.

In the conventional approach, see for example the review paper [2], components of the transformation matrix can be determined by using the irreducible representation matrices. It is much simpler for the one-dimensional representations A_1 and A_2, because the irreducible representation matrices are indeed the characters that can be read off from the character table in Table 7.1.

For the boundary nodes corresponding to A_1 representation, its transformation matrix $\bar{\mathbf{T}}_{\mathrm{b}}^{\mathrm{A}_1}$ without normalization is

$$\bar{\mathbf{T}}_{\mathrm{b}}^{\mathrm{A}_1} = \left(\begin{array}{cccc|cccc|cc} 1 & 1 & \dots & 1 & 1 & 1 & \dots & 1 & 0 & 0 \end{array}\right). \tag{7.17}$$

Following the numbering of nodes, the first N entries, which are equal to 1, of $\bar{\mathbf{T}}_{\mathrm{b}}^{\mathrm{A}_1}$ are the characters (irreducible representation matrices) of $\mathbf{A}_1$ representation corresponding to the N-fold cyclic rotations C_N^i $(i = 0, 1, \dots, N-1)$ about z-axis, and the next N entries, which are also equal to 1, are those corresponding to the two-fold rotations $\mathrm{C}_{2,i}$ $(i = 0, 1, \dots, N-1)$; the last two components are zero because they correspond to transformation of center nodes.

Similar to $\bar{\mathbf{T}}_{\mathrm{b}}^{\mathrm{A}_1}$, the transformation matrix $\bar{\mathbf{T}}_{\mathrm{c}}^{\mathrm{A}_2}$ without normalization of the center nodes for the A_2 representation is given as

$$\bar{\mathbf{T}}_{\mathrm{c}}^{\mathrm{A}_2} = \left(\begin{array}{cccc|cccc|cc} 0 & 0 & \dots & 0 & 0 & 0 & \dots & 0 & 1 & -1 \end{array}\right), \tag{7.18}$$

where the $(2N+1)$th entry, which is 1, is the character of $\mathbf{A}_2$ representation corresponding to the N-fold cyclic rotations of the center nodes, and the last $2N$th entry, which is -1, is the one corresponding to the two-fold rotations.

In a similar way, the transformation matrix $\bar{\mathbf{T}}$ without normalization of the boundary nodes and center nodes corresponding to the one-dimensional representations A_1 and A_2 can be summarized as follows by using Eq. (7.16):

$$\bar{\mathbf{T}} = \left(\begin{array}{cccc|cccc|cc} 1 & 1 & \dots & 1 & 1 & 1 & \dots & 1 & 0 & 0 \\ 0 & 0 & \dots & 0 & 0 & 0 & \dots & 0 & 1 & 1 \\ 1 & 1 & \dots & 1 & -1 & -1 & \dots & -1 & 0 & 0 \\ 0 & 0 & \dots & 0 & 0 & 0 & \dots & 0 & 1 & -1 \end{array}\right). \tag{7.19}$$

Moreover, the normalized version $\mathbf{T}$ of the transformation matrix $\bar{\mathbf{T}}$ used in Eq. (7.15) is

$$\mathbf{T} = \mathbf{N}\bar{\mathbf{T}}, \tag{7.20}$$

where the matrix $\mathbf{N}$ is

$$\mathbf{N} = \frac{1}{\sqrt{2N}} \begin{pmatrix} 1 & & & \\ & \sqrt{N} & & \\ & & 1 & \\ & & & \sqrt{N} \end{pmatrix}. \tag{7.21}$$

Applying transformation matrix $\mathbf{T}$ to the force density matrix $\mathbf{E}$ as in Eq. (7.15), we derive the A_1 and A_2 blocks $\tilde{\mathbf{E}}^{\mathrm{A}_1}$ and $\tilde{\mathbf{E}}^{\mathrm{A}_2}$ of the symmetry-adapted force density matrix $\tilde{\mathbf{E}}$ as

$$\begin{aligned} \tilde{\mathbf{E}}^{\mathrm{A}_1} &= \begin{pmatrix} q_{\mathrm{r}} & -\sqrt{N}q_{\mathrm{r}} \\ -\sqrt{N}q_{\mathrm{r}} & Nq_{\mathrm{r}} \end{pmatrix}, \\ \tilde{\mathbf{E}}^{\mathrm{A}_2} &= \begin{pmatrix} 2(q_{\mathrm{v}} + q_{\mathrm{s}}) + q_{\mathrm{r}} & -\sqrt{N}q_{\mathrm{r}} \\ -\sqrt{N}q_{\mathrm{r}} & Nq_{\mathrm{r}} \end{pmatrix}. \end{aligned} \tag{7.22}$$

Since the center nodes are involved only in A_1 and A_2 representations, for the symmetry-adapted blocks corresponding to other representations, the direct formulation in Eq. (7.14) for the boundary nodes, which belong to a regular orbit of dihedral group, is applicable.

When N is even, the one-dimensional representations B_1 and B_2 exist; from Table 7.2, their unique irreducible representation matrices (characters) are

$$\begin{aligned} \mathrm{B}_1 : \mathbf{R}_0^{\mathrm{B}_1} &= 1, \ \mathbf{R}_N^{\mathrm{B}_1} = 1, \quad \mathbf{R}_{N+v}^{\mathrm{B}_1} = (-1)^v, \\ \mathrm{B}_2 : \mathbf{R}_0^{\mathrm{B}_2} &= 1, \ \mathbf{R}_N^{\mathrm{B}_2} = -1, \ \mathbf{R}_{N+v}^{\mathrm{B}_2} = (-1)^{v+1}. \end{aligned} \tag{7.23}$$

Substituting Eq. (7.23) into Eq. (7.14), the blocks $\tilde{\mathbf{E}}^{\mathrm{B}_1}$ and $\tilde{\mathbf{E}}^{\mathrm{B}_2}$ corresponding to representations B_1 and B_2 can be written as

$$\begin{aligned} \tilde{\mathbf{E}}^{\mathrm{B}_1} &= (q_{\mathrm{v}} + q_{\mathrm{r}} + q_{\mathrm{s}})\mathbf{R}_0^{\mathrm{B}_1} - q_{\mathrm{s}}\mathbf{R}_N^{\mathrm{B}_1} - q_{\mathrm{v}}\mathbf{R}_{N+v}^{\mathrm{B}_1} \\ &= q_{\mathrm{r}} - q_{\mathrm{s}} + (-1)^{v+1}q_{\mathrm{v}}, \\ \tilde{\mathbf{E}}^{\mathrm{B}_2} &= (q_{\mathrm{v}} + q_{\mathrm{r}} + q_{\mathrm{s}})\mathbf{R}_0^{\mathrm{B}_2} - q_{\mathrm{s}}\mathbf{R}_N^{\mathrm{B}_2} - q_{\mathrm{v}}\mathbf{R}_{N+v}^{\mathrm{B}_2} \\ &= q_{\mathrm{r}} + q_{\mathrm{s}} + (-1)^{v}q_{\mathrm{v}}. \end{aligned} \tag{7.24}$$

The irreducible representation matrices $\mathbf{R}_0^{\mathrm{E}_k}$, $\mathbf{R}_N^{\mathrm{E}_k}$, and $\mathbf{R}_{N+v}^{\mathrm{E}_k}$ for the two-dimensional representations E_k ($k = 1, 2, \ldots, p$) are

$$\mathbf{R}_0^{\mathrm{E}_k} = \begin{pmatrix} 1 & 0 \\ 0 & 1 \end{pmatrix},$$

$$\mathbf{R}_N^{\mathrm{E}_k} = \begin{pmatrix} 1 & 0 \\ 0 & -1 \end{pmatrix},$$

$$\mathbf{R}_{N+v}^{\mathrm{E}_k} = \begin{pmatrix} C_{vk} & S_{vk} \\ S_{vk} & -C_{vk} \end{pmatrix}. \tag{7.25}$$

From Eq. (7.14), $\tilde{\mathbf{E}}^{\mathrm{E}_k}$ corresponding to the two-dimensional representations $\mathbf{E}_k$, which appears twice in $\tilde{\mathbf{E}}$, can be written as

$$\begin{aligned}\tilde{\mathbf{E}}^{\mathrm{E}_k} &= (q_\mathrm{v} + q_\mathrm{r} + q_\mathrm{s})\mathbf{R}_0^{\mathrm{E}_k} - q_\mathrm{s}\mathbf{R}_N^{\mathrm{E}_k} - q_\mathrm{v}\mathbf{R}_{N+v}^{\mathrm{E}_k} \\ &= \begin{pmatrix} q_\mathrm{r} + q_\mathrm{v}(1 - C_{kv}) & -q_\mathrm{v} S_{kv} \\ -q_\mathrm{v} S_{kv} & q_\mathrm{r} + q_\mathrm{v}(1 - C_{kv}) + 2q_\mathrm{s} \end{pmatrix}.\end{aligned} \tag{7.26}$$

7.3 Self-equilibrium Conditions

In order to have a non-degenerate geometry realization, the force density matrix $\mathbf{E}$, or its symmetry-adapted form $\tilde{\mathbf{E}}$, of a three-dimensional tensegrity structure should have at least four zero eigenvalues. This is the non-degeneracy condition for three-dimensional tensegrity structures given in Sect. 2.5.

The A_1 block $\tilde{\mathbf{E}}^{\mathrm{A}_1}$ of the symmetry-adapted force density matrix $\tilde{\mathbf{E}}$ is a 2-by-2 matrix as presented in Eq. (7.22), and its two eigenvalues $\lambda_1^{\mathrm{A}_1}$ and $\lambda_2^{\mathrm{A}_1}$ are easy to calculate as

$$\begin{aligned}\lambda_1^{\mathrm{A}_1} &= 0, \\ \lambda_2^{\mathrm{A}_1} &= (N+1)q_\mathrm{r} \ (>0).\end{aligned} \tag{7.27}$$

$\lambda_2^{\mathrm{A}_1}$ is positive because we assume that the (radial) cables carry tension, such that q_r is positive.

As indicated in Table 7.2, the A_2 representation corresponds to z-coordinates of the structure with dihedral symmetry. Thus, the A_2 block $\tilde{\mathbf{E}}^{\mathrm{A}_2}$ should be singular with zero determinant:

$$|\tilde{\mathbf{E}}^{\mathrm{A}_2}| = 2Nq_\mathrm{r}(q_\mathrm{v} + q_\mathrm{s}) = 0. \tag{7.28}$$

Therefore, we have the relation between the force densities of the vertical cables q_v and the struts q_s as follows from Eq. (7.28):

$$q_\mathrm{v} = -q_\mathrm{s}, \tag{7.29}$$

because q_r is positive.

Using Eq. (7.29), the A_1 and A_2 blocks can be simplified as

$$\tilde{\mathbf{E}}^{A_1} = \tilde{\mathbf{E}}^{A_2} = \begin{pmatrix} q_r & -\sqrt{N}q_r \\ -\sqrt{N}q_r & Nq_r \end{pmatrix}. \tag{7.30}$$

The two eigenvalues $\lambda_1^{A_2}$ and $\lambda_2^{A_2}$ of the A_2 block $\tilde{\mathbf{E}}^{A_2}$ of the symmetry-adapted force density matrix $\tilde{\mathbf{E}}$ are the same as those of $\tilde{\mathbf{E}}^{A_1}$:

$$\begin{aligned} \lambda_1^{A_2} &= 0, \\ \lambda_2^{A_2} &= (N+1)q_r > 0. \end{aligned} \tag{7.31}$$

Moreover, the other blocks can be simplified as follows by using Eq. (7.29):

$$\begin{aligned} \tilde{\mathbf{E}}^{B_1} &= q_r + (-1)^{\nu+3}q_v = q_r + (-1)^{\nu+1}q_v, \\ \tilde{\mathbf{E}}^{B_2} &= q_r + (-1)^{\nu+1}q_v, \\ \tilde{\mathbf{E}}^{E_k} &= \begin{pmatrix} q_r + q_v(1 - C_{k\nu}) & -q_v S_{k\nu} \\ -q_v S_{k\nu} & q_r - q_v(1 + C_{k\nu}) \end{pmatrix}. \end{aligned} \tag{7.32}$$

In Eqs. (7.27) and (7.31), we have already derived two zero eigenvalues in the force density matrix, one in the A_1 block and the other in the A_2 block. Therefore, the remaining two zero eigenvalues for satisfying non-degeneracy condition should lie in only one two-dimensional block $\tilde{\mathbf{E}}^{E_k}$, because there exist two copies of the block in the symmetry-adapted force density matrix $\tilde{\mathbf{E}}$. This two-dimensional block might be the one $\tilde{\mathbf{E}}^{E_1}$ corresponding to the two-dimensional representation E_1, because it corresponds to *xy*-coordinates as shown in Table 7.2 for dihedral groups. From Eq. (7.32), $\tilde{\mathbf{E}}^{E_1}$ is

$$\tilde{\mathbf{E}}^{E_1} = \begin{pmatrix} q_r + q_v(1 - C_\nu) & -q_v S_\nu \\ -q_v S_\nu & q_r - q_v(1 + C_\nu) \end{pmatrix}, \tag{7.33}$$

which is singular with zero determinant; i.e.,

$$\begin{aligned} |\tilde{\mathbf{E}}^{E_1}| &= q_r^2 - 2q_v^2(1 - C_\nu) \\ &= 0. \end{aligned} \tag{7.34}$$

Thus, we have the following relation between the force densities of vertical cables q_v and radial cables q_r:

$$q_r = q_v\sqrt{2(1 - C_\nu)}, \tag{7.35}$$

where only the positive solution is adopted because both q_r and q_v are positive.

To summarize, the force densities of the star-shaped structures $\mathbf{D}_N^\nu$ with $\mathbf{D}_N$ symmetry in the self-equilibrium state are

Relations among the force densities of struts q_s, radial cables q_r, and vertical cables q_v of the symmetric star-shaped tensegrity structure $\mathbf{D}_N^v$:

$$q_s = -q_v,$$
$$q_r = q_v\sqrt{2(1 - C_v)}. \tag{7.36}$$

It is notable that the relations in Eq. (7.36) are identical to those in Eqs. (3.77) and (3.79) previously derived in Sect. 3.4 by using force equilibrium analysis. Moreover, coordinates of the reference node 0 can be obtained from the null-space of the symmetry-adapted blocks $\tilde{\mathbf{E}}^{E_1}$ and $\tilde{\mathbf{E}}^{A_2}$, which respectively correspond to xy- and z-coordinates. The coordinates of the representative node Eq. (3.84), and those of the other nodes can be determined by using their corresponding symmetry operations.

7.4 Stability Conditions

In the stability investigation, the divisible structures that can be *mechanically* separated into several identical substructures[1] are excluded, because they should be considered in the cases with lower symmetry.

7.4.1 Divisibility Conditions

A structure is said to be *divisible* if the members and nodes can be mechanically separated into several identical substructures. The substructures are pinned to the common center nodes. A divisible structure has finite mechanisms and is intrinsically unstable, because rotation of one substructure about z-axis has no mechanical influence on the others.

Example 7.4 Divisibility of the star-shaped tensegrity structure $\mathbf{D}_8^2$ as shown in Fig. 7.2a.

The structure $\mathbf{D}_8^2$ as shown in Fig. 7.2a has dihedral symmetry $\mathbf{D}_8$, and consists of eight struts, 16 radial cables, and eight vertical cables. Due to connectivity of the vertical cables, it is mechanically separated into two identical structures $\mathbf{D}_4^1$ as shown in Fig. 7.2b, c. The substructures have lower dihedral symmetry $\mathbf{D}_4$ than the original structure; each of the substructures consists of four struts, eight radial cables, and four vertical cables.

[1] The divisible star-shaped structures might not be physically separated into serval substructures because they share the common center nodes.

The struts and vertical cables in each substructure $\mathbf{D}_4^1$ connect one to another to form a closed circuit, so that the substructures are indivisible. It is obvious that the substructures $\mathbf{D}_4^1$ are self-equilibrated, and they are not super-stable from the conditions for super-stability presented later in this chapter.

Because the boundary nodes are connected to the common center nodes by radial cables, divisibility of the structure is only related to the connectivity of struts and vertical cables.

From the connectivity defined in Sect. 7.1 for the star-shaped structure $\mathbf{D}_N^v$, we know that node i ($i = 0, 1, \ldots, N-1$) on the upper plane is connected to node $N + i + v$ on the lower plane by a vertical cable, node $N + i + v$ is connected to node $i + v$ on the upper plane by a strut, node $i + v$ connects node $N + i + 2v$ on the lower plane through a vertical cable, and so on. Eventually, we must return to the starting node i in the trip. If we stop when the trip returns to the starting node i for the first time, the boundary nodes on the upper plane in the linkage can be listed as follows:

$$i \to i + v \to i + 2v \to \cdots \to i + jv - kN \; (= i), \tag{7.37}$$

where j and k are the smallest positive integers satisfying the following condition:

$$0 \leq i + jv - kN \leq N - 1. \tag{7.38}$$

The parameters j and k indicate the number of boundary nodes on the upper plane that have been visited and the number of rounds about z-axis, respectively.

From $i + jv - kN = i$ for returning to the starting node i, we have

$$jv = kN. \tag{7.39}$$

If the structure is indivisible, we should have visited N boundary nodes on the upper plane. Thus, we have $j = N$, and therefore, $v = k$, which can happen if and only if v and N have no common divisor except 1.

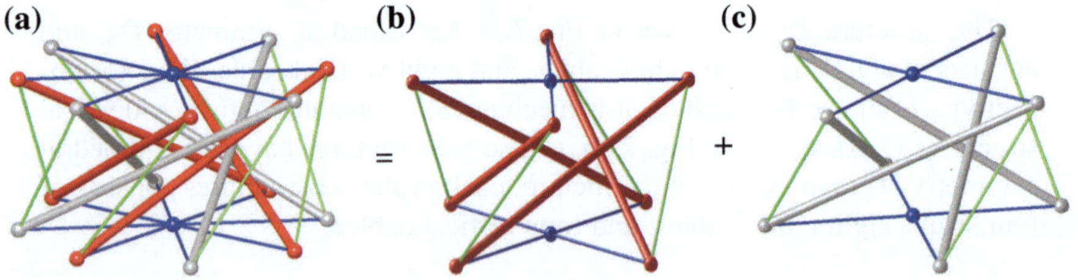

Fig. 7.2 Divisible star-shaped tensegrity structure $\mathbf{D}_8^2$. The structure $\mathbf{D}_8^2$ in (**a**) can be divided into two identical substructures $\mathbf{D}_4^1$ in (**b**) and (**c**)

Hence, we can have the following condition for the indivisibility of star-shaped structures:

The necessary and sufficient indivisibility condition for a star-shaped tensegrity structure $\mathbf{D}_N^{\nu}$:
The star-shaped tensegrity structure $\mathbf{D}_N^{\nu}$ is indivisible if and only if the connectivity parameter ν for vertical cable and the number N of struts have no common divisor except 1.

7.4.2 Super-stability Conditions

In Lemma 4.7, we presented the sufficient conditions for super-stability of a three-dimensional tensegrity structure:

1. The geometry matrix $\mathbf{G}$ defined in Eq. (4.110) has (full) rank of six.
2. The force density matrix $\mathbf{E}$, or its symmetry-adapted form $\tilde{\mathbf{E}}$, has rank deficiency of four;
3. The force density matrix $\mathbf{E}$, or its symmetry-adapted form $\tilde{\mathbf{E}}$, is positive semi-definite;

The first condition will not be discussed in the following discussions on stability, because it is usually satisfied for indivisible star-shaped structures, while divisible structures are unstable. Therefore, only the last two conditions concerning signs of the force density matrix are considered to present the conditions for super-stability of the star-shaped structures.

$\tilde{\mathbf{E}}^{\mathrm{A}_1}$ and $\tilde{\mathbf{E}}^{\mathrm{A}_2}$ are positive semi-definite because their eigenvalues are

$$\begin{aligned} \lambda_1^{\mathrm{A}_1} &= 0, \\ \lambda_2^{\mathrm{A}_1} &= (N+1)q_{\mathrm{r}} > 0, \\ \lambda_1^{\mathrm{A}_2} &= 0, \\ \lambda_2^{\mathrm{A}_2} &= (N+1)q_{\mathrm{r}} > 0, \end{aligned} \tag{7.40}$$

in which the force density q_{r} of the radial cables carrying tension is positive.

Because B_1 and B_2 representations exist only if N is even, ν should not be even; otherwise, N and ν have common divisor except 1 so that the structure is divisible. Therefore, the eigenvalues λ^{B_1} and λ^{B_2} of the one-dimensional blocks $\tilde{\mathbf{E}}^{\mathrm{B}_1}$ and $\tilde{\mathbf{E}}^{\mathrm{B}_2}$ for ν odd are written as follows by using Eq. (7.32):

$$\lambda^{\mathrm{B}_1} = \lambda^{\mathrm{B}_2} = q_{\mathrm{r}} + q_{\mathrm{v}} > 0. \tag{7.41}$$

Because both q_{r} and q_{v} are positive (for cables), both λ^{B_1} and λ^{B_2} are positive.

From Eqs. (7.40) and (7.41), the one-dimensional blocks are positive semi-definite if the structure is indivisible, with two zero eigenvalues in $\tilde{\mathbf{E}}^{A_1}$ and $\tilde{\mathbf{E}}^{A_2}$. If all the two-dimensional blocks $\tilde{\mathbf{E}}^{E_k}$ $(k \neq 1)$ are positive definite, while $\tilde{\mathbf{E}}^{E_1}$ is positive semi-definite, the last two conditions for super-stability of tensegrity structures are satisfied, and therefore, the structure is super-stable. Thus, the problem of finding the super-stable structures becomes that of finding the conditions for $\tilde{\mathbf{E}}^{E_k}$ $(k \neq 1)$ being positive definite and $\tilde{\mathbf{E}}^{E_1}$ being positive semi-definite.

The eigenvalues of the two-dimensional blocks $\tilde{\mathbf{E}}^{E_k}$ are

$$\begin{aligned}\frac{\lambda_1^{E_k}}{q_v} &= \sqrt{2(1-C_v)} + \sqrt{2(1-C_{kv})} > 0,\\ \frac{\lambda_2^{E_k}}{q_v} &= \sqrt{2(1-C_v)} - \sqrt{2(1-C_{kv})}.\end{aligned} \tag{7.42}$$

From Eq. (7.36), $\lambda_2^{E_1} = 0$ for $k = 1$ such that the block $\tilde{\mathbf{E}}^{E_1}$ is positive semi-definite. In order to ensure that $\tilde{\mathbf{E}}^{E_k}$ $(k \neq 1)$ is positive definite, the following inequality has to be satisfied from Eq. (7.42):

$$C_{kv} > C_v. \tag{7.43}$$

For Eq. (7.43), we need to consider the cosine values for different v and kv. This is clearly seen from the following example for the structures with $\mathbf{D}_{11}$ symmetry.

Example 7.5 The conditions for satisfying Eq. (7.43) for the structures with $\mathbf{D}_{11}$ symmetry.

The cosine values for different v (=1, 2, ..., 5) for the star-shaped structures with $\mathbf{D}_{11}$ symmetry are shown in Fig. 7.3.

C_v has the minimum value when $v = 5$, which is the maximum possible value for v.

It is obvious from the above example that C_{kv} $(k = 0, 1, \ldots, N-1)$ are cyclic functions of kv, and they are of reflection symmetry with respect to $kv = N/2$; i.e., $C_{kv} = C_{N-kv}$. C_v has the smallest value when v is nearest to $N/2$, but cannot be $N/2$ because N is odd such that the structure is indivisible.

Consider the boundary nodes on the upper plane. If the structure is indivisible, we will eventually stop at node 0 after starting from node v travelling along N sets of strut and vertical cable. In the trip, every node on the plane is visited exactly once.

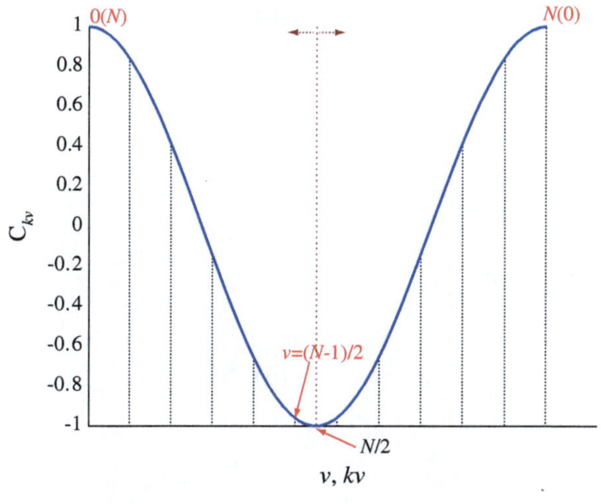

Fig. 7.3 Value of C_{kv} corresponding to the connectivity of vertical cables v ($N = 11$)

Example 7.6 Travelling along the struts and vertical cables for the structures with $\mathbf{D}_9$ symmetry.

Different connectivities of the nine nodes on the upper plane of the indivisible structures with $\mathbf{D}_9$ symmetry are shown in Fig. 7.4a, b, and d. If the structure is divisible, at least one node is visited more than once within the N steps, e.g., only three nodes of the divisible structure in Fig. 7.4c are visited three times in nine steps.

If the structure is divisible, some nodes with $C_{kv} = C_v$ for their cosines are visited more than once, but some of them are not visited. Hence, it will introduce additional zero eigenvalues in the force density matrix. The only exception is $\mathbf{D}_4^2$: there are only two values for the cosines, 1 and -1, and hence, they would be visited exactly once in the trip; however, it is unstable because it is divisible.

Because number of the two-dimensional representations of a dihedral group cannot exceed $N/2$, we can have the cosine values exactly once for C_{vk} if the structure is indivisible. Hence, in order to satisfy Eq. (7.43), the structure should be indivisible and C_v should have the minimum cosine value. When $v = N/2$ with N even, C_v has the minimum value -1; however, the structure is divisible in this case. Hence, the case $v = (N-1)/2$ with N odd is the only possibility that Eq. (7.43) can be satisfied for an indivisible star-shaped structure.

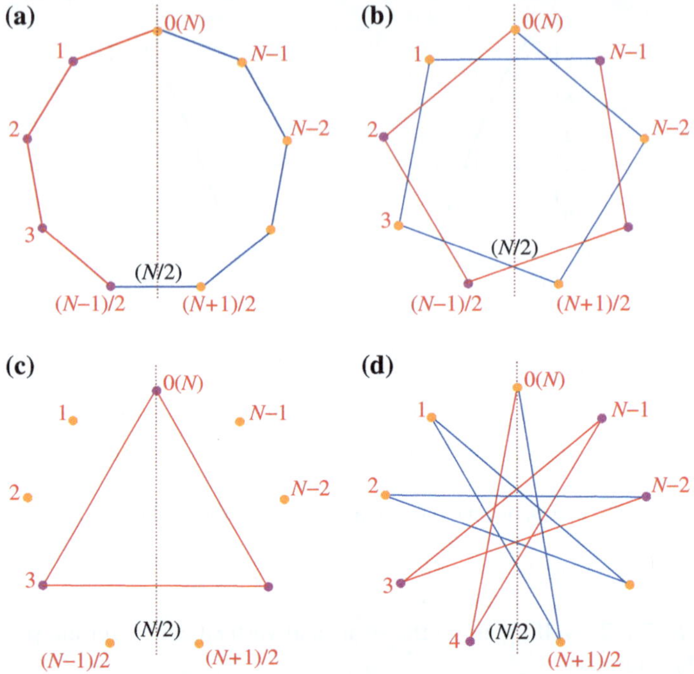

Fig. 7.4 Connectivity of boundary nodes on the same plane through struts and vertical cables ($N = 9$). It shows the idea, for super-stability condition for the star-shaped tensegrity structures, that $C_{kv} > C_v$ holds for any k only if $v = (N-1)/2$. **a** $v = 1$ (indivisible), **b** $v = 2$ (indivisible), **c** $v = 3$ (divisible), **d** $v = 4$ (indivisible)

From the above discussions, we may make the conclusions for super-stability of star-shaped structures:

Super-stability conditions for the star-shaped tensegrity structure $\mathbf{D}_N^v$:
A star-shaped tensegrity structure $\mathbf{D}_N^v$ with dihedral symmetry $\mathbf{D}_N$ is super-stable if the following two conditions are satisfied:

1. The structure has odd number of struts; i.e., N is odd.
2. The struts are as close to each other as possible, or in another term, the connectivity v of the vertical cables is $(N-1)/2$.

Example 7.7 Examples of super-stable star-shaped tensegrity structures.

According to the super-stability conditions, the star-shaped structures $\mathbf{D}_N^v$ cannot be super-stable when N is even. For the structures with $\mathbf{D}_3$, $\mathbf{D}_5$, and

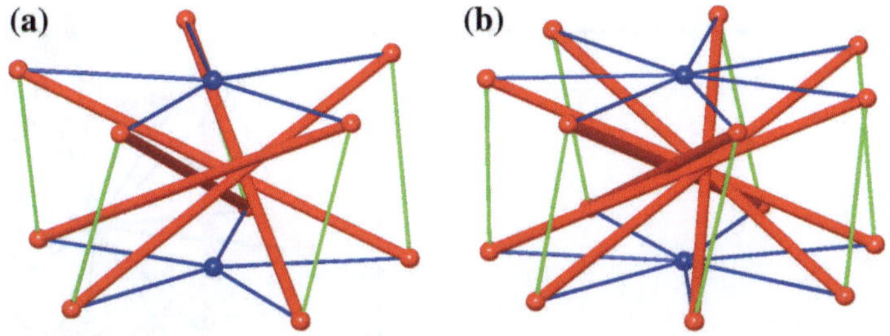

Fig. 7.5 Star-shaped tensegrity structures that are super-stable. All of them have odd number of struts, and the struts have shortest possible distances to each other by definition of vertical cables. **a** $\mathbf{D}_5^2$, **b** $\mathbf{D}_7^3$

$\mathbf{D}_7$ symmetries, they are super-stable when the connectivity parameter v for vertical cables are

$$\begin{aligned}\mathbf{D}_3:\ v&=\frac{N-1}{2}=\frac{3-1}{2}=1,\\ \mathbf{D}_5:\ v&=\frac{N-1}{2}=\frac{5-1}{2}=2,\\ \mathbf{D}_7:\ v&=\frac{N-1}{2}=\frac{7-1}{2}=3.\end{aligned}\tag{7.44}$$

Therefore, the structure $\mathbf{D}_3^1$ in Fig. 7.1b and the structures $\mathbf{D}_5^2$ and $\mathbf{D}_7^3$ in Fig. 7.5 are super-stable.

7.4.3 Prestress-stability

The star-shaped structures that are not super-stable might be prestress-stable; i.e., the structures are stable with positive definite tangent stiffness matrix when the level of prestresses is low enough, or the axial stiffness of its members is high enough. In this subsection, we further demonstrate that their prestress-stability is affected by the height/radius ratio.

Radius of the parallel circles where the boundary nodes are located is denoted by R, and the distance between the two parallel circle is denoted by H indicating height of a star-shaped tensegrity structure.

Example 7.8 Influence of height/radius ratio to prestress-stability of the structures with $\mathbf{D}_7$ symmetry.

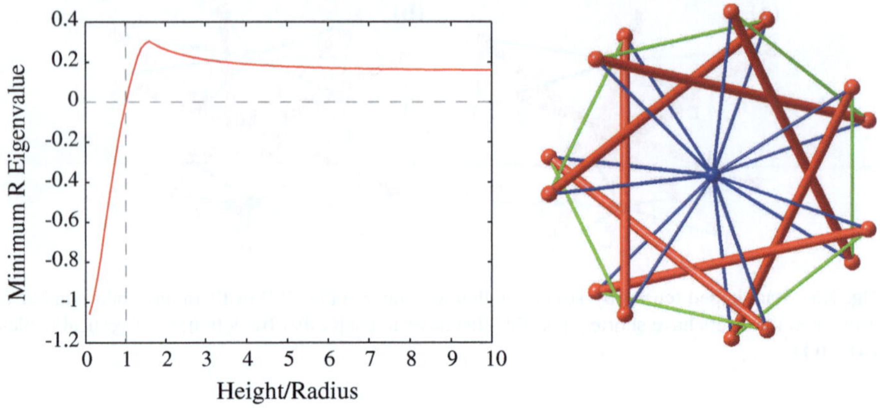

Fig. 7.6 Star-shaped tensegrity structure $\mathbf{D}_7^1$. It is prestress-stable when its height/radius ratio H/R is large enough; i.e., $H/R > 1.0$ in this case

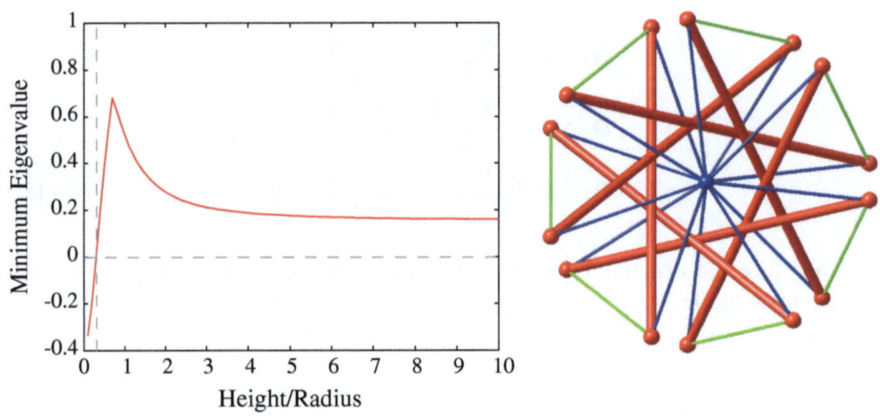

Fig. 7.7 Star-shaped tensegrity structure $\mathbf{D}_7^2$. It is prestress-stable when its height/radius ratio H/R is large enough; i.e., $H/R > 0.3$ in this case

We consider the star-shaped tensegrity structures $\mathbf{D}_7^1$, $\mathbf{D}_7^2$, and $\mathbf{D}_7^3$ with dihedral symmetry $\mathbf{D}_7$, which are respectively shown in right-hand sides in Figs. 7.6, 7.7, and 7.8.

On the left-hand sides of these figures, the minimum eigenvalues of the quadratic form $\mathbf{Q}_\mathrm{G}$ of the geometrical stiffness matrix with respect to the mechanisms defined in Eq. (4.78) are plotted against the height/radius ratio of the structure. The structure is prestress-stable if the minimum eigenvalue of $\mathbf{Q}_\mathrm{G}$ is positive.

It is obvious that $\mathbf{D}_7^3$ is always prestress-stable because it is super-stable. We can also observe from Figs. 7.6 and 7.7 that the structures $\mathbf{D}_7^1$ and $\mathbf{D}_7^2$ are prestress-stable if the height/radius ratio is large enough: the structure $\mathbf{D}_7^1$ is prestress-stable when $H/R > 1.0$, and the structure $\mathbf{D}_7^2$ when $H/R > 0.3$.

Based on the numerical study on stability of the star-shaped tensegrity structures up to $N = 10$, we have their stability properties as listed in Table 7.3. Note that in Table 7.3, 'P', 'S', and 'D' denote prestress-stability, super-stability, and divisibility, respectively.

From the numerical investigations and Table 7.3, we may have the following conclusion (conjecture) for prestress-stability of an indivisible star-shaped structure:

Conjecture on prestress-stability of an indivisible star-shaped tensegrity structure $\mathbf{D}_N^{\nu}$:

An indivisible star-shaped tensegrity structure $\mathbf{D}_N^{\nu}$ that is not super-stable is prestress-stable when its height/radius is large enough.

The verification of this conjecture is out of scope of this book.

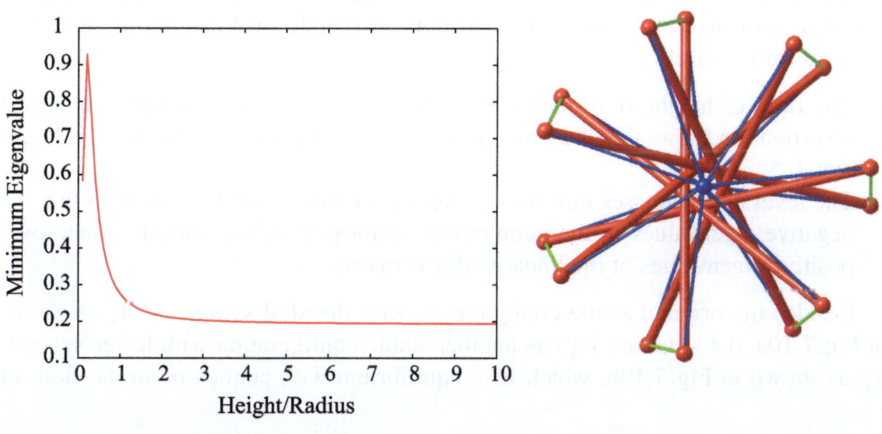

Fig. 7.8 Star-shaped tensegrity structure $\mathbf{D}_7^3$. The structure is super-stable, and therefore, always prestress-stable

Table 7.3 Stability of star-shaped tensegrity structures $\mathbf{D}_N^{\nu}$

$\nu \backslash N$	3	4	5	6	7	8	9	10
1	S	P	P	P	P	P	P	P
2		D	S	D	P	D	P	D
3				D	S	P	N	P
4						D	S	D
5								D

P: prestress-stability, S: super-stability, D: divisibility

7.5 Multi-stable Star-shaped Structure

In the previous section, we found that the star-shaped structures that are not super-stable might be prestress-stable; interestingly, these structures may have several 'stable' configurations when they are subjected to large deformation. The structures having more than one stable configurations are called *multi-stable* structures. In this section, we do not try to present all possible multi-stable star-shaped structures, instead, we study only the example structure $\mathbf{D}_4^1$ in detail to demonstrate its multi-stable behaviour.

7.5.1 Preliminary Study

The star-shaped structure $\mathbf{D}_4^1$ that is of dihedral symmetry $\mathbf{D}_4$ is shown in Fig. 7.9. The structure consists of eight boundary nodes and two center nodes, moreover, there are 16 members including four struts, eight radial cables, and four vertical cables.

We know from the super-stability conditions for a star-shaped tensegrity structure presented in Sect. 7.4.2 that the structure $\mathbf{D}_4^1$ is not super-stable, because it has even number of struts. The structure is locally stable with positive definite tangent stiffness matrix as indicated in Table 7.3. Furthermore, two additional conditions should be satisfied for its stability:

1. The ratio of height H to radius R of the structure is large enough; numerical investigation shows that the structure is prestress-stable only when H/R is larger than 0.5.
2. The level of prestresses introduced into the members is not too high so that the negative eigenvalues in the geometrical stiffness matrix do not dominate over positive eigenvalues of the linear stiffness matrix.

Besides the original stable configuration with dihedral symmetry $\mathbf{D}_4$ as shown in Fig. 7.10a, the structure $\mathbf{D}_4^1$ has another stable configuration with lower symmetry as shown in Fig. 7.10b, which is at equilibrium with contact of struts. Both of

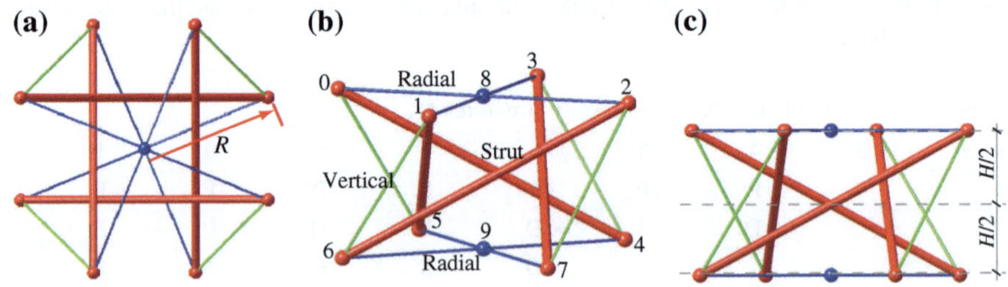

Fig. 7.9 Initial stable configuration of the star-shaped tensegrity structure $\mathbf{D}_4^1$. **a** Top, **b** diagonal, **c** side

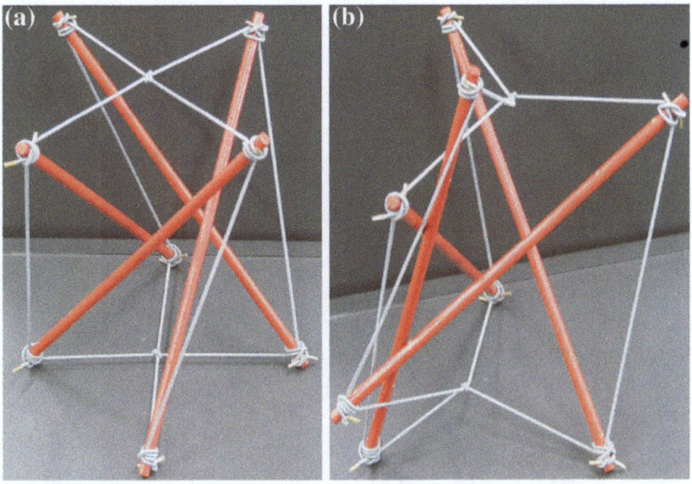

Fig. 7.10 Physical model of the multi-stable star-shaped tensegrity structure $\mathbf{D}_4^1$. The initial stable configuration with dihedral symmetry $\mathbf{D}_4$ is shown in (**a**), and the other stable configuration with lower symmetry is shown in (**b**). The two stable configurations can be switched to each other by proper external loads

the configurations are stable in the sense that they will recover the original shape after release of small enforced deformations. Moreover, the two stable states can be switched to each other by applying a large enforced deformation—the movement of any node around the principal z-axis of the symmetry.

7.5.2 Multi-stable Equilibrium Path

In this subsection, we conduct structural analysis to confirm multi-stable behavior of the structure $\mathbf{D}_4^1$.

For the initial settings of structural analysis, both height H and radius R of the structure are set as 1.0 m, hence, it is prestress-stable because the ratio H/R (=1.0) is larger than the necessary value 0.5. The stiffness parameters $A_i E_i$ ($i = 1, 2, \ldots, 16$), which is the product of cross-sectional area A_i and Young's modulus E_i, of struts and cables are 1.0×10^6 N and 1.0×10^2 N, respectively. According to the self-equilibrium condition presented in Sect. 7.3, the force densities of the struts, vertical cables, and radial cables in the state of self-equilibrium are -1.0 N, 1.0 N, and $\sqrt{2}$ N, respectively.

Numerical investigation shows that the structure associated with the highly symmetric configuration in Fig. 7.9 is stable, since its tangent stiffness matrix is positive semi-definite while the zero eigenvalues correspond to the rigid-body motions.

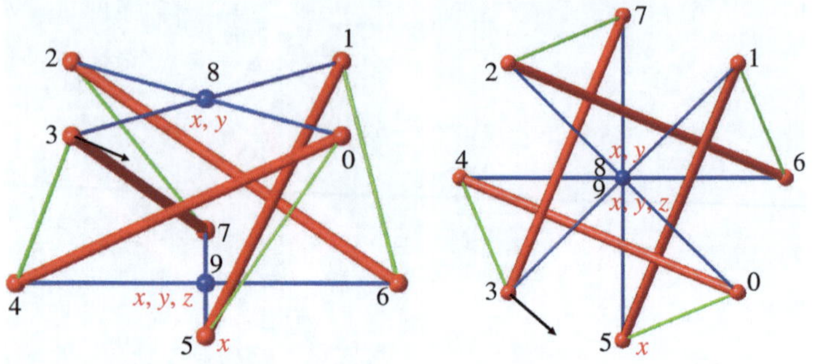

Fig. 7.11 Constraints and external load applied at the structure $\mathbf{D}_4^1$

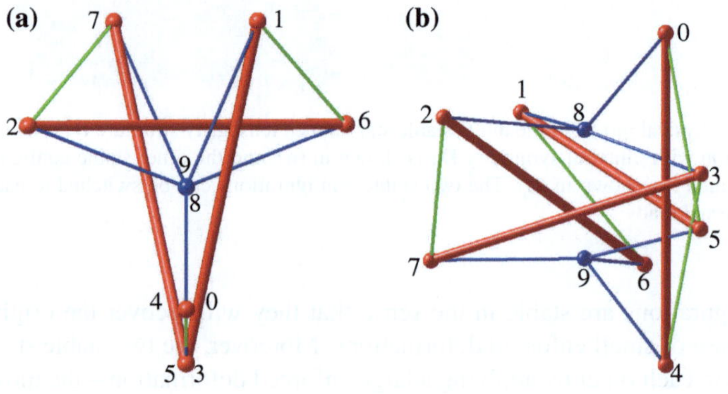

Fig. 7.12 The other stable configuration of the star-shaped tensegrity structure $\mathbf{D}_4^1$ with less symmetry. The structure can switch between this configuration and the initial configuration as shown in Fig. 7.9 through large deformation. **a** Top view, **b** side view

To constrain the rigid-body motions for structural analysis, the center node 9 on the bottom plane is fixed in all directions, while the x-displacement of node 5 and xy-displacements of the center node 8 on the upper plane are constrained as indicated in Fig. 7.11. Moreover, enforced rotation of node 3 about z-axis is applied. Node 3 is rotated counter-clockwise by $\pi/4$, and finally arrives at the position as shown in Fig. 7.12, which is the other stable configuration of the structure with contacts between struts. It can be observed from Fig. 7.12 that nodes 0, 3, 4, and 5 fall on the same plane at the final stable configuration.

Multi-stability of the structure can be understood much easier by looking at the change of its strain energy during the enforced deformation. Denote the axial force, member length, stress, and strain of member i by s_i, l_i, σ_i, and ε_i, respectively.

Assuming that

$$\sigma_i = E_i \varepsilon_i, \tag{7.45}$$

and using the linear relation between axial force s_i and stress σ_i; i.e.,

$$s_i = A_i \sigma_i, \tag{7.46}$$

the strain energy Π_{E} stored in the structure can be calculated as the sum of those stored in each member:

$$\Pi_{\mathrm{E}} = \sum_{i=1}^{16} \int \frac{1}{2} \sigma_i \varepsilon_i dV_i = \frac{1}{2} \sum_{i=1}^{16} A_i l_i \frac{\sigma_i^2}{E_i} = \frac{1}{2} \sum_{i=1}^{16} \frac{s_i^2 l_i}{A_i E_i}$$
$$= \frac{1}{2} \sum_{i=1}^{16} \frac{q_i^2 l_i^3}{A_i E_i}, \tag{7.47}$$

where force density $q_i = s_i / l_i$ and volume $V_i = A_i l_i$ of member i have been used.

Displacement control is utilized in the structural analysis so as to capture the detailed behaviour of the structure during the enforced rotation. The enforced rotation, expressed by the angle θ, is divided into consecutive 20 steps. All displacement components of the nodes are functions of θ. Therefore, the strain energy $\Pi_{\mathrm{E}}(\theta)$ is a function of θ only. The stain energy at each step is plotted in Fig. 7.13, and the deformed configurations of the structure at each step during the structural analysis are shown in Fig. 7.14. Moreover, the following facts can be observed from Fig. 7.13 concerning the rotation angle θ of node 3 about z-axis:

A: At the initial position $\theta = 0$:
The strain energy takes a local minimum in the neighborhood. Therefore, the structure at this position is in the state of self-equilibrium with zero gradient of strain energy, and moreover, it is stable with positive change in strain energy subjected to any small disturbance.

B: Between the position with $\theta = 0$ and that with peak strain energy $\Pi_{\max}$:
The strain energy increases along with the increase of enforced rotation angle θ.

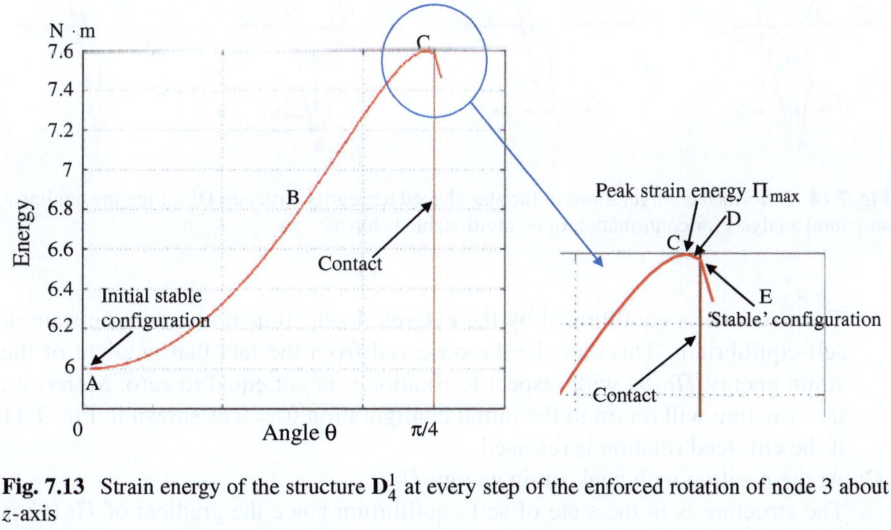

Fig. 7.13 Strain energy of the structure $\mathbf{D}_4^1$ at every step of the enforced rotation of node 3 about z-axis

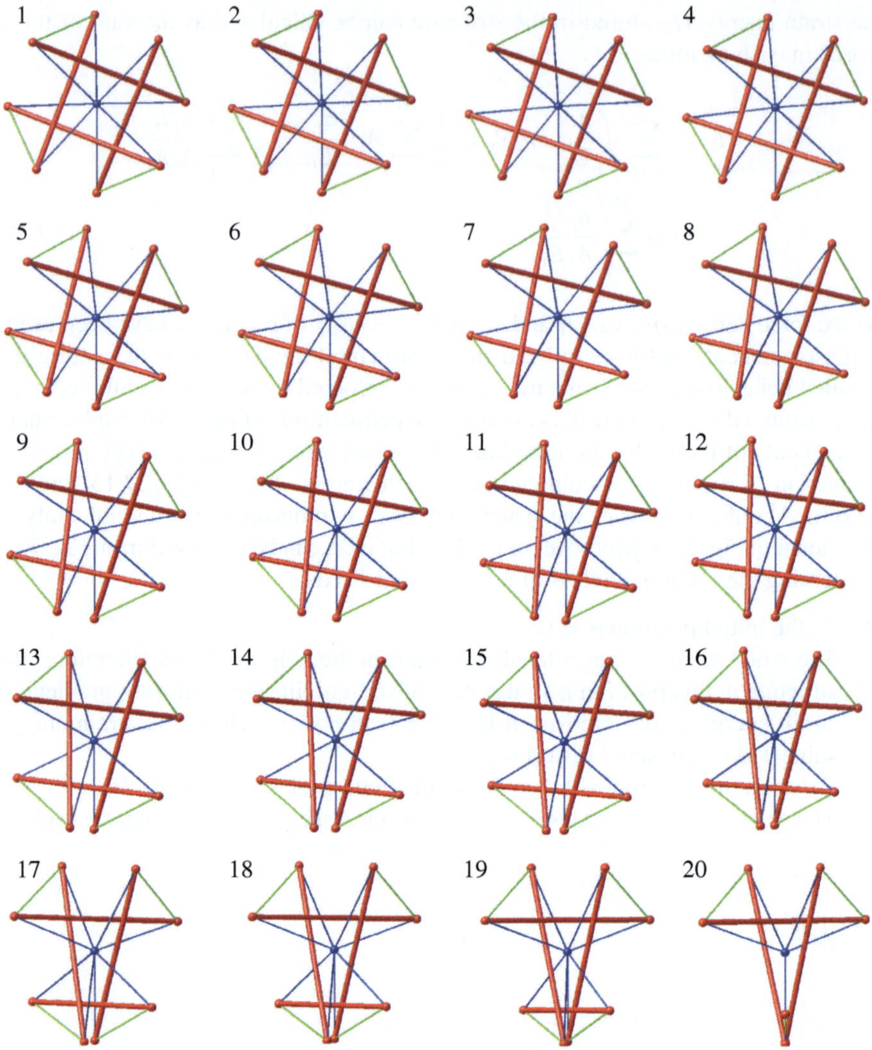

Fig. 7.14 Deformed configurations of the star-shaped tensegrity structure $\mathbf{D}_4^1$ during the nonlinear structural analysis for confirmation of its multi-stable behavior

The structure is equilibrated by the external loads, thus it is not in the state of self-equilibrium. This can also be observed from the fact that gradient of the strain energy $\Pi_{\mathrm{E}}(\theta)$ with respect to rotation θ is not equal to zero. Moreover, the structure will return to the initial configuration $\theta = 0$ as shown in Fig. 7.11 if the enforced rotation is released.

C: At the position with peak strain energy $\Pi_{\max}$:
The structure is in the state of self-equilibrium since the gradient of $\Pi_{\mathrm{E}}(\theta)$ is

zero, but it is unstable because the corresponding strain energy is maximum such that any disturbance leads to decrease of strain energy. It will move back to the initial stable configuration as shown in Fig. 7.11, or it will move forward to the next stable configuration as shown in Fig. 7.12, depending on direction of the infinitesimal disturbance.

D: Beyond $\Pi_{\max}$:
This is not the stable nor the self-equilibrated configuration as in the second stage 'B' in Fig. 7.13.

E: At the final position $\theta = \pi/4$:
Although the structure is not in the state of stability nor self-equilibrium in the conventional (mathematical) sense from the viewpoint of energy, further deformation is prevented by the contact of struts, and hence, deformation stops at this configuration and forms another 'stable' configuration.

7.6 Remarks

In this chapter, self-equilibrium analysis as well as stability investigation of the star-shaped tensegrity structures with dihedral symmetry have been analytically conducted, following the same procedure presented in Chap. 6 for prismatic tensegrity structures.

It has been proved that the star-shaped structures are indivisible, if the parameter N describing symmetry $\mathbf{D}_N$ of the structures and the parameter ν describing connectivity of the vertical cables do not have common divisor except 1.

Moreover, it has been proved that this class of structures are super-stable if (a) there exist odd number of struts; i.e., N is odd; and (b) the struts are arranged to be as close to each other as possible by the connectivity of vertical cables; i.e., $\nu = (N - 1)/2$.

For the indivisible structures, numerical investigations show that they can be prestress-stable, if not super-stable, when the height/radius ratios are large enough.

Furthermore, some of the prestress-stable structures that are not super-stable, for example the structure $\mathbf{D}_4^1$ studied this chapter, might have several stable configurations with lower level of symmetry than the initial configuration.This multi-stable behavior has been confirmed by numerical non-linear analysis as well as physical models.

References

1. Fowler, P. W., & Guest, S. D. (2000). A symmetry extension of Maxwell's rule for rigidity of frames. *International Journal of Solids and Structures*, *37*(12), 1793–1804.
2. Kangwai, R. D., Guest, S. D., & Pellegrino, S. (1999). An introduction to the analysis of symmetric structures. *Computers and Structures*, *71*(6), 671–688.
3. Kettle, S. F. A. (2007). *Symmetry and structure: readable group theory for chemists*. New York: Wiley.

Chapter 8
Regular Truncated Tetrahedral Structures

Abstract In this chapter, we present the analytical conditions for self-equilibrium and super-stability of the regular truncated tetrahedral tensegrity structures. These structures have tetrahedral symmetry, and their nodes have one-to-one correspondence to the symmetry operations of a tetrahedral group. The analytical conditions in terms of force densities are derived by investigating the symmetry-adapted force density matrix, using an extension of the direct formulation for dihedral symmetry in Appendix D.5.

Keywords Truncated tetrahedron · Tetrahedral symmetry · Self-equilibrium condition · Super-stability condition

8.1 Preliminary Study

In this chapter, we study the tensegrity structures with tetrahedarl symmetry; in particular, we concentrate on the regular truncated tetrahedral structures, the nodes of which have one-to-one correspondence to the symmetry operations of a tetrahedral group. The nodes of such structures belong to a regular orbit: any of the nodes can be moved to the position of another node by only one proper symmetry operation of the group. An example of regular truncated tensegrity structures is shown in Fig. 8.1a, which is generated from the regular truncated tetrahedron in Fig. 8.1b. This kind of structure was first introduced by Fuller [2], when he tried to make novel tensegrity structures utilizing the well-known geometries.

The cables of a tetrahedral structure lie along the edges of a truncated tetrahedron as shown in Fig. 8.1b: Type-1 cables are generated by replacing the original edges of the regular tetrahedron, and Type-2 cables are generated by replacing the new edges of the truncated tetrahedron after cutting off the vertices of the tetrahedron. Moreover, the struts are the diagonals connecting the vertices of the truncated tetrahedron. Hence, a tetrahedral structure consists of 12 nodes and 24 members, including 6 struts in compression and 18 cables in tension; i.e., $n = 12$, $m = 24$.

According to the modified Maxwell's rule presented in Eq. (2.61) in Chap. 2 for free-standing structures, we have the following relation for the number n^{s} of

J.Y. Zhang and M. Ohsaki, *Tensegrity Structures*, Mathematics for Industry 6,
DOI 10.1007/978-4-431-54813-3_8

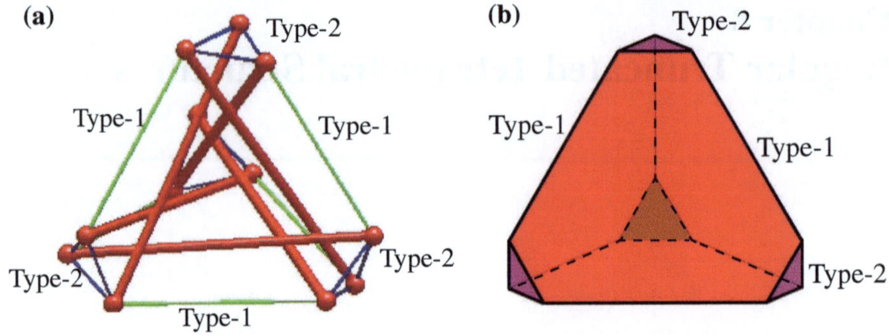

Fig. 8.1 An example of regular truncated tetrahedral tensegrity structure in (**a**), connectivity of which is generated from the regular truncated tetrahedron in (**b**)

independent prestress modes as well as the number n^{m} of infinitesimal mechanisms for the three-dimensional ($d = 3$) structure in Fig. 8.1:

$$\begin{aligned} n^{\mathrm{s}} - n^{\mathrm{m}} &= m - dn + n^{\mathrm{b}} \\ &= 24 - 3 \times 12 + 6 \\ &= -6, \end{aligned} \tag{8.1}$$

where $n^{\mathrm{b}}(=6)$ is the number of rigid-body motions of the structure in the three-dimensional space. The negative value in Eq. (8.1) indicates that the structure is *unstable*, if the structure is in the unstressed state. The structure in Fig. 8.1 has only one mode of prestress; i.e., $n^{\mathrm{s}} = 1$, and hence, there exists in total seven infinitesimal mechanisms; i.e., $n^{\mathrm{m}} = 7$; however, as will be presented later in this chapter, the structure in Fig. 8.1 is stable, and furthermore, it is *super-stable* if it has proper distribution of prestresses. It is this single prestress mode that stiffens the seven infinitesimal mechanisms to guarantee stability of the structure.

For the self-equilibrium of regular truncated tetrahedral tensegrity structures, we have presented the analytical condition in Sect. 3.5, by considering the self-equilibrium of only one representative node. However, (super-)stability of these structures has not been investigated. In some other studies, for example those in [4, 5], (super-)stability of the structures with tetrahedral symmetry has been studied *numerically*; theoretical proof was not provided.

Moreover, in the previous two chapters, we have studied self-equilibrium as well as (super-)stability of the prismatic and star-shaped tensegrity structures that have dihedral symmetry, by using the *analytical* formulation for block-diagonalization of force density matrices as presented in Appendix D.5. These studies demonstrate that high symmetry of the structures can be extensively utilized via group representation theory to significantly simplify the problems, and furthermore, derivation of analytical conditions becomes possible.

In this chapter, we extend the symmetry-adapted approach for the structures with dihedral symmetry to those with tetrahedral symmetry, and then present analytical conditions for their self-equilibrium as well as super-stability in a similar way.

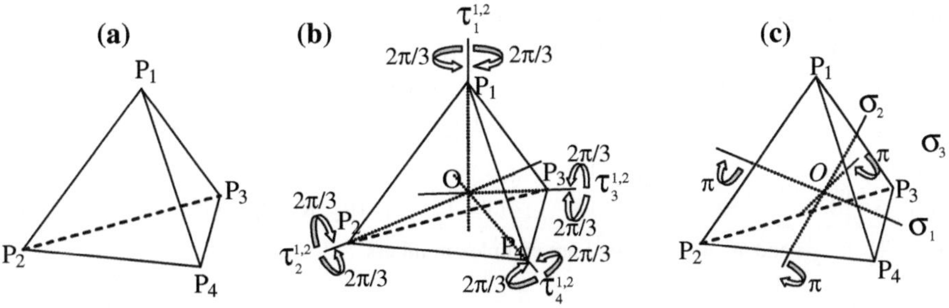

Fig. 8.2 The twelve symmetry operations for a regular tetrahedron. **a** Identity (E), **b** three-fold rotations ($\tau_j^{1,2}$), **c** two-fold rotations (σ_k)

8.2 Tetrahedral Symmetry

In Chaps. 6 and 7, we have demonstrated that symmetry of a structure can be systematically dealt with by using group representation theory. In particular, the theory on irreducible representation matrices is important for us to derive the symmetry-adapted force density matrix, with the independent blocks (sub-matrices) lying on its diagonal. Investigation on positive definiteness of these independent blocks enables us to present the analytical conditions for self-equilibrium analysis as well as super-stability of the structures, because the eigenvalues of the blocks can be easily computed in an analytical manner. In this section, we first give a brief introduction to tetrahedral group and its irreducible representation matrices.

A tetrahedron is composed of four vertices, four faces, and six edges. Moreover, all edges of a *regular tetrahedron* have the same length, an example of which is shown in Fig. 8.2a. A regular tetrahedron will have the same appearance by applying any of the following twelve symmetry operations:

- The identity operation E doing nothing as indicated in Fig. 8.2a;
- The three-fold rotations τ_j^1 and τ_j^2 ($j = 1, 2, 3, 4$) of the angles $2\pi/3$ and $4\pi/3$, respectively about the j-axis going through the vertex j and the origin 'O' (central point) of the tetrahedron as indicated in Fig. 8.2b; and
- The two-fold rotations σ_k ($k = 1, 2, 3$) of the angle π about the axes going through the middle of the edges as well as the origin as indicated in Fig. 8.2c.

For a tetrahedral group, there are three one-dimensional irreducible representations[1]: A, E_1, and E_2; and there is only one three-dimensional irreducible representation T. The characters corresponding to irreducible representations as well as symmetry operations of a tetrahedral group [3] are listed in Table 8.1, where $\mathrm{i} = \sqrt{-1}$. The symbols x, y, and z in the last column of the table respectively stand for x-, y-, and z-coordinates, and R_x, R_y, and R_z stand for rotations about these axes.

[1] Readers should not be confused by the two-dimensional representation denoted by E_k for a dihedral group in Chaps. 6 and 7.

Table 8.1 The characters of the tetrahedral group

	E	$4C_3(\tau_j^1)$	$4C_3^2(\tau_j^2)$	$3C_2(\sigma_k)$	$j = 1, 2, 3, 4;\ k = 1, 2, 3$
A	1	1	1	1	$x^2 + y^2 + z^2$
E_1	1	ω	ω^2	1	$\left(\frac{1}{\sqrt{6}}[2z^2 - x^2 - y^2], \frac{1}{\sqrt{2}}[x^2 - y^2]\right)$
E_2	1	ω^2	ω	1	
T	3	0	0	-1	$(T_x, T_y, T_z),\ (R_x, R_y, R_z),\ (x, y, z),\ (xy, yz, zx)$

$\omega = -\frac{1}{2} + \frac{\sqrt{3}}{2}\mathrm{i}$ and $\omega^2 = -\frac{1}{2} - \frac{\sqrt{3}}{2}\mathrm{i}$ are complex numbers

Let $\mathbf{R}^\mu$ denote the irreducible representation matrix of the representation μ. From Table 8.1, the (one-by-one) representation matrices, which turn out to be scalars, corresponding to representation $(\mu{=})A$ are all equal to one; i.e.,

$$\begin{aligned} &\mathbf{R}^{\mathrm{A}}(\mathrm{E}) = 1; \\ &\mathbf{R}^{\mathrm{A}}(\tau_j^1) = \mathbf{R}^{\mathrm{A}}(\tau_j^2) = 1, \quad (j = 1, 2, 3, 4); \\ &\mathbf{R}^{\mathrm{A}}(\sigma_k) = 1, \quad (k = 1, 2, 3). \end{aligned} \tag{8.2}$$

Similarly, for the one-dimensional representations E_1 and E_2, their (one-by-one) representation matrices corresponding to each symmetry operation can be directly read off from Table 8.1 as follows:

$$\begin{aligned} &\mathbf{R}^{\mathrm{E}_1}(\mathrm{E}) = 1, \\ &\mathbf{R}^{\mathrm{E}_1}(\tau_j^1) = -\frac{1}{2} + \frac{\sqrt{3}}{2}\mathrm{i}, \quad (j = 1, 2, 3, 4); \\ &\mathbf{R}^{\mathrm{E}_1}(\tau_j^2) = -\frac{1}{2} - \frac{\sqrt{3}}{2}\mathrm{i}, \quad (j = 1, 2, 3, 4); \\ &\mathbf{R}^{\mathrm{E}_1}(\sigma_k) = 1, \quad (k = 1, 2, 3); \end{aligned} \tag{8.3}$$

and

$$\begin{aligned} &\mathbf{R}^{\mathrm{E}_2}(\mathrm{E}) = 1, \\ &\mathbf{R}^{\mathrm{E}_2}(\tau_j^1) = -\frac{1}{2} - \frac{\sqrt{3}}{2}\mathrm{i}, \quad (j = 1, 2, 3, 4); \\ &\mathbf{R}^{\mathrm{E}_2}(\tau_j^2) = -\frac{1}{2} + \frac{\sqrt{3}}{2}\mathrm{i}, \quad (j = 1, 2, 3, 4); \\ &\mathbf{R}^{\mathrm{E}_2}(\sigma_k) = 1, \quad (k = 1, 2, 3). \end{aligned} \tag{8.4}$$

Unlike the one-dimensional irreducible representation matrices, the three-dimensional irreducible representation matrices are not unique; i.e., there can be many choices as long as their characters (traces) satisfy the character table. In particular, the three-dimensional irreducible representation matrices $\mathbf{R}^{\mathrm{T}}$ are in fact the

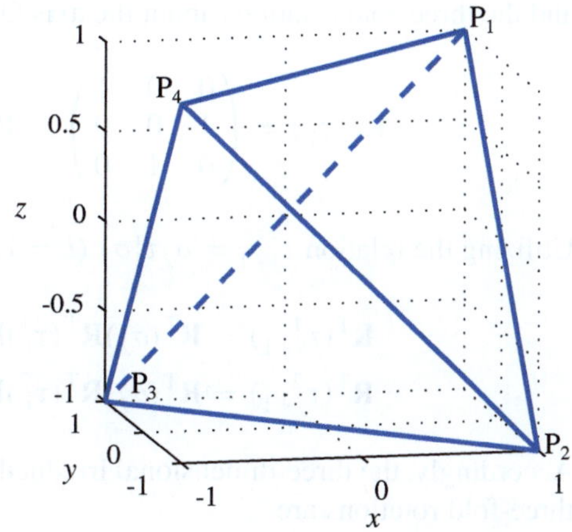

Fig. 8.3 One possible choice of coordinate system (x, y, z) for a regular tetrahedron

symmetry (rotation) operations of a regular tetrahedron, as indicated by (R_x, R_y, R_z) in the last column of Table 8.1. Different representation matrices come from different selections of coordinate system.

Figure 8.3 shows a possible choice of coordinate system (x, y, z) adapted here. The coordinates of the four vertices P_i $(i = 1, 2, 3, 4)$ of a regular tetrahedron in this coordinate system are given as

$$P_1 : \begin{pmatrix} 1 \\ 1 \\ 1 \end{pmatrix}, \quad P_2 : \begin{pmatrix} 1 \\ -1 \\ -1 \end{pmatrix}, \quad P_3 : \begin{pmatrix} -1 \\ 1 \\ -1 \end{pmatrix}, \quad P_4 : \begin{pmatrix} -1 \\ -1 \\ 1 \end{pmatrix}. \tag{8.5}$$

The irreducible representation matrix corresponding to the symmetry operation σ_1 of the three-dimensional representation $(\mu=)$T, which exchanges the vertices P_1 and P_2, and P_3 and P_4; i.e., $(1, 2)(3, 4)$, in this coordinate system is

$$\mathbf{R}^{\mathrm{T}}(\sigma_1) = \begin{pmatrix} 1 & 0 & 0 \\ 0 & -1 & 0 \\ 0 & 0 & -1 \end{pmatrix}, \tag{8.6}$$

with the trace equal to -1 satisfying the character table of the tetrahedral group.

Corresponding to the adopted coordinate system in Fig. 8.3, the three-dimensional irreducible representation matrices $\mathbf{R}^{\mathrm{T}}(\sigma_2)$ and $\mathbf{R}^{\mathrm{T}}(\sigma_3)$ for the two-fold rotations $\sigma_2 = (1\ 3)(2\ 4)$, exchanging the vertices P_1 and P_3, and P_2 and P_4, and $\sigma_3 = (1\ 4)(2\ 3)$, exchanging the vertices P_1 and P_4, and P_2 and P_3, are defined as

$$\mathbf{R}^{\mathrm{T}}(\sigma_2) = \begin{pmatrix} -1 & 0 & 0 \\ 0 & 1 & 0 \\ 0 & 0 & -1 \end{pmatrix}, \quad \mathbf{R}^{\mathrm{T}}(\sigma_3) = \begin{pmatrix} -1 & 0 & 0 \\ 0 & -1 & 0 \\ 0 & 0 & 1 \end{pmatrix}, \tag{8.7}$$

and the three-fold rotations about the axis OP_1 are

$$\mathbf{R}^{\mathrm{T}}(\tau_1^1) = \begin{pmatrix} 0 & 0 & 1 \\ 1 & 0 & 0 \\ 0 & 1 & 0 \end{pmatrix}, \quad \mathbf{R}^{\mathrm{T}}(\tau_1^2) = \begin{pmatrix} 0 & 1 & 0 \\ 0 & 0 & 1 \\ 1 & 0 & 0 \end{pmatrix}. \tag{8.8}$$

Utilizing the relation $\tau_{j+1}^i = \sigma_j \tau_1^i \sigma_j$ $(i = 1, 2;\ j = 1, 2, 3)$, we have

$$\begin{aligned} \mathbf{R}^{\mathrm{T}}(\tau_{j+1}^1) &= \mathbf{R}^{\mathrm{T}}(\sigma_j)\mathbf{R}^{\mathrm{T}}(\tau_1^1)\mathbf{R}^{\mathrm{T}}(\sigma_j), \\ \mathbf{R}^{\mathrm{T}}(\tau_{j+1}^2) &= \mathbf{R}^{\mathrm{T}}(\sigma_j)\mathbf{R}^{\mathrm{T}}(\tau_1^2)\mathbf{R}^{\mathrm{T}}(\sigma_j), \quad (j = 1, 2, 3). \end{aligned} \tag{8.9}$$

Accordingly, the three-dimensional irreducible representation matrices for the other three-fold rotations are

$$\mathbf{R}^{\mathrm{T}}(\tau_2^1) = \begin{pmatrix} 0 & 0 & -1 \\ -1 & 0 & 0 \\ 0 & 1 & 0 \end{pmatrix}, \quad \mathbf{R}^{\mathrm{T}}(\tau_2^2) = \begin{pmatrix} 0 & -1 & 0 \\ 0 & 0 & 1 \\ -1 & 0 & 0 \end{pmatrix}, \tag{8.10}$$

$$\mathbf{R}^{\mathrm{T}}(\tau_3^1) = \begin{pmatrix} 0 & 0 & 1 \\ -1 & 0 & 0 \\ 0 & -1 & 0 \end{pmatrix}, \quad \mathbf{R}^{\mathrm{T}}(\tau_3^2) = \begin{pmatrix} 0 & -1 & 0 \\ 0 & 0 & -1 \\ 1 & 0 & 0 \end{pmatrix}, \tag{8.11}$$

$$\mathbf{R}^{\mathrm{T}}(\tau_4^1) = \begin{pmatrix} 0 & 0 & -1 \\ 1 & 0 & 0 \\ 0 & -1 & 0 \end{pmatrix}, \quad \mathbf{R}^{\mathrm{T}}(\tau_4^2) = \begin{pmatrix} 0 & 1 & 0 \\ 0 & 0 & -1 \\ -1 & 0 & 0 \end{pmatrix}. \tag{8.12}$$

It is easy to verify that traces of these matrices are equal to the characters of corresponding symmetry operations for the three-dimensional irreducible representations listed in Table 8.1:

$$\begin{aligned} &\mathrm{trace}(\mathbf{R}^{\mathrm{T}}(\sigma_k)) = -1, && (k = 1, 2, 3); \\ &\mathrm{trace}(\mathbf{R}^{\mathrm{T}}(\tau_j^1)) = \mathrm{trace}(\mathbf{R}^{\mathrm{T}}(\tau_j^1)) = 0, && (j = 1, 2, 3, 4). \end{aligned}$$

8.3 Symmetry-adapted Force Density Matrix

Self-equilibrium and super-stability of a tensegrity structure can be verified by respectively investigating the nullities and positive-definiteness of its force density matrix. These investigations are ideally conducted on the symmetry-adapted force density matrix, because the independent blocks or sub-matrices of this form are of much smaller sizes such that they can be easily handled even by hand calculations.

For such purpose, we present the analytical block-diagonalization of the force density matrix for the regular truncated tetrahedral structures in this section, which is extended from the formulation for the structures with dihedral symmetry presented in Appendix D.5.

8.3.1 Structure of Symmetry-adapted Force Density Matrix

Structure of the symmetry-adapted force density matrix can be determined by considering permutation representation of the nodes, written in terms of sum of the irreducible representations [1, 3].

Any symmetry operation of a regular tetrahedron, except for the identity operation E, will exchange positions of the nodes. Thus, the numbers of nodes that will not change their positions by the corresponding symmetry operations, or equivalently the reducible representation $\Gamma(\mathrm{N})$ of the nodes, are

Symmetry operation:	E	$4C_3$	$4C_3^2$	$3C_2$
$\Gamma(\mathrm{N})$	12	0	0	0

(8.13)

From characters of the irreducible representations of tetrahedral group as listed in Table 8.1, we have

	E	$4C_3$	$4C_3^2$	$3C_2$
A	1	1	1	1
$+E_1$	1	ω	ω^2	1
$+E_2$	1	ω^2	ω	1
+3T	9	0	0	-3
=	12	0	0	0

(8.14)

where $\omega + \omega^2 = -1$ has been used. Therefore, the reducible representation $\Gamma(\mathrm{N})$ of the nodes in Eq. (8.13) can be written as a linear combination of the irreducible representations in a general form as follows:

$$\Gamma(\mathrm{N}) = \mathrm{A} + \mathrm{E}_1 + \mathrm{E}_2 + 3\mathrm{T}. \tag{8.15}$$

The linear combination of irreducible representations for nodes in Eq. (8.15) characterizes structure of the symmetry-adapted force density matrix $\tilde{\mathbf{E}}$, where $\tilde{(\cdot)}$ is used to denote the symmetry-adapted form of a matrix: there are

- one one-dimensional block $\tilde{\mathbf{E}}^{\mathrm{A}}$ corresponding to the one-dimensional representation A,
- one one-dimensional block $\tilde{\mathbf{E}}^{\mathrm{E}_1}$ corresponding to the one-dimensional representation E_1,

- one one-dimensional block $\tilde{\mathbf{E}}^{\mathrm{E}_2}$ corresponding to the one-dimensional representation E_2, and
- three copies of the three-dimensional block $\tilde{\mathbf{E}}^{\mathrm{T}}$ corresponding to the three-dimensional representation T.

Hence, structure of the symmetry-adapted force density matrix $\tilde{\mathbf{E}}$ can be summarized as follows, with the independent sub-matrices $\tilde{\mathbf{E}}^{\mu}$ lying on the leading diagonal:

$$\underset{12\times 12}{\tilde{\mathbf{E}}} = \begin{pmatrix} \underset{1\times 1}{\tilde{\mathbf{E}}^{\mathrm{A}}} & & & & & \\ & \underset{1\times 1}{\tilde{\mathbf{E}}^{\mathrm{E}_1}} & & & \mathbf{O} & \\ & & \underset{1\times 1}{\tilde{\mathbf{E}}^{\mathrm{E}_2}} & & & \\ & & & \underset{3\times 3}{\tilde{\mathbf{E}}^{\mathrm{T}}} & & \\ & \mathbf{O} & & & \underset{3\times 3}{\tilde{\mathbf{E}}^{\mathrm{T}}} & \\ & & & & & \underset{3\times 3}{\tilde{\mathbf{E}}^{\mathrm{T}}} \end{pmatrix}, \tag{8.16}$$

or it can be written in the form of direct sum as follows:

$$\tilde{\mathbf{E}} = \tilde{\mathbf{E}}^{\mathrm{A}} \oplus \tilde{\mathbf{E}}^{\mathrm{E}_1} \oplus \tilde{\mathbf{E}}^{\mathrm{E}_2} \oplus 3\tilde{\mathbf{E}}^{\mathrm{T}}. \tag{8.17}$$

8.3.2 Blocks of Symmetry-adapted Force Density Matrix

The block $\tilde{\mathbf{E}}^{\mu}$ corresponding to representation μ of the symmetry-adapted force density matrix $\tilde{\mathbf{E}}$ can be obtained by applying (coordinate) transformation matrices on both sides of the original force density matrix [4]. However, numerical approaches might not have the opportunity to have thorough understanding of stability of the structure. This subsection presents the blocks of the symmetry-adapted force density matrix in an analytical manner. The formulation is extended from that for the structures with dihedral symmetry as in Chaps. 6 and 7. A strict proof for this extension is not provided here for compactness.

Due to symmetry, the members of each type have the same length, and contain the same prestress, and therefore, they have the same force density. The force densities of Type-1 cables, Type-2 cables, and struts are denoted by q_{v}, q_{h}, and q_{s}, respectively.

Each node of a regular truncated tetrahedral tensegrity structure is connected by only one Type-1 cable. One node can be moved to the other end of the Type-1 cable by an appropriate two-fold rotation σ_j $(j = 1, 2, 3)$ around the axis perpendicular to the Type-1 cable. Similarly, the two nodes connecting to a strut can exchange their positions by a two-fold rotation σ_k $(k = 1, 2, 3)$ about the axis perpendicular to the strut. Note here that $j \neq k$.

On the other hand, each node of the structure is connected by two Type-2 cables. One node can be moved to the other two nodes connected by the Type-2 cables by the three-fold rotations τ_i^1 and τ_i^2 $(i = 1, 2, 3, 4)$.

Similar to the direct formulation in Eq. (D.20) for the structures with dihedral symmetry, the blocks corresponding to each distinct representation μ of a tetrahedral group are formulated as sum of the multiplication of irreducible representation matrices and force densities connected to the reference node. Therefore, the blocks $\tilde{\mathbf{E}}^{\mu}$ corresponding to representation μ can be written in a general form as

$$\tilde{\mathbf{E}}^{\mu} = (2q_{\mathrm{h}} + q_{\mathrm{v}} + q_{\mathrm{s}})\mathbf{I} - q_{\mathrm{h}}\left[\mathbf{R}^{\mu}(\tau_i^1) + \mathbf{R}^{\mu}(\tau_i^2)\right] - q_{\mathrm{v}}\mathbf{R}^{\mu}(\sigma_j) - q_{\mathrm{s}}\mathbf{R}^{\mu}(\sigma_k),$$
$$(i = 1, 2, 3, 4;\ j, k = 1, 2, 3;\ j \neq k), \tag{8.18}$$

where $\mathbf{I}$ is an identity matrix: it is a one-by-one matrix $\mathbf{I}_1$ if μ is a one-dimensional representation (A, E_1, or E_2), or it is a three-by-three matrix $\mathbf{I}_3$ if μ is the three-dimensional representation T.

The parameters (i, j, k) define connectivity pattern of the members. It is obvious from Table 8.1 that values of (i, j, k) will not affect the sub-matrix $\tilde{\mathbf{E}}^{\mu}$ for the one-dimensional representations $\mu = \mathrm{A}, \mathrm{E}_1, \mathrm{E}_2$.

According to Eq. (8.18) and Table 8.1, we have the block $\tilde{\mathbf{E}}^{\mathrm{A}}$ for the one-dimensional representation A as

$$\begin{aligned}\tilde{\mathbf{E}}^{\mathrm{A}} &= (2q_{\mathrm{h}} + q_{\mathrm{v}} + q_{\mathrm{s}}) - q_{\mathrm{h}}(1 + 1) - q_{\mathrm{v}} - q_{\mathrm{s}} \\ &= 0,\end{aligned} \tag{8.19}$$

such that $\tilde{\mathbf{E}}^{\mathrm{A}}$ is always zero.

For the one-dimensional representations E_1 and E_2, we have

$$\begin{aligned}\tilde{\mathbf{E}}^{\mathrm{E}_1} = \tilde{\mathbf{E}}^{\mathrm{E}_2} &= (2q_{\mathrm{h}} + q_{\mathrm{v}} + q_{\mathrm{s}}) - q_{\mathrm{h}}(\omega + \omega^2) - q_{\mathrm{v}} - q_{\mathrm{s}} \\ &= 3q_{\mathrm{h}}\ (> 0).\end{aligned} \tag{8.20}$$

Because q_{h} is the force density of Type-2 cables carrying tension and it is positive, the blocks $\tilde{\mathbf{E}}^{\mathrm{E}_1}$ and $\tilde{\mathbf{E}}^{\mathrm{E}_2}$ are positive (definite).

Different values of (i, j, k) will result in different entries in the three-dimensional representation matrices, but importantly, they will not affect the eigenvalues of the three-dimensional block $\tilde{\mathbf{E}}^{\mathrm{T}}$.

Example 8.1 The three-dimensional block $\tilde{\mathbf{E}}^{\mathrm{T}}$ corresponding to different connectivity of a regular truncated tetrahedral structure defined by the parameters (i, j, k).

For the case of $i = 1$, $j = 2$, and $k = 3$, we have

$$\tilde{\mathbf{E}}^{\mathrm{T}} = (2q_{\mathrm{h}} + q_{\mathrm{v}} + q_{\mathrm{s}})\mathbf{I}_3 - q_{\mathrm{h}}\left[\mathbf{R}^{\mathrm{T}}(\tau_1^1) + \mathbf{R}^{\mathrm{T}}(\tau_1^2)\right] - q_{\mathrm{v}}\mathbf{R}^{\mathrm{T}}(\sigma_2) - q_{\mathrm{s}}\mathbf{R}^{\mathrm{T}}(\sigma_3)$$
$$= \begin{pmatrix} 2(q_{\mathrm{h}} + q_{\mathrm{v}} + q_{\mathrm{s}}) & -q_{\mathrm{h}} & -q_{\mathrm{h}} \\ -q_{\mathrm{h}} & 2(q_{\mathrm{h}} + q_{\mathrm{s}}) & -q_{\mathrm{h}} \\ -q_{\mathrm{h}} & -q_{\mathrm{h}} & 2(q_{\mathrm{h}} + q_{\mathrm{v}}) \end{pmatrix}. \tag{8.21}$$

For another case with different values, for instance $i = 2$, $j = 1$, and $k = 2$, we have

$$\tilde{\mathbf{E}}^{\mathrm{T}} = (2q_{\mathrm{h}} + q_{\mathrm{v}} + q_{\mathrm{s}})\mathbf{I}_3 - q_{\mathrm{h}}\left[\mathbf{R}^{\mathrm{T}}(\tau_2^1) + \mathbf{R}^{\mathrm{T}}(\tau_2^2)\right] - q_{\mathrm{v}}\mathbf{R}^{\mathrm{T}}(\sigma_1) - q_{\mathrm{s}}\mathbf{R}^{\mathrm{T}}(\sigma_2)$$
$$= \begin{pmatrix} 2(q_{\mathrm{h}} + q_{\mathrm{s}}) & q_{\mathrm{h}} & q_{\mathrm{h}} \\ q_{\mathrm{h}} & 2(q_{\mathrm{h}} + q_{\mathrm{v}}) & -q_{\mathrm{h}} \\ q_{\mathrm{h}} & -q_{\mathrm{h}} & 2(q_{\mathrm{h}} + q_{\mathrm{v}} + q_{\mathrm{s}}) \end{pmatrix}. \tag{8.22}$$

It is obvious from Example 8.1 that the two three-dimensional blocks $\tilde{\mathbf{E}}^{\mathrm{T}}$ corresponding to different connectivity patterns defined by the parameters (i, j, k) have the same eigenvalues, although arrangement of the leading diagonal entries is different.

8.4 Self-equilibrium Conditions

In order to guarantee a non-degenerate geometry realization for a three-dimensional tensegrity structure, the (symmetry-adapted) force density matrix should have at least four zero eigenvalues as discussed in Sect. 2.5. We already know from Eq. (8.19) that there exists one zero eigenvalue in the one-dimensional block $\tilde{\mathbf{E}}^{\mathrm{A}}$, while the other two one-dimensional blocks $\tilde{\mathbf{E}}^{\mathrm{E}_1}$ and $\tilde{\mathbf{E}}^{\mathrm{E}_2}$ are positive (definite). Therefore, the remaining three zero eigenvalues have to be in the three copies of the three-dimensional block $\tilde{\mathbf{E}}^{\mathrm{T}}$; i.e., it is sufficient that there exists one zero eigenvalue in $\tilde{\mathbf{E}}^{\mathrm{T}}$.

The eigenvalues λ of $\tilde{\mathbf{E}}^{\mathrm{T}}$ can be obtained by solving the following equation

$$\det(\tilde{\mathbf{E}}^{\mathrm{T}} - \lambda\mathbf{I}) = 0. \tag{8.23}$$

Using $\tilde{\mathbf{E}}^{\mathrm{T}}$ either in Eq. (8.21) or in Eq. (8.22), the characteristic polynomial in Eq. (8.23) can be explicitly written as

$$\begin{aligned} &\det(\tilde{\mathbf{E}}^{\mathrm{T}} - \lambda\mathbf{I}) \\ &= \lambda[-\lambda^2 + 2\lambda(3q_{\mathrm{h}} + 2q_{\mathrm{v}} + 2q_{\mathrm{s}}) - 9q_{\mathrm{h}}^2 - 16q_{\mathrm{h}}q_{\mathrm{v}} - 16q_{\mathrm{h}}q_{\mathrm{s}} - 4q_{\mathrm{v}}^2 - 4q_{\mathrm{s}}^2 - 12q_{\mathrm{v}}q_{\mathrm{s}}] \\ &\quad + 4(3q_{\mathrm{h}}^2q_{\mathrm{v}} + 3q_{\mathrm{h}}^2q_{\mathrm{s}} + 2q_{\mathrm{h}}q_{\mathrm{v}}^2 + 2q_{\mathrm{v}}^2q_{\mathrm{s}} + 2q_{\mathrm{h}}q_{\mathrm{s}}^2 + 2q_{\mathrm{v}}q_{\mathrm{s}}^2 + 6q_{\mathrm{h}}q_{\mathrm{v}}q_{\mathrm{s}}) \\ &= 0. \end{aligned} \tag{8.24}$$

To ensure a zero eigenvalue; i.e., $\lambda_1 = 0$ in Eq. (8.24), we have

$$3q_h^2 q_v + 3q_h^2 q_s + 2q_h q_v^2 + 2q_v^2 q_s + 2q_h q_s^2 + 2q_v q_s^2 + 6q_h q_v q_s = 0, \tag{8.25}$$

which yields the same result in self-equilibrium analysis of the structure in Eq. (3.101) in Chap. 3. Furthermore, the other two eigenvalues λ_2 and λ_3 are calculated from the following equation:

$$\lambda^2 - 2\lambda(3q_h + 2q_v + 2q_s) + 9q_h^2 + 16q_h q_v + 16q_h q_s + 4q_v^2 + 4q_s^2 + 12q_v q_s = 0, \tag{8.26}$$

which leads to

$$\lambda_{2,3} = \xi \pm 2\sqrt{\eta}, \tag{8.27}$$

where

$$\begin{aligned} \xi &= 3q_h + 2q_v + 2q_s, \\ \eta &= -(q_h q_v + q_h q_s + q_v q_s). \end{aligned} \tag{8.28}$$

Solving Eq. (8.25), we obtain the two solutions q_{h_1} and q_{h_2} in Eq. (8.29) for the force density of Type-2 cables q_h in the state of self-equilibrium.

Self-equilibrated force densities for the regular truncated tetrahedral tensegrity structures:

$$\begin{aligned} q_{h_1} &= \frac{\phi^2 - 5\psi^2 + \sqrt{\psi^4 + 14\psi^2\phi^2 + \phi^4}}{6\psi}, \quad \text{with } \psi > 0, \\ q_{h_2} &= \frac{\phi^2 - 5\psi^2 - \sqrt{\psi^4 + 14\psi^2\phi^2 + \phi^4}}{6\psi}, \quad \text{with } \psi < 0, \end{aligned} \tag{8.29}$$

where

$$q_v = \psi + \phi \ (> 0) \quad \text{and} \quad q_s = \psi - \phi \ (< 0). \tag{8.30}$$

More details on the solutions, including discussions on the sign of ψ, can be found in Sect. 3.5. It it notable that the solutions in Eq. (8.29) coincide with the results in Eqs. (3.115) and (3.117) derived by force equilibrium analysis of the representative node.

8.5 Super-stability Conditions

Using the relationship among the force densities q_{v}, q_{h}, and q_{s} of the three types of members in Eq. (8.29) derived in the self-equilibrium analysis, this section presents the super-stability conditions, also in terms of force densities, for the regular truncated tetrahedral tensegrity structures.

8.5.1 Eigenvalues of the Three-dimensional Block

As has been discussed previously, the independent blocks corresponding to the one-dimensional representations are positive semi-definite: $\tilde{\mathbf{E}}^{\mathrm{A}}$ is always zero, and $\tilde{\mathbf{E}}^{\mathrm{E}_1} = \tilde{\mathbf{E}}^{\mathrm{E}_2} = 3q_{\mathrm{h}}$ are positive because the force density q_{h} of Type-2 cables must be positive. Moreover, there exists (at least) one zero eigenvalue ($\lambda_1 = 0$) in the three-dimensional block $\tilde{\mathbf{E}}^{\mathrm{T}}$, which is to satisfy the non-degeneracy condition for a tensegrity structure in three-dimensional space.

To guarantee super-stability of the regular truncated tetrahedral structures, the remaining two non-zero eigenvalues λ_2 and λ_3 of the three-dimensional block $\tilde{\mathbf{E}}^{\mathrm{T}}$ in Eq. (8.27) have to be positive.

First, we show that λ_2 and λ_3 are real, for which we need to prove that

$$\begin{aligned}\eta &= -(q_{\mathrm{h}}q_{\mathrm{v}} + q_{\mathrm{h}}q_{\mathrm{s}} + q_{\mathrm{v}}q_{\mathrm{s}}) \\ &= -2q_{\mathrm{h}}\psi - (\psi^2 - \phi^2) \\ &= \frac{2(\psi^2+\phi^2) \mp \sqrt{\psi^4 + 14\psi^2\phi^2 + \phi^4}}{3} \\ &> 0,\end{aligned} \tag{8.31}$$

where the two non-trivial solutions for q_{h} in Eq. (8.29) have been incorporated. From the relations

$$\begin{aligned}[2(\psi^2+\phi^2)]^2 - (\psi^4 + 14\psi^2\phi^2 + \phi^4) &= 3(\psi^2 - \phi^2)^2 \\ &> 0,\end{aligned} \tag{8.32}$$

as well as

$$\psi^2 + \phi^2 > 0, \tag{8.33}$$

we have

$$2(\psi^2+\phi^2) \mp \sqrt{\psi^4 + 14\psi^2\phi^2 + \phi^4} > 0, \tag{8.34}$$

which leads to $\eta > 0$ and coincides with the fact that the force density matrix is symmetric and thus its eigenvalues are real.

To have positive values for the eigenvalues λ_2 and λ_3 in Eq. (8.27), the following relation has to be satisfied

$$\xi_1 \pm 2\sqrt{\eta} > 0 \quad \text{or} \quad \xi_2 \pm 2\sqrt{\eta} > 0, \tag{8.35}$$

where

$$\begin{aligned}\xi_{1,2} &= 3q_{\mathrm{h}_{1,2}} + 2q_{\mathrm{v}} + 2q_{\mathrm{s}} \\ &= \frac{3\psi^2 + \phi^2 \pm \sqrt{\psi^4 + \phi^4 + 14\psi^2\phi^2}}{2\psi}.\end{aligned} \tag{8.36}$$

In the following subsections, we will prove that only the first solution q_{h_1} can guarantee a super-stable structure, while the second solution q_{h_2} never leads to a super-stable structure because there always exists a negative eigenvalue.

8.5.2 Super-stability Condition for the First Solution q_{h_1}

For the first solution q_{h_1}, we have concluded that

$$\psi = \frac{1}{2}(q_{\mathrm{v}} + q_{\mathrm{s}}) > 0, \tag{8.37}$$

as discussed in Sect. 3.5, and furthermore, the inequality

$$\phi = \frac{1}{2}(q_{\mathrm{v}} - q_{\mathrm{s}}) > 0 \tag{8.38}$$

is always satisfied because $q_{\mathrm{v}} > 0$ for Type 1 cables and $q_{\mathrm{s}} < 0$ for struts. Therefore, the following inequality obviously holds for ξ_1 corresponding to q_{h_1}

$$\begin{aligned}\xi_1 &= \frac{3\psi^2 + \phi^2 + \sqrt{\psi^4 + \phi^4 + 14\psi^2\phi^2}}{2\psi} \\ &> 0.\end{aligned} \tag{8.39}$$

Furthermore, we have

$$\begin{aligned}&\xi_1^2 - (2\sqrt{\eta})^2 \\ &= \frac{1}{12\psi^2}\Big(-2\psi^4 + 6\phi^4 + 28\psi^2\phi^2 \\ &\qquad + 34\psi^2\sqrt{\psi^4 + \phi^4 + 14\psi^2\phi^2} + 6\phi^2\sqrt{\psi^4 + \phi^4 + 14\psi^2\phi^2}\Big)\end{aligned}$$

$$
\begin{aligned}
&> \frac{1}{12\psi^2}\left(32\psi^4 + 6\phi^4 + 28\psi^2\phi^2 + 6\phi^2\sqrt{\psi^4 + \phi^4 + 14\psi^2\phi^2}\right) \\
&> 0,
\end{aligned} \tag{8.40}
$$

for which

$$
\sqrt{\psi^4 + \phi^4 + 14\psi^2\phi^2} > \psi^2 \tag{8.41}
$$

has been used. Therefore, we have proved that

$$
\begin{aligned}
\lambda_{2,3} &= \xi_1 \pm 2\sqrt{\eta} \\
&> 0.
\end{aligned} \tag{8.42}
$$

From the discussions in this subsection, we have the following conclusion for super-stability of regular truncated tetrahedral tensegrity structures.

Super-stability condition for regular truncated tetrahedral tensegrity structures:

The structure is super-stable with the solution q_{h_1} of the Type-2 cables in Eq. (8.29), when

$$
q_v > -q_s > 0. \tag{8.43}
$$

When the solution q_{h_1} of the Type-2 cables in Eq. (8.29) is adopted for self-equilibrium of the structures with Eq. (8.43) being satisfied, the (symmetry-adapted) force density matrix is positive semi-definite because

- the one-dimensional block $\tilde{\mathbf{E}}^{\mathrm{A}}$ is zero from Eq. (8.19);
- the one-dimensional blocks $\tilde{\mathbf{E}}^{\mathrm{E}_1}$ and $\tilde{\mathbf{E}}^{\mathrm{E}_2}$ are positive with positive q_h from Eq. (8.20);
- the three-dimensional block $\tilde{\mathbf{E}}^{\mathrm{T}}$ is positive semi-definite, with one zero eigenvalue for self-equilibrium while the remaining two eigenvalues are positive from Eq. (8.42).

8.5.3 Super-stability Condition for the Second Solution q_{h_2}

For the second solution q_{h_2} of the Type-2 cables in Eq. (8.29), the relation that $\psi < 0$ is satisfied as indicated in Eq. (8.29). Thus, the following inequality holds

$$
\begin{aligned}
\xi_2 &= \frac{3\psi^2 + \phi^2 - \sqrt{\psi^4 + \phi^4 + 14\psi^2\phi^2}}{2\psi} \\
&< 0,
\end{aligned} \tag{8.44}
$$

because

$$\psi < 0 \quad \text{and} \quad 3\psi^2 + \phi^2 - \sqrt{\psi^4 + \phi^4 + 14\psi^2\phi^2} > 0. \tag{8.45}$$

The latter inequality in Eq. (8.45) holds because

$$3\psi^2 + \phi^2 > 0, \tag{8.46}$$

and

$$\begin{aligned}(3\psi^2 + \phi^2)^2 - (\psi^4 + \phi^4 + 14\psi^2\phi^2) &= 8\psi^2(\psi^2 - \phi^2) \\ &= 64\psi^2 q_v^2 q_s^2 \\ &> 0.\end{aligned} \tag{8.47}$$

Therefore, there always exists one negative eigenvalue for the second solution

$$\lambda_2 = \xi_2 - 2\sqrt{\eta} < 0, \tag{8.48}$$

such that *the regular truncated tetrahedral tenesgrity structure can never be super-stable for the second solution* q_{h_2} *of the force density of Type-2 cables in Eq.* (8.29).

To illustrate the above-mentioned discussions on super-stability, the two non-zero eigenvalues λ_2 and λ_3 of the three-dimensional block $\tilde{\mathbf{E}}^T$ are plotted in Fig. 8.4 corresponding to the two solutions q_{h_1} and q_{h_2} of the Type-2 cables with respect to the force density ratio q_s/q_v. Because the cables have positive force densities and struts negative, only the negative region of q_s/q_v is possible; moreover, a structure

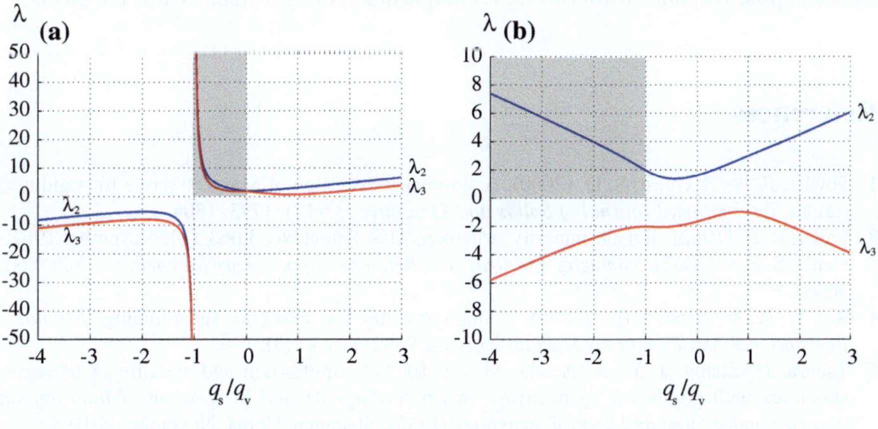

Fig. 8.4 Eigenvalues of $\tilde{\mathbf{E}}^T$ corresponding to the solutions q_{h_1} and q_{h_2} in Eq. (8.29). The zero eigenvalues have been omitted for clarity in the figures. **a** Solution q_{h_1}, **b** solution q_{h_2}

can be super-stable only if the eigenvalues λ_2 and λ_3 are both positive, as indicated by the shaded regions in the figure.

It can be observed from Fig. 8.4a that the two eigenvalues fall into the shaded region, and hence they are always positive for the first solution q_{h_1} when $q_v > -q_s > 0$. As a result, the structure is super-stable in this region. On the other hand, there is always one negative eigenvalue outside of the shaded region in Fig. 8.4b for the second solution q_{h_2}; thus, the structure can never be super-stable for this solution.

8.6 Remarks

In this chapter, we analytically presented the conditions of self-equilibrium as well as super-stability for the regular truncated tetrahedral structures, where the nodes have one-to-one correspondence to the symmetry operations of the tetrahedral group.

It has been demonstrated that the analytical formulation for the structures with dihedral symmetry in Appendix D.5 can also be directly applied to the regular truncated tetrahedral structures: the force density matrix is analytically decomposed (block-diagonalized) to the symmetry-adapted form with three one-dimensional blocks and three copies of the three-dimensional block.

The conditions for self-equilibrium were derived by enforcing the three-dimensional block of the symmetry-adapted force density matrix to be singular. The conditions for super-stability were derived by ensuring positive semi-definiteness of its three-dimensional block.

Unlike the prismatic and star-shaped structures with dihedral symmetry, the super-stability of which is related to the connectivity patterns, the structures with tetrahedral symmetry are super-stable if and only if the force density of Type-2 cables is positive, and furthermore, its magnitude is larger than that of the struts.

References

1. Fowler, P. W., & Guest, S. D. (2000). A symmetry extension of Maxwell's rule for rigidity of frames. *International Journal of Solids and Structures*, *37*(12), 1793–1804.
2. Fuller, R. B. (1962). Tensile-integrity structures. U.S. Patent No. 3,063,521, November 1962.
3. Kettle, S. F. A. (2007). *Symmetry and structure: Readable group theory for chemists*. New York: Wiley.
4. Raj, R. P., & Guest, S. D. (2006). Using symmetry for tensegrity form-finding. *Journal of International Association for Shell and Spatial Structures*, *47*(3), 1–8.
5. Tsuura, F., Zhang, J. Y., & Ohsaki, M. (2010). Self-equilibrium and stability of tensegrity structures with polyhedral symmetries. In *Proceedings Annual Symposium of International Association for Shell and Spatial Structures (IASS)*. Shanghai, China, November 2010.

Appendix A
Linear Algebra

Abstract This chapter summarizes some basic knowledge on linear algebra, which is necessary for the study in this book.

A.1 Introduction

Linear algebra is a basic and important branch of mathematics concerning vector spaces as well as linear mappings between such spaces. Its initial development was motivated by a system of linear equations containing several unknowns, which are naturally expressed by using matrices and vectors.

A system of linear equations is a collection of m equations with respect to the n variable quantities x_1, x_2, ..., x_n in a linear form as follows:

$$\begin{aligned} a_{11}x_1 + a_{12}x_2 + a_{13}x_3 + \cdots + a_{1n}x_n &= b_1, \\ a_{21}x_1 + a_{22}x_2 + a_{23}x_3 + \cdots + a_{2n}x_n &= b_2, \\ &\vdots \\ a_{m1}x_1 + a_{m2}x_2 + a_{m3}x_3 + \cdots + a_{mn}x_n &= b_m, \end{aligned} \tag{A.1}$$

where a_{ij} and b_j $(i = 1, 2, \ldots, m;\ j = 1, 2, \ldots, n)$ are constant coefficients in the equations. Moreover, we restrict ourselves in this book that a_{ij}, b_j, and x_j are real values, although in general they can be complex numbers.

Suppose that the n real values $\hat{x}_1, \hat{x}_2, \ldots, \hat{x}_n$ are the *solutions* to the system of linear equations in Eq. (A.1). Therefore, all equations in Eq. (A.1) are true simultaneously, if we substitute $\hat{x}_j$ to x_j $(j = 1, 2, \ldots, n)$. The *set of solutions* of a system of linear equations is the set which contains every solution to the system. The number of solutions in the set is one or infinite, otherwise, there is no (exact) solution, depending on number of variables as well as that of the linear independent equations. Linear independence of equations indicates that any of the equations cannot be written as a linear combination of the other equations. More discussions on solutions of a system of linear equations will be given in Sect. A.4.

J.Y. Zhang and M. Ohsaki, *Tensegrity Structures*, Mathematics for Industry 6,
DOI 10.1007/978-4-431-54813-3

Example A.1 Linearly dependent and independent equations.

The following linear equations are linearly dependent

$$\begin{aligned} f_1(x_1, x_2, x_3) &: \ x_1 + 2x_2 + 3x_3 = 2, \\ f_2(x_1, x_2, x_3) &: 2x_1 + \ x_2 \qquad\ \ = 1, \\ f_3(x_1, x_2, x_3) &: \ x_1 + \ x_2 + \ x_3 = 1, \end{aligned} \tag{A.2}$$

because the third equation can be written in a *linear* combination of the first two equations:

$$f_3 = \frac{1}{3}(f_1 + f_2). \tag{A.3}$$

Therefore, there are in total two independent equations in this system, and the set of solution for Eq. (A.2) is

$$\begin{aligned} x_1 &= x_3, \\ x_2 &= 1 + 2x_3. \end{aligned} \tag{A.4}$$

There are in fact infinite number of solutions for x_1, x_2, x_3 as defined in Eq. (A.4), because the system will have a unique solution only when any one of them is assigned.

By contrast, the following three linear equations are linearly independent to each other

$$\begin{aligned} f_1(x_1, x_2, x_3) &: \ x_1 + 2x_2 + 3x_3 = 2, \\ \bar{f}_2(x_1, x_2, x_3) &: 2x_1 + \ x_2 + \ x_3 = 1, \\ f_3(x_1, x_2, x_3) &: \ x_1 + \ x_2 + \ x_3 = 1, \end{aligned} \tag{A.5}$$

because they cannot be written in a linear form. Thus, the number of independent equations is equal to that of the variables, and the only solution is

$$\begin{aligned} x_1 &= 0, \\ x_2 &= 1, \\ x_3 &= 0. \end{aligned} \tag{A.6}$$

Note that this is in fact a special case for the solutions in Eq. (A.4).

However, there exists no exact solution for the following system of linear equations

$$\begin{aligned} f_1(x_1, x_2, x_3) &: \quad x_1 + 2x_2 + 3x_3 = 2, \\ \bar{f}_2(x_1, x_2, x_3) &: 2x_1 + \quad x_2 + \quad x_3 = 1, \\ f_3(x_1, x_2, x_3) &: \quad x_1 + \quad x_2 + \quad x_3 = 1, \\ f_4(x_1, x_2, x_3) &: 3x_1 + 2x_2 + \quad x_3 = 1, \end{aligned} \tag{A.7}$$

because these four equations are linearly independent, which is larger than the number of unknown parameters x_1, x_2, x_3.

A.2 Vector and Matrix

A *vector* is an array of (real) values, while a *matrix* is a rectangular array of values arranged in its rows and columns. Vector is the special form of matrix: a matrix with only one column is called column vector, and a matrix with only one row is called row vector. In this book, a vector always refers to a column vector if not specified in order to avoid any confusion in notations. Moreover, we use upper-case characters for matrices and lower-case characters for vectors in general.

A matrix $\mathbf{A}$ with m rows and n columns is denoted by $\mathbf{A} \in \mathbb{R}^{m \times n}$; the individual item in the ith row and the jth column of $\mathbf{A}$ is called the (i, j)th entry denoted by a_{ij} or $A_{(i,j)}$.

The system of linear equations in Eq. (A.1) can be summarized in a matrix form as

$$\mathbf{Ax} = \mathbf{b}, \tag{A.8}$$

where

$$\mathbf{A} = \begin{pmatrix} a_{11} & a_{12} & \dots & a_{1j} & \dots & a_{1n} \\ a_{21} & a_{22} & \dots & a_{2j} & \dots & a_{2n} \\ \vdots & \vdots & \vdots & \vdots & \vdots & \vdots \\ a_{i1} & a_{i2} & \dots & a_{ij} & \dots & a_{in} \\ \vdots & \vdots & \vdots & \vdots & \vdots & \vdots \\ a_{m1} & a_{m2} & \dots & a_{mj} & \dots & a_{mn} \end{pmatrix},$$

$$\mathbf{x} = \begin{pmatrix} x_1 \\ x_2 \\ \vdots \\ x_n \end{pmatrix}, \quad \mathbf{b} = \begin{pmatrix} b_1 \\ b_2 \\ \vdots \\ b_m \end{pmatrix}. \tag{A.9}$$

The law of multiplication of a matrix and a vector will be presented later in Eq. (A.25).

Example A.2 Matrix form of the systems of linear equations in Example A.1.

The system of linear equations in Eq. (A.2) can be summarized in a matrix form as follows by using $\mathbf{A}_1 \in \mathbb{R}^{3\times 3}$, $\mathbf{b}_1 \in \mathbb{R}^3$, and $\mathbf{x} \in \mathbb{R}^3$:

$$\mathbf{A}_1 = \begin{pmatrix} 1 & 2 & 3 \\ 2 & 1 & 0 \\ 1 & 1 & 1 \end{pmatrix}, \quad \mathbf{b}_1 = \begin{pmatrix} 2 \\ 1 \\ 1 \end{pmatrix}, \quad \mathbf{x} = \begin{pmatrix} x_1 \\ x_2 \\ x_3 \end{pmatrix}. \tag{A.10}$$

Similarly, the system of equations in Eq. (A.5) can be summarized as

$$\mathbf{A}_2 = \begin{pmatrix} 1 & 2 & 3 \\ 2 & 1 & 1 \\ 1 & 1 & 1 \end{pmatrix}, \quad \mathbf{b}_2 = \begin{pmatrix} 2 \\ 1 \\ 1 \end{pmatrix}, \tag{A.11}$$

and that in Eq. (A.7) is

$$\mathbf{A}_3 = \begin{pmatrix} 1 & 2 & 3 \\ 2 & 1 & 1 \\ 1 & 1 & 1 \\ 3 & 2 & 1 \end{pmatrix}, \quad \mathbf{b}_3 = \begin{pmatrix} 2 \\ 1 \\ 1 \\ 1 \end{pmatrix}. \tag{A.12}$$

It should be noted that the system of linear equations in Eq. (A.1) and that in Eq. (A.8) are equivalent to each other, although they may look very different.

A.2.1 Addition of Matrices and Multiplication with Scalars

Addition of matrices is possible only when they have the same size; i.e., the same numbers of rows and columns. If the matrix $\mathbf{C} \in \mathbb{R}^{m\times n}$ is the addition of two matrices $\mathbf{A}$ and $\mathbf{B}$ ($\in \mathbb{R}^{m\times n}$); i.e.,

$$\mathbf{C} = \mathbf{A} + \mathbf{B}, \tag{A.13}$$

then the (i, j)th entry c_{ij} of $\mathbf{C}$ is calculated as

$$c_{ij} = a_{ij} + b_{ij}, \quad (i = 1, 2, \ldots, m; \quad j = 1, 2, \ldots, n). \tag{A.14}$$

If the matrix $\mathbf{C} \in \mathbb{R}^{m\times n}$ is the multiplication of the matrix $\mathbf{A} \in \mathbb{R}^{m\times n}$ with a scalar k (arbitrary value); i.e.,

$$\mathbf{C} = k\mathbf{A}, \tag{A.15}$$

then the (i, j)th entry c_{ij} of $\mathbf{C}$ is calculated as

$$c_{ij} = ka_{ij}, \quad (i = 1, 2, \ldots, m; \quad j = 1, 2, \ldots, n). \tag{A.16}$$

From the definitions of addition and multiplication of the matrices and scalars, we have the following laws, where **A**, **B**, and **C** ($\in \mathbb{R}^{m\times n}$) are matrices and k and p are scalars:

- Associativity of addition

$$\mathbf{A} + (\mathbf{B} + \mathbf{C}) = (\mathbf{A} + \mathbf{B}) + \mathbf{C}. \tag{A.17}$$

- Commutativity of addition

$$\mathbf{A} + \mathbf{B} = \mathbf{B} + \mathbf{A}. \tag{A.18}$$

- Identity element of addition

$$\mathbf{A} + \mathbf{O} = \mathbf{A}, \tag{A.19}$$

  where $\mathbf{O} \in \mathbb{R}^{m\times n}$ is the zero matrix, with all entries equal to zero.
- Inverse elements of addition

$$\mathbf{A} + (-\mathbf{A}) = \mathbf{O}. \tag{A.20}$$

- Distributivity of scalar multiplication with matrix addition

$$k(\mathbf{A} + \mathbf{B}) = k\mathbf{A} + k\mathbf{B}. \tag{A.21}$$

- Distributivity of scalar multiplication with matrix

$$(k + p)\mathbf{A} = k\mathbf{A} + p\mathbf{A}. \tag{A.22}$$

- Compatibility of scalar multiplication with scalar multiplication

$$k(p\mathbf{A}) = (kp)\mathbf{A}. \tag{A.23}$$

- Identity element of scalar multiplication

$$1\mathbf{A} = \mathbf{A}. \tag{A.24}$$

A.2.2 Multiplication of Matrices

The *matrix-vector multiplication* of a matrix $\mathbf{A} \in \mathbb{R}^{m\times n}$ and a (column) vector $\mathbf{x} \in \mathbb{R}^n$ as in Eq. (A.8) is written as follows using their entries:

$$b_i = \sum_{j=1}^{n} a_{ij} x_j, \quad (i = 1, 2, \ldots, m). \tag{A.25}$$

Matrix multiplication of two matrices $\mathbf{A} \in \mathbb{R}^{m\times p}$ and $\mathbf{B} \in \mathbb{R}^{p\times n}$ is defined as follows by using the matrix $\mathbf{C} \in \mathbb{R}^{m\times n}$:

$$\mathbf{C} = \mathbf{AB}, \tag{A.26}$$

where the (i, j)th entry c_{ij} of $\mathbf{C}$ is obtained by

$$c_{ij} = \sum_{k=1}^{p} a_{ik} b_{kj}, \quad (i = 1, 2, \ldots, m; \quad j = 1, 2, \ldots, n). \tag{A.27}$$

Note that the matrix multiplication is applicable only if the number of columns of $\mathbf{A}$ is equal to the number of rows of $\mathbf{B}$.

For the matrices $\mathbf{A}$, $\mathbf{B}$, and $\mathbf{C}$ of appropriate sizes and a scalar k, we have the following four properties for the matrix multiplication:

$$(\mathbf{AB})\mathbf{C} = \mathbf{A}(\mathbf{BC}), \tag{A.28a}$$
$$\mathbf{A}(\mathbf{B} + \mathbf{C}) = \mathbf{AB} + \mathbf{AC}, \tag{A.28b}$$
$$(\mathbf{B} + \mathbf{C})\mathbf{A} = \mathbf{BA} + \mathbf{CA}, \tag{A.28c}$$
$$k(\mathbf{AB}) = (k\mathbf{A})\mathbf{B} = \mathbf{A}(k\mathbf{B}). \tag{A.28d}$$

Tensor product, denoted by $\otimes$, of two matrices (or vectors) $\mathbf{A} \in \mathbb{R}^{m_1\times n_1}$ and $\mathbf{B} \in \mathbb{R}^{m_2\times n_2}$ is defined as

$$\mathbf{C} = \mathbf{A} \otimes \mathbf{B}, \tag{A.29}$$

where the (i, j)th sub-matrix $\mathbf{C}_{(i,j)} \in \mathbb{R}^{m_2\times n_2}$ of $\mathbf{C} \in \mathbb{R}^{m_1 m_2\times n_1 n_2}$ is

$$\mathbf{C}_{(i,j)} = a_{ij}\mathbf{B}, \quad (i = 1, 2, \ldots, m_1; \ j = 1, 2, \ldots, n_1). \tag{A.30}$$

Tensor product is very useful for simplification of formulations, for example, the formulation of geometry stiffness matrix $\mathbf{K}_\mathrm{G}$ as presented in Sect. 4.2.2 by using the force density matrix $\mathbf{E}$ as components.

A.2.3 Useful Matrices

The *transpose* of a matrix $\mathbf{A}$ is denoted by $\mathbf{A}^\top$. If the matrix $\mathbf{B} \in \mathbb{R}^{m\times n}$ is the transpose of the matrix $\mathbf{A} \in \mathbb{R}^{n\times m}$; i.e., $\mathbf{B} = \mathbf{A}^\top$, then their entries have the following relationship

$$b_{ij} = a_{ji}, \quad (i = 1, 2, \ldots, m; \quad j = 1, 2, \ldots, n). \tag{A.31}$$

If the number of rows and the number of columns of a matrix are the same, then the matrix is said to be a *square matrix*. A square matrix with n rows and n columns is said to be order n and is called an n-square matrix.

Trace of an n-square matrix $\mathbf{A}$ is defined as the sum of the entries lying on its diagonal:

$$\text{trace}(\mathbf{A}) = \sum_{i=1}^{n} a_{ii}. \tag{A.32}$$

If the diagonal entries of an n-square matrix are all equal to one and the other entries are zero, it is called the *identity matrix*, and is usually denoted by $\mathbf{I}$ or $\mathbf{I}_n$ to indicate its size.

A square matrix $\mathbf{A}$ is said to be invertible if there exists a matrix $\mathbf{B}$ satisfying

$$\mathbf{AB} = \mathbf{BA} = \mathbf{I}. \tag{A.33}$$

Such a matrix $\mathbf{B}$ is unique and it is called the *inverse matrix* of $\mathbf{A}$ and denoted by $\mathbf{A}^{-1}$. Moreover, $\mathbf{B}$ is the inverse of $\mathbf{A}$ if and only if $\mathbf{A}$ is the inverse matrix of $\mathbf{B}$. If a square matrix is not invertible, then it is said to be *singular*.

Determinant of a square matrix $\mathbf{A}$ is denoted by $\det(\mathbf{A})$ or $|\mathbf{A}|$. It can be computed from the entries of the matrix by a specific arithmetic expression. Determinant of the coefficient matrix provides important information on solutions of a system of linear equations. The system has a unique solution when the determinant is non-zero, because the coefficient matrix is invertible. On the other hand, there are either no solution or many solutions, when the determinant is zero.

It is important to know that a square matrix is invertible if and only if its determinant is non-zero. In the self-equilibrium analysis of symmetric tensegrity structures, we have used zero determinant of the (reduced) force density matrix to ensure its singularity.

If the transpose $\mathbf{A}^\top$ of a square matrix $\mathbf{A} \in \mathbb{R}^{n\times n}$ is equal to itself:

$$\mathbf{A} = \mathbf{A}^\top, \tag{A.34}$$

then it is a *symmetric matrix*. Many matrices in structural engineering are symmetric, for example the force density matrix $\mathbf{E}$ in Chap. 2 and the stiffness matrices $\mathbf{K}_\text{E}$, $\mathbf{K}_\text{G}$, and $\mathbf{K}$ in Chap. 4. The *eigenvalues of a symmetric matrix must be real values*, which is of extreme importance to us in stability investigation of tensegrity structures.

The maximum number of linearly independent rows in a matrix $\mathbf{A}$ is called the *row rank* of $\mathbf{A}$, and the maximum number of linearly independent columns in $\mathbf{A}$ is called the column rank of $\mathbf{A}$. A result of fundamental importance is that the column rank and the row rank are always the same. The rank of $\mathbf{A}$ is denoted by $\text{rank}(\mathbf{A})$ or r^A. An n-square matrix $\mathbf{A} \in \mathbb{R}^{n\times n}$ is said to be *full-rank*, when its rank is n; i.e., $\text{rank}(\mathbf{A}) = n$. A square matrix is invertible only if it is full-rank.

A.2.4 Partial Derivative of Vectors and Matrices

Partial derivative of a vector $\mathbf{a} \in \mathbb{R}^n$ with respect to a variable x is defined as

$$\frac{\partial \mathbf{a}}{\partial x} = \begin{pmatrix} \frac{\partial a_1}{\partial x} \\ \frac{\partial a_2}{\partial x} \\ \vdots \\ \frac{\partial a_i}{\partial x} \\ \vdots \\ \frac{\partial a_n}{\partial x} \end{pmatrix}. \tag{A.35}$$

Partial derivative of a matrix $\mathbf{A} \in \mathbb{R}^{m \times n}$ with respect to a variable x is defined as

$$\frac{\partial \mathbf{A}}{\partial x} = \begin{pmatrix} \frac{\partial a_{11}}{\partial x} & \frac{\partial a_{12}}{\partial x} & \cdots & \frac{\partial a_{1j}}{\partial x} & \cdots & \frac{\partial a_{1n}}{\partial x} \\ \frac{\partial a_{21}}{\partial x} & \frac{\partial a_{22}}{\partial x} & \cdots & \frac{\partial a_{2j}}{\partial x} & \cdots & \frac{\partial a_{2n}}{\partial x} \\ \vdots & \vdots & \vdots & \vdots & \vdots & \vdots \\ \frac{\partial a_{i1}}{\partial x} & \frac{\partial a_{i2}}{\partial x} & \cdots & \frac{\partial a_{ij}}{\partial x} & \cdots & \frac{\partial a_{in}}{\partial x} \\ \vdots & \vdots & \vdots & \vdots & \vdots & \vdots \\ \frac{\partial a_{m1}}{\partial x} & \frac{\partial a_{m2}}{\partial x} & \cdots & \frac{\partial a_{mj}}{\partial x} & \cdots & \frac{\partial a_{mn}}{\partial x} \end{pmatrix}. \tag{A.36}$$

Moreover, partial derivative of transpose $\mathbf{y}^\top$ of a (column) vector $\mathbf{y} \in \mathbb{R}^m$ with respect to a (column) variable vector $\mathbf{x} \in \mathbb{R}^n$ is defined as

$$\frac{\partial \mathbf{y}^\top}{\partial \mathbf{x}} = \begin{pmatrix} \frac{\partial y_1}{\partial x_1} & \frac{\partial y_2}{\partial x_1} & \cdots & \frac{\partial y_j}{\partial x_1} & \cdots & \frac{\partial y_m}{\partial x_1} \\ \frac{\partial y_1}{\partial x_2} & \frac{\partial y_2}{\partial x_2} & \cdots & \frac{\partial y_j}{\partial x_2} & \cdots & \frac{\partial y_m}{\partial x_2} \\ \vdots & \vdots & \vdots & \vdots & \vdots & \vdots \\ \frac{\partial y_1}{\partial x_i} & \frac{\partial y_2}{\partial x_i} & \cdots & \frac{\partial y_j}{\partial x_i} & \cdots & \frac{\partial y_m}{\partial x_i} \\ \vdots & \vdots & \vdots & \vdots & \vdots & \vdots \\ \frac{\partial y_1}{\partial x_n} & \frac{\partial y_m}{\partial x_n} & \cdots & \frac{\partial y_j}{\partial x_n} & \cdots & \frac{\partial y_m}{\partial x_n} \end{pmatrix}. \tag{A.37}$$

A.3 Eigenvalues, Eigenvectors, and Spectral Decomposition

An *eigenvector* $\boldsymbol{\phi}$ of a square matrix $\mathbf{A}$ is a non-zero vector that matrix multiplication of $\mathbf{A}$ and $\boldsymbol{\phi}$ is equal to multiplication of $\boldsymbol{\phi}$ with a specific scalar λ:

$$\mathbf{A}\boldsymbol{\phi} = \lambda\boldsymbol{\phi}, \tag{A.38}$$

where λ is called the *eigenvalue* of $\mathbf{A}$. When $\mathbf{A}$ is a symmetric matrix, λ must be a real value.

Eigenvalues of an n-square matrix $\mathbf{A}$ can be determined by solving the following characteristic equation with respect to λ

$$\det(\mathbf{A} - \lambda\mathbf{I}_n) = 0. \tag{A.39}$$

Because Eq. (A.39) is of degree n for λ, an n-square matrix has n eigenvalues, which are usually arranged in a non-decreasing order as

$$\lambda_1 \leq \lambda_2 \leq \cdots \leq \lambda_n. \tag{A.40}$$

The sum of eigenvalues of $\mathbf{A}$ is equal to its trace trace($\mathbf{A}$); i.e.,

$$\sum_{i=1}^{n} \lambda_i = \text{trace}(\mathbf{A}) = \sum_{i=1}^{n} a_{ii}. \tag{A.41}$$

Moreover, the product of the n eigenvalues λ_i $(i = 1, 2, \ldots, n)$ of $\mathbf{A}$ is equal to its determinant; i.e.,

$$\prod_{i=1}^{n} \lambda_i = \det(\mathbf{A}). \tag{A.42}$$

Using the eigenvalues and eigenvectors of $\mathbf{A}$, it can be decomposed as follows [1]:

$$\mathbf{A} = \boldsymbol{\Phi}\boldsymbol{\Lambda}\boldsymbol{\Phi}^\top, \tag{A.43}$$

where the diagonal entries of the diagonal matrix $\boldsymbol{\Lambda}$ are the eigenvalues λ_i $(i = 1, 2, \ldots, n)$ of $\mathbf{A}$. The ith column $\boldsymbol{\phi}_i$ of $\boldsymbol{\Phi}$ is the eigenvector corresponding to the ith eigenvalue λ_i; i.e.,

$$\mathbf{A}\boldsymbol{\phi}_i = \lambda_i\boldsymbol{\phi}_i, \tag{A.44}$$

and $\boldsymbol{\phi}_i$ is orthor-nomalized as

$$(\boldsymbol{\phi}_i)^\top \boldsymbol{\phi}_j = \delta_{ij}, \tag{A.45}$$

where δ_{ij} is the Kronecker delta:

$$\delta_{ij} = \begin{cases} 0, & \text{if } i \neq j, \\ 1, & \text{if } i = j. \end{cases} \tag{A.46}$$

A.4 Least Square Solution

Consider a system of linear equations described in a matrix form as in Eq. (A.8), where $\mathbf{A} \in \mathbb{R}^{m \times n}$, $\mathbf{b} \in \mathbb{R}^m$, and $\mathbf{x} \in \mathbb{R}^n$. Denote the matrix $\hat{\mathbf{A}} \in \mathbb{R}^{m \times (n+1)}$ as

$$\hat{\mathbf{A}} = (\mathbf{A},\ \mathbf{b})\,. \tag{A.47}$$

Ranks of the matrices $\mathbf{A}$ and $\hat{\mathbf{A}}$ are respectively denoted by r^{A} and $r^{\hat{\mathrm{A}}}$.

In the following discussions on possible solutions of the system, we consider only the case that $\mathbf{A}$ is full-rank; i.e., $r^{\mathrm{A}} = m$ if $m \leq n$, or $r^{\mathrm{A}} = n$ if $m \geq n$. The linearly dependent equations can be eliminated from the system to make the coefficient matrix $\mathbf{A}$ full-rank, if its original version is not. Therefore, considering only full-rank $\mathbf{A}$ does not loss any generality.

Furthermore, we have only two possible cases for r^{A} and $r^{\hat{\mathrm{A}}}$: $r^{\hat{\mathrm{A}}} = r^{\mathrm{A}}$ or $r^{\hat{\mathrm{A}}} = r^{\mathrm{A}} + 1$. Existence of exact solutions for the system of linear equations depends on relation between r^{A} and $r^{\hat{\mathrm{A}}}$.

- $r^{\hat{\mathrm{A}}} = r^{\mathrm{A}}$: There exist an exact solution for the system.
  - If the coefficient matrix $\mathbf{A}$ is a square matrix; i.e., $m = n$, then it is invertible, because we have $r^{\mathrm{A}} = m = n$ from our assumption that $\mathbf{A}$ is full-rank. There exists only one solution $\mathbf{x}$ for the system of linear equations:

$$\mathbf{x} = \mathbf{A}^{-1}\mathbf{b}. \tag{A.48}$$

  - If $m > n$ with $r^{\mathrm{A}} = m$, a similar form to Eq. (A.48) is also applicable to present the only one exact solution, with $m - n$ linearly independent equations moved out from the system.
  - For the case $m < n$ with $r^{\mathrm{A}} = n$, where the number of equations is less than the number of variables, (exact) solutions can be written in a general form as follows:

$$\mathbf{x} = \hat{\mathbf{x}} + \sum_{i=1}^{n-m} \alpha_i \boldsymbol{\phi}_i, \tag{A.49}$$

  where $\hat{\mathbf{x}}$ is a special solution satisfying $\mathbf{A}\hat{\mathbf{x}} = \mathbf{b}$, and $\boldsymbol{\phi}_i$ is a vector lying in the null-space of $\mathbf{A}$ with an arbitrary coefficient α_i; i.e.,

$$\mathbf{A}\boldsymbol{\phi}_i = \mathbf{0}. \tag{A.50}$$

- $r^{\hat{A}} = r^A + 1$: There exists no exact solution for the system, and approximate (least square) solution is sometimes found to be useful. The approximate solution of the system, denoted by $\bar{\mathbf{x}}$, is written in a general form as

$$\bar{\mathbf{x}} = \mathbf{A}^{-}\mathbf{b}, \tag{A.51}$$

where $\mathbf{A}^{-}$ ic called Moore-Penrose generalized inverse matrix of $\mathbf{A}$. More details on formulation of $\mathbf{A}^{-}$ will be given later in this section. It should be noted that Eq. (A.51) is the least square solution for the case $m > n$ with $r^A = n$ and $r^{\hat{A}} = n + 1$, while it is the approximate solution with the least norm for the case $m < n$ with $r^A = m$ and $r^{\hat{A}} = m + 1$.

Example A.3 Solutions of the systems of linear equations in Example A.1.

The coefficient matrices $\mathbf{A}_1 \in \mathbb{R}^{3\times 3}$ and $\mathbf{A}_2 \in \mathbb{R}^{3\times 3}$ are square matrices. However, rank of $\mathbf{A}_1$ is 2 such that it is not invertible. On the other hand, $\mathbf{A}_2$ is a full-rank matrix, and its inverse matrix $\mathbf{A}^{-1}$ is

$$\mathbf{A}_2^{-1} = \begin{pmatrix} 0 & 1 & -1 \\ -1 & -2 & 5 \\ 1 & 1 & -3 \end{pmatrix}. \tag{A.52}$$

From Eq. (A.48), the (exact) solution of the system of equations is

$$\begin{aligned} \mathbf{x}_2 = \mathbf{A}_2^{-1}\mathbf{b}_2 &= \begin{pmatrix} 0 & 1 & -1 \\ -1 & -2 & 5 \\ 1 & 1 & -3 \end{pmatrix} \begin{pmatrix} 2 \\ 1 \\ 1 \end{pmatrix} \\ &= \begin{pmatrix} 0 \\ 1 \\ 0 \end{pmatrix}, \end{aligned} \tag{A.53}$$

which coincides with the solution in Eq. (A.6) derived in another way.

Moreover, $\mathbf{A}_3 \in \mathbb{R}^{4\times 3}$ is not a square matrix, and therefore, it is not invertible. This system of linear equations may have approximate (least square) solutions by using the Moore-Penrose generalized inverse matrix $\mathbf{A}^{-}$.

For the case $m > n$ with $r^A = n$ and $r^{\hat{A}} = n + 1$, there exists no exact solution for the system, and error $\boldsymbol{\varepsilon} \in \mathbb{R}^m$ of the approximate solution $\bar{\mathbf{x}}$ is defined as

$$\boldsymbol{\varepsilon} = \mathbf{b} - \mathbf{A}\bar{\mathbf{x}}. \tag{A.54}$$

The square of this error can be written as

$$|\boldsymbol{\varepsilon}|^2 = \boldsymbol{\varepsilon}^\top \boldsymbol{\varepsilon} = (\mathbf{b} - \mathbf{A}\bar{\mathbf{x}})^\top (\mathbf{b} - \mathbf{A}\bar{\mathbf{x}}). \tag{A.55}$$

The stationary condition of the square of error $|\boldsymbol{\varepsilon}|^2$ with respect to $\bar{\mathbf{x}}$ leads to

$$\mathbf{A}^\top \mathbf{A}\bar{\mathbf{x}} - \mathbf{A}^\top \mathbf{b} = \mathbf{0}. \tag{A.56}$$

Since we are discussing the case that $m > n$ and $r^{\mathrm{A}} = n$, we know that rank of $\mathbf{A}^\top \mathbf{A}$ is n, and hence, it is invertible. Therefore, the least square solution of Eq. (A.8) can be written as

$$\bar{\mathbf{x}} = (\mathbf{A}^\top \mathbf{A})^{-1} \mathbf{A}^\top \mathbf{b}. \tag{A.57}$$

Using the following definition of Moore-Penrose generalized inverse matrix $\mathbf{A}^-$ for the case $m > n$

$$\mathbf{A}^- = (\mathbf{A}^\top \mathbf{A})^{-1} \mathbf{A}, \tag{A.58}$$

Equation (A.57) is rewritten in a simpler form as in Eq. (A.51). Note here that $\mathbf{A}^-$ satisfies the following laws of for a Moore-Penrose generalized inverse matrix:

$$\begin{aligned} (\mathbf{A}\mathbf{A}^-)^\top &= \mathbf{A}\mathbf{A}^-, \\ (\mathbf{A}^-\mathbf{A})^\top &= \mathbf{A}^-\mathbf{A}, \\ \mathbf{A}\mathbf{A}^-\mathbf{A} &= \mathbf{A}, \\ \mathbf{A}^-\mathbf{A}\mathbf{A}^- &= \mathbf{A}^-. \end{aligned} \tag{A.59}$$

Example A.4 Solutions of the third system of linear equations in Example A.1.

Moore-Penrose generalized inverse matrix $\mathbf{A}_3^-$ of the coefficient matrix $\mathbf{A}_3$ of the third system of linear equations is

$$\mathbf{A}_3^- = \begin{pmatrix} -0.2222 & 1.0000 & -0.1111 & -0.2222 \\ 0.1667 & -2.0000 & 0.3333 & 1.1667 \\ 0.2778 & 1.0000 & -0.1111 & -0.7222 \end{pmatrix}, \tag{A.60}$$

and its least square solutions are

$$\bar{\mathbf{x}}_3 = \mathbf{A}_3 \mathbf{b} = \begin{pmatrix} 0.2222 \\ -0.1667 \\ 0.7222 \end{pmatrix}. \tag{A.61}$$

Errors $\boldsymbol{\varepsilon}$ of the solutions is

$$\boldsymbol{\varepsilon} = \mathbf{b} - \mathbf{A}_3\bar{\mathbf{x}}_3 = \begin{pmatrix} -0.0556 \\ 0.0000 \\ 0.2222 \\ -0.0556 \end{pmatrix}. \tag{A.62}$$

A.5 Reduced Row-Echelon Form (RREF)

Any (possibly not square) matrix $\mathbf{A}$ with finite sizes can be reduced by a finite sequence of linear elementary row operations $\mathbf{B}_1, \mathbf{B}_2, \ldots, \mathbf{B}_k$ to a *Reduced Row-Echelon Form* (RREF):

$$\mathbf{U} := \mathbf{B}_k \cdots \mathbf{B}_2\mathbf{B}_1\mathbf{A}, \tag{A.63}$$

where each of the row operations $\mathbf{B}_i$ $(i = 1, 2, \ldots, k)$ is invertible. Moreover, the RREF of $\mathbf{A}$ is characterized by the following three properties:

a. The first non-zero entry in any non-zero row is 1.
b. The leading 1 of each non-zero row appears in a column of which all the other entries are 0.
c. Each such leading 1 comes in a column after every preceding row's leading zeros.

Matrix $\mathbf{A}$ determines its RREF uniquely, even though $\mathbf{A}$ does not determine uniquely the sequences of elementary row operations that reduce $\mathbf{A}$ to $\mathbf{U}$.

Moreover, rank of $\mathbf{A}$ will not be changed by multiplication with invertible matrices corresponding to elementary row (or column) operations, hence, we have rank $(\mathbf{A}) = \text{rank}(\mathbf{U})$.

Example A.5 Reduced row-echelon form of an example matrix.

The RREF of a matrix $\mathbf{A}$ given as

$$\mathbf{A} = \begin{pmatrix} 16 & 2 & 3 & 13 \\ 5 & 11 & 10 & 8 \\ 9 & 7 & 6 & 12 \\ 4 & 14 & 15 & 1 \end{pmatrix} \tag{A.64}$$

is

$$\mathbf{U} = \left(\begin{array}{ccc|c} 1 & 0 & 0 & 1 \\ 0 & 1 & 0 & 3 \\ 0 & 0 & 1 & -3 \\ \hline 0 & 0 & 0 & 0 \end{array}\right). \tag{A.65}$$

It is obvious that rank of **U** is 3, because the fourth row is a zero (row) vector after elementary row operations of the matrix **A**. Since we have applied only row operations to **A** to obtain **U**, we may learn from **U** given in Eq. (A.65) that the first three columns of **A** are mutually independent, because the first three columns of **U** are obviously independent to each other.

Reference

1. Lay, D. C. (2011). *Linear algebra and its applications* (4th ed.). London: Pearson.

Appendix B
Affine Motions and Rigidity Condition

Abstract In Chap. 2, we proved that the force density matrix of a three-dimensional free-standing structure should have at least four zero eigenvalues, so that the structure does not degenerate to a space of lower dimensions. Accordingly, the geometrical stiffness matrix of a non-degenerate structure has at least twelve zero eigenvalues. Since tensegrity structures are free-standing, six of these zero eigenvalues correspond to the rigid-body motions. In this chapter, we will show that remaining six eigenvalues correspond to the non-trivial affine motions. These motions are used to present the sufficient conditions for super-stability of a tensegrity structure in Chap. 4. Furthermore, we will prove that investigation of stability of a tensegrity structure using rank of its geometry matrix as discussed in Chap. 4 is equivalent to the condition in the field of structural rigidity in mathematics.

B.1 Affine Motions

An *affine motion* is a motion that preserves colinearity and ratios of distances; i.e., all points lying on one line are transformed to points on another line, and ratios of the distances between any pairs of the points on the line are preserved [2]. However, an affine motion does not necessarily preserve angles or lengths. Hence, a triangle can be transformed into any other triangle by an affine motion.

There are $d^2 + d$ independent affine motions in d-dimensional space. In general, an affine motion can be a linear combination of *rotation*, *translation*, *dilation*, and *shear*. The rotation and translation of the structure are rigid-body motions of a structure, because they always preserve the member lengths (distances between the nodes) as well as angles between the members. The other two types of motions—dilation and shear—are called *non-trivial affine motions*, which do not preserve angles and lengths. Hence, half of the affine motions of a free-standing structure are the rigid-body motions, and the other half are the non-trivial affine motions.

J.Y. Zhang and M. Ohsaki, *Tensegrity Structures*, Mathematics for Industry 6,
DOI 10.1007/978-4-431-54813-3

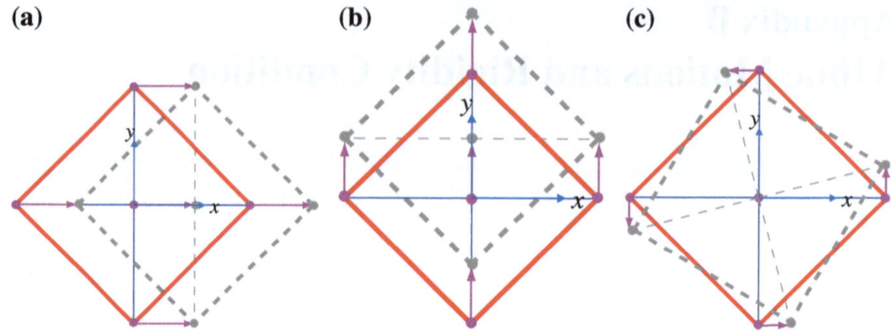

Fig. B.1 Rigid-body motions of a two-dimensional tensegrity structure. **a** Translation in x-direction, **b** translation in y-direction, **c** rotation about z-axis

It is shown later in this section that the rigid-body motions lie in the null-spaces of both the linear and geometrical stiffness matrices, and the non-trivial affine motions lie in the null-space of the geometrical stiffness matrix if the structure is non-degenerate.

Example B.1 Affine motions of the two-dimensional structure as shown in Figs. B.1 and B.2.

For the two-dimensional tensegrity structure as shown in Figs. B.1 and B.2, there exist six affine motions: three rigid-body motions as shown in Fig. B.1 and three non-trivial affine motions as shown in Fig. B.2.

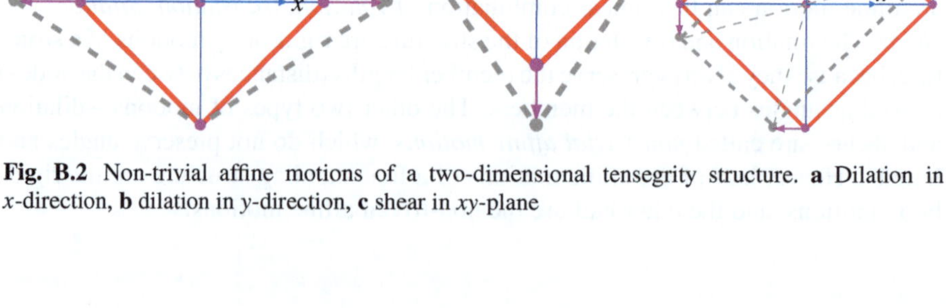

Fig. B.2 Non-trivial affine motions of a two-dimensional tensegrity structure. **a** Dilation in x-direction, **b** dilation in y-direction, **c** shear in xy-plane

The solid and dashed lines in the figures denote the members before and after application of affine motions, respectively.

Subjected to rigid-body motions, the structure is moved to a new position without any deformation; i.e., the distance of any pairs of points in the structure keeps unchanged.

B.1.1 Rigid-Body Motions

A d-dimensional free-standing structure has $(d^2 + d)/2$ independent rigid-body motions. In a Cartesian coordinate system, these rigid-body motions include (a) translation in each direction, and (b) rotation about each axis. With respect to the rigid-body motions, the quadratic form Q_{K} of the tangent stiffness matrix $\mathbf{K}$ is always equal to zero irrespective of the geometrical and mechanical properties of the structure. This is because that the member lengths are not changed by these motions (displacements), and therefore, the strain energy stored in the structure does not change.

B.1.1.1 Translation for Linear and Geometrical Stiffness

Suppose that the structure consists of n (free) nodes and m members. The translation vectors $\mathbf{d}_{\mathrm{t}}^{x}$, $\mathbf{d}_{\mathrm{t}}^{y}$, and $\mathbf{d}_{\mathrm{t}}^{z}$ in x-, y-, and z-directions are defined as

$$\mathbf{d}_{\mathrm{t}}^{x} = \begin{pmatrix} \mathbf{i} \\ \mathbf{0} \\ \mathbf{0} \end{pmatrix}, \ \mathbf{d}_{\mathrm{t}}^{y} = \begin{pmatrix} \mathbf{0} \\ \mathbf{i} \\ \mathbf{0} \end{pmatrix}, \ \mathbf{d}_{\mathrm{t}}^{z} = \begin{pmatrix} \mathbf{0} \\ \mathbf{0} \\ \mathbf{i} \end{pmatrix}, \tag{B.1}$$

where all the entries of the vector $\mathbf{i} \in \mathbb{R}^n$ are 1. It is easy to show that $\mathbf{d}_{\mathrm{t}}^{x}$, $\mathbf{d}_{\mathrm{t}}^{y}$, and $\mathbf{d}_{\mathrm{t}}^{z}$ are mutually orthogonal, and therefore, independent to each other:

$$(\mathbf{d}_{\mathrm{t}}^{x})^{\top}\mathbf{d}_{\mathrm{t}}^{y} = (\mathbf{d}_{\mathrm{t}}^{x})^{\top}\mathbf{d}_{\mathrm{t}}^{z} = (\mathbf{d}_{\mathrm{t}}^{y})^{\top}\mathbf{d}_{\mathrm{t}}^{z} = 0. \tag{B.2}$$

From the definition of connectivity matrix $\mathbf{C}$ of a structure in Eq. (2.7), we have

$$\mathbf{C}\mathbf{i} = \mathbf{0}; \tag{B.3}$$

and furthermore, from the definition of x-component $\mathbf{D}^x$ of the equilibrium matrix in Eq. (2.30), we have

$$\begin{aligned}(\mathbf{D}^x)^\top \mathbf{i} &= \left(\mathbf{C}^\top \mathbf{U}\mathbf{L}^{-1}\right)^\top \mathbf{i} = \mathbf{L}^{-1}\mathbf{U}\mathbf{C}\mathbf{i} \\ &= \mathbf{0}. \end{aligned} \tag{B.4}$$

Moreover, the above equation is also true for the y- and z-components, and therefore, we have the following equation from the formulation of the equilibrium matrix $\mathbf{D}$ in Eq. (2.34)

$$\mathbf{D}^\top \mathbf{d}_t^i = \mathbf{0}, \tag{B.5}$$

where i stands for x, y, and z. Furthermore, from the formulation of the linear stiffness matrix $\mathbf{K}_E$ by using $\mathbf{D}$ in Eq. (4.55), we have

$$\begin{aligned}\mathbf{K}_E \mathbf{d}_t^i &= \mathbf{D}\bar{\mathbf{K}}\mathbf{D}^\top \mathbf{d}_t^i = \mathbf{D}\bar{\mathbf{K}}\mathbf{0} \\ &= \mathbf{0}, \end{aligned} \tag{B.6}$$

where $\bar{\mathbf{K}}$ is the member stiffness matrix.

From the definition of the force density matrix $\mathbf{E}$ in Eq. (2.104), we have

$$\begin{aligned}\mathbf{E}\mathbf{i} &= \mathbf{C}^\top \mathbf{Q}\mathbf{C}\mathbf{i} = \mathbf{C}^\top \mathbf{Q}\mathbf{0} \\ &= \mathbf{0}, \end{aligned} \tag{B.7}$$

where $\mathbf{Q}$ is the diagonal version of the force density vector. Because d copies of $\mathbf{E}$ are the independent components lying on diagonal of the geometrical stiffness matrix $\mathbf{K}_G$ as indicated in Eq. (4.55), and moreover, the vector $\mathbf{i}$ is the only non-zero component in $\mathbf{d}_t^i$, we have

$$\mathbf{K}_G \mathbf{d}_t^i = \mathbf{0}. \tag{B.8}$$

Suppose that a translation $\mathbf{d}_t$ is a weighted sum of $\mathbf{d}_t^i$ through the arbitrary coefficients α_i:

$$\mathbf{d}_t = \sum_{i \in \{x,y,z\}} \alpha_i \mathbf{d}_t^i. \tag{B.9}$$

From Eqs. (B.6) and (B.8), the quadratic form Q_K of the tangent stiffness matrix $\mathbf{K}$ with respect to the translation motion $\mathbf{d}_t$ is

$$\begin{aligned} Q_K &= \mathbf{d}_t^\top \mathbf{K}\mathbf{d}_t = \mathbf{d}_t^\top (\mathbf{K}_E + \mathbf{K}_G)\mathbf{d}_t \\ &= \mathbf{d}_t^\top \left(\sum_{i \in \{x,y,z\}} (\alpha_i \mathbf{K}_E \mathbf{d}_t^i + \alpha_i \mathbf{K}_G \mathbf{d}_t^i) \right) \\ &= 0. \end{aligned} \tag{B.10}$$

This indicates that any translation motions are rigid-body motions resulting in no change of the strain energy.

B.1.2 Rotation for Geometrical Stiffness

In order to show that rotations about the axes are also rigid-body motions, the quadratic forms $Q_{\rm G}$ and $Q_{\rm E}$ of the geometrical and linear stiffness matrices are considered separately. For simplicity, only the rotation about z-axis is considered in the following. The formulation can be easily extended to the rotations about x- and y-axes.

Let $\bar{\mathbf{X}}$ and $\mathbf{X}$ $(\in \mathbb{R}^{3n})$ denote the new and old generalized coordinate vectors, respectively:

$$\begin{aligned}\bar{\mathbf{X}} &= \left(\bar{\mathbf{x}}^\top, \bar{\mathbf{y}}^\top, \bar{\mathbf{z}}^\top\right)^\top, \\ \mathbf{X} &= \left(\mathbf{x}^\top, \mathbf{y}^\top, \mathbf{z}^\top\right)^\top, \end{aligned} \tag{B.11}$$

where $\bar{\mathbf{x}}$ and $\mathbf{x}$, $\bar{\mathbf{y}}$ and $\mathbf{z}$, and $\bar{\mathbf{z}}$ and $\mathbf{z}$ are the new and old coordinate vectors in x-, y-, and z-directions, respectively. Suppose that the structure is rotated about z-axis by an arbitrary angle θ. Thus, $\bar{\mathbf{X}}$ and $\mathbf{X}$ are related by a rotation matrix $\mathbf{R}$

$$\bar{\mathbf{X}} = \mathbf{R}\mathbf{X}, \tag{B.12}$$

and $\mathbf{R}$ is defined as

$$\mathbf{R} = \begin{pmatrix} c\mathbf{I}_n & -s\mathbf{I}_n & \\ s\mathbf{I}_n & c\mathbf{I}_n & \\ & & \mathbf{I}_n \end{pmatrix}, \tag{B.13}$$

where $c = \cos\theta$, $s = \sin\theta$, and $\mathbf{I}_n \in \mathbb{R}^{n\times n}$ is an identity matrix.

From the self-equilibrium equations given in Eq. (2.113) and formulation of the geometrical stiffness matrix $\mathbf{K}_{\rm G}$ in Eq. (4.55), we have

$$\mathbf{K}_{\rm G}\mathbf{X} = \mathbf{0}. \tag{B.14}$$

Because

$$\begin{aligned}\mathbf{R}\mathbf{K}_{\rm G} &= \begin{pmatrix} c\mathbf{I}_n & -s\mathbf{I}_n & \\ s\mathbf{I}_n & c\mathbf{I}_n & \\ & & \mathbf{I}_n \end{pmatrix}\begin{pmatrix} \mathbf{E} & & \\ & \mathbf{E} & \\ & & \mathbf{E} \end{pmatrix} \\ &= \begin{pmatrix} c\mathbf{E} & -s\mathbf{E} & \\ s\mathbf{E} & c\mathbf{E} & \\ & & \mathbf{E} \end{pmatrix} \\ &= \begin{pmatrix} \mathbf{E} & & \\ & \mathbf{E} & \\ & & \mathbf{E} \end{pmatrix}\begin{pmatrix} c\mathbf{I}_n & -s\mathbf{I}_n & \\ s\mathbf{I}_n & c\mathbf{I}_n & \\ & & \mathbf{I}_n \end{pmatrix} \\ &= \mathbf{K}_{\rm G}\mathbf{R}, \end{aligned} \tag{B.15}$$

and from Eq. (B.14), we can further have the following relation for $\bar{\mathbf{X}}$ associated with $\mathbf{K}_\mathrm{G}$:

$$\begin{aligned}\mathbf{K}_\mathrm{G}\bar{\mathbf{X}} &= \mathbf{K}_\mathrm{G}\mathbf{R}\mathbf{X} = \mathbf{R}\mathbf{K}_\mathrm{G}\mathbf{X} \\ &= \mathbf{0}.\end{aligned} \tag{B.16}$$

The displacement $\mathbf{d}_\mathrm{r}$ of a structure is equal to the nodal coordinate differences $\bar{\mathbf{X}} - \mathbf{X}$ after and before the transformation; i.e., $\mathbf{d}_\mathrm{r} = \bar{\mathbf{X}} - \mathbf{X}$. From Eqs. (B.14) and (B.16), we have the following relation for the displacement $\mathbf{d}_\mathrm{r}$

$$\begin{aligned}\mathbf{K}_\mathrm{G}\mathbf{d}_\mathrm{r} &= \mathbf{K}_\mathrm{G}(\bar{\mathbf{X}} - \mathbf{X}) = \mathbf{K}_\mathrm{G}\bar{\mathbf{X}} - \mathbf{K}_\mathrm{G}\mathbf{X} \\ &= \mathbf{0}.\end{aligned} \tag{B.17}$$

Hence, the quadratic form Q_G of $\mathbf{K}_\mathrm{G}$ with respect to $\mathbf{d}_\mathrm{r}$ vanishes.

B.1.3 Rotation for Linear Stiffness

Let $\bar{x}_i$, $\bar{y}_i$, and $\bar{z}_i$ denote the new coordinates of node i as a result of rotation about z-axis by an arbitrary angle θ. The relation between the new coordinate vector $\bar{\mathbf{X}}_i$ and the old coordinate vector $\mathbf{X}_i$ of node i can be written as

$$\bar{\mathbf{X}}_i = \begin{pmatrix} \bar{x}_i \\ \bar{y}_i \\ \bar{z}_i \end{pmatrix} = \mathbf{R}_i\mathbf{X}_i = \mathbf{R}_i \begin{pmatrix} x_i \\ y_i \\ z_i \end{pmatrix}, \tag{B.18}$$

where the rotation matrix $\mathbf{R}_i$ for node i is

$$\mathbf{R}_i = \begin{pmatrix} c & -s & 0 \\ s & c & 0 \\ 0 & 0 & 1 \end{pmatrix}. \tag{B.19}$$

Consider member k connecting nodes i and j. The member length l_k before rotation can be expressed as follows by using the generalized coordinate vectors: $\mathbf{X}_i$ and $\mathbf{X}_j$ of nodes i and j

$$l_k^2 = \left(\mathbf{X}_i - \mathbf{X}_j\right)^\top \left(\mathbf{X}_i - \mathbf{X}_j\right), \tag{B.20}$$

and the member length $\bar{l}_k$ after rotation is

$$\bar{l}_k^2 = \left(\bar{\mathbf{X}}_i - \bar{\mathbf{X}}_j\right)^\top \left(\bar{\mathbf{X}}_i - \bar{\mathbf{X}}_j\right)$$
$$= \left(\mathbf{X}_i - \mathbf{X}_j\right)^\top \mathbf{R}_i^\top \mathbf{R}_i \left(\mathbf{X}_i - \mathbf{X}_j\right)$$
$$= l_k^2, \qquad \text{(B.21)}$$

because

$$\mathbf{R}_i^\top \mathbf{R}_i = \mathbf{I}_3, \qquad \text{(B.22)}$$

where $\mathbf{I}_3 \in \mathbb{R}^{3\times 3}$ is an identity matrix. Hence, extension e_k of member k due to rotation of the structure is zero:

$$e_k = \bar{l}_k - l_k = 0. \qquad \text{(B.23)}$$

From Eq. (B.23), the zero member length extension Eq. (2.82) is satisfied for all members ($k = 1, 2, \ldots, m$); i.e., $\mathbf{D}^\top \mathbf{d}_\mathrm{r} = \mathbf{0}$. Therefore, the following equation holds for the rotation $\mathbf{d}_\mathrm{r}$ about z-axis by an arbitrary angle θ:

$$\mathbf{K}_\mathrm{E}\mathbf{d}_\mathrm{r} = \mathbf{D}\bar{\mathbf{K}}\mathbf{D}^\top \mathbf{d}_\mathrm{r} = \mathbf{D}\bar{\mathbf{K}}\mathbf{0}$$
$$= \mathbf{0}, \qquad \text{(B.24)}$$

and furthermore the quadratic form Q_E of $\mathbf{K}_\mathrm{E}$ with respect to $\mathbf{d}_\mathrm{r}$ is

$$Q_\mathrm{E} = \mathbf{d}_\mathrm{r}^\top \mathbf{K}_\mathrm{E}\mathbf{d}_\mathrm{r} = \mathbf{d}_\mathrm{r}^\top \mathbf{0}$$
$$= 0. \qquad \text{(B.25)}$$

Thus, the quadratic form of $\mathbf{K}_\mathrm{E}$ with respect to $\mathbf{d}_\mathrm{r}$ also vanishes.

Because both of the quadratic forms Q_G and Q_E of $\mathbf{K}_\mathrm{G}$ and $\mathbf{K}_\mathrm{E}$ with respect to the rotation about z-axis are zero, this rotation is a rigid-body motion. Similar approach can be used to verify that the rotations of the whole structure about x- and y-axes are also rigid-body motions.

So far, we have demonstrated that translations and rotations in the affine motions are the rigid-body motions. A linear combination of these motions is certainly a rigid-body motion.

B.1.4 Non-trivial Affine Motions

By applying dilation motions, the structure expands or contracts in one direction and remains unchanged in the perpendicular directions. Directions of these motions $\mathbf{d}_\mathrm{a}^x$, $\mathbf{d}_\mathrm{a}^y$, and $\mathbf{d}_\mathrm{a}^z$ ($\in \mathbb{R}^{dn}$) of a structure in the three-dimensional ($d = 3$) space can be written as follows by using the nodal coordinate vectors:

$$\mathbf{d}_\mathrm{a}^x = \begin{pmatrix} \mathbf{x} \\ \mathbf{0} \\ \mathbf{0} \end{pmatrix}, \ \mathbf{d}_\mathrm{a}^y = \begin{pmatrix} \mathbf{0} \\ \mathbf{y} \\ \mathbf{0} \end{pmatrix}, \ \mathbf{d}_\mathrm{a}^z = \begin{pmatrix} \mathbf{0} \\ \mathbf{0} \\ \mathbf{z} \end{pmatrix}. \tag{B.26}$$

Figure B.2a, b show the two dilations $\mathbf{d}_\mathrm{a}^x$ and $\mathbf{d}_\mathrm{a}^y$ in x- and y-directions of a two-dimensional structure, respectively.

The shear motions $\mathbf{d}_\mathrm{a}^{xy}$, $\mathbf{d}_\mathrm{a}^{xz}$, and $\mathbf{d}_\mathrm{a}^{yz}$ ($\in \mathbb{R}^{dn}$) on xy-, xz-, and yz-planes can be defined as

$$\mathbf{d}_\mathrm{a}^{xy} = \begin{pmatrix} \mathbf{y} \\ \mathbf{x} \\ \mathbf{0} \end{pmatrix}, \ \mathbf{d}_\mathrm{a}^{xz} = \begin{pmatrix} \mathbf{z} \\ \mathbf{0} \\ \mathbf{x} \end{pmatrix}, \ \mathbf{d}_\mathrm{a}^{yz} = \begin{pmatrix} \mathbf{0} \\ \mathbf{z} \\ \mathbf{y} \end{pmatrix}. \tag{B.27}$$

In the shears $\mathbf{d}_\mathrm{a}^{ij}$ ($i, j \in \{x, y, z\}$ and $i \neq j$), the motion in i-direction is proportional to the nodal coordinates in j-direction, and vice versa. We have only one shear motion $\mathbf{d}_\mathrm{a}^{xy}$ for the two-dimensional case as shown in Fig. B.2c.

Lemma B.1 *The non-trivial affine motions* $\mathbf{d}_\mathrm{a}^{i(j)}$ *defined in Eqs. (B.26) and (B.27) lie in the null-space of* $\mathbf{K}_\mathrm{G}$.

Proof It is obvious from the self-equilibrium equations with respect to the nodal coordinates in Eq. (2.113) and definition of the geometrical stiffness matrix $\mathbf{K}_\mathrm{G}$ in Eq. (4.55) that

$$\mathbf{K}_\mathrm{G}\mathbf{d}_\mathrm{a}^{i(j)} = \mathbf{0}. \tag{B.28}$$

Therefore, $\mathbf{d}_\mathrm{a}^{i(j)}$ are the eigenvectors of $\mathbf{K}_\mathrm{G}$ corresponding to its zero eigenvalues, which proves the lemma. □

Because the non-trivial affine motions defined above are dependent on the nodal coordinates, while the rigid-body motions are not, the non-trivial affine motions and the rigid-body motions are linearly independent. Furthermore, the following lemma shows that the non-trivial affine motions are linearly independent.

Lemma B.2 *The non-trivial affine motions defined in Eqs. (B.26) and (B.27) of a non-degenerate tensegrity structure in d-dimensional space are linearly independent.*

Proof Consider the three-dimensional case. Let an arbitrary affine motion $\mathbf{d}_\mathrm{a} \in \mathbb{R}^{dn}$ be given as a linear combination of the affine motions defined in Eqs. (B.26) and (B.27) by the arbitrary coefficients β_k ($k = 1, 2, \ldots, 6$) as follows:

$$\mathbf{d}_\mathrm{a} = \beta_1 \mathbf{d}_\mathrm{a}^x + \beta_2 \mathbf{d}_\mathrm{a}^y + \beta_3 \mathbf{d}_\mathrm{a}^z + \beta_4 \mathbf{d}_\mathrm{a}^{xy} + \beta_5 \mathbf{d}_\mathrm{a}^{xz} + \beta_6 \mathbf{d}_\mathrm{a}^{yz}. \tag{B.29}$$

By incorporating Eqs. (B.26) and (B.27), $\mathbf{d}_\mathrm{a}$ can be divided as

$$\mathbf{d}_\mathrm{a} = \left((\mathbf{d}_\mathrm{a}^1)^\top, (\mathbf{d}_\mathrm{a}^2)^\top, (\mathbf{d}_\mathrm{a}^3)^\top \right)^\top, \tag{B.30}$$

where

$$\begin{aligned} \mathbf{d}_\mathrm{a}^1 &= \beta_1 \mathbf{x} + \beta_4 \mathbf{y} + \beta_5 \mathbf{z}, \\ \mathbf{d}_\mathrm{a}^2 &= \beta_4 \mathbf{x} + \beta_2 \mathbf{y} + \beta_6 \mathbf{z}, \\ \mathbf{d}_\mathrm{a}^3 &= \beta_5 \mathbf{x} + \beta_6 \mathbf{y} + \beta_3 \mathbf{z}. \end{aligned} \tag{B.31}$$

From Lemma 2.1, we know that the coordinate vectors $\mathbf{x}$, $\mathbf{y}$, and $\mathbf{z}$ of a non-degenerate (free-standing) structure in three-dimensional space are linearly independent. Thus, $\mathbf{d}_\mathrm{a}^1 = \mathbf{d}_\mathrm{a}^2 = \mathbf{d}_\mathrm{a}^3 = \mathbf{0}$ is satisfied if and only if

$$\begin{aligned} \beta_1 = \beta_4 = \beta_5 = 0, \\ \beta_4 = \beta_2 = \beta_6 = 0, \\ \beta_5 = \beta_6 = \beta_3 = 0. \end{aligned} \tag{B.32}$$

Hence, $\mathbf{d}_\mathrm{a} = \mathbf{0}$ is satisfied if and only if

$$\beta_k = 0, \quad (k = 1, 2, \ldots, 6). \tag{B.33}$$

Therefore, the non-trivial affine motions are linearly independent. Linear independence can also be shown for the two-dimensional case, which concludes the proof. □

Furthermore, we have the following lemma clarifying components of null-space of the geometrical stiffness matrix $\mathbf{K}_\mathrm{G}$.

Lemma B.3 *The affine motions, including the rigid-body motions and the non-trivial affine motions, span the whole null-space of the geometrical stiffness matrix* $\mathbf{K}_\mathrm{G}$ *of a d-dimensional tensegrity structure, when it has the minimum rank deficiency* $d^2 + d$ *to satisfy the non-degeneracy condition.*

Proof When the non-degeneracy condition is satisfied by a prestressed pin-jointed structure in d-dimensional space, the rank deficiency of the geometrical stiffness matrix is at least $d^2 + d$.

From Lemmas B.1 and B.2, we know that the $(d^2+d)/2$ non-trivial affine motions are linearly independent, and moreover, these motions lie in the null-space of the geometrical stiffness matrix. Therefore, together with the $(d^2 + d)/2$ linearly independent rigid-body motions, they span the null-space of the geometrical stiffness matrix, when it has the minimum rank deficiency of $d^2 + d$. □

B.2 Equivalent Stability Condition in Rigidity

In the field of structural rigidity in mathematics, a necessary condition for stability of tensegrity structures was presented by Connelly [1]:

> **Stability condition in structural rigidity:**
>
> If a structure is stable, then its member direction vectors do not lie on the same conic at infinity.

Denote the coordinates of node i as $\mathbf{X}_i = (x_i, y_i, z_i)^\top \in \mathbb{R}^d$. By applying the affine motion defined by the transformation matrix $\mathbf{T} \in \mathbb{R}^{d\times d}$ and translation vector $\mathbf{t} \in \mathbb{R}^d$, $\mathbf{X}_i$ is transformed to $\bar{\mathbf{X}}_i$ as

$$\bar{\mathbf{X}}_i = \mathbf{T}\mathbf{X}_i + \mathbf{t}. \tag{B.34}$$

Suppose that nodes i and j $(i < j)$ are connected by member k. The *member direction* $\mathbf{d}_k \in \mathbb{R}^d$ of member k is given as

$$\mathbf{d}_k = \mathbf{X}_i - \mathbf{X}_j. \tag{B.35}$$

If the member directions in the d-dimensional space lie on a *conic at infinity* denoted by C, then C can be defined as follows by using $\mathbf{d}_k$ and a non-trivial symmetric matrix $\mathbf{N} \in \mathbb{R}^{d\times d}$:

$$C = \{\mathbf{d} \mid \mathbf{d}_k^\top \mathbf{N}\mathbf{d}_k = 0\}. \tag{B.36}$$

If the structure has a non-trivial motion preserving the lengths of all members, the (strain) energy of the structure does not change; therefore, the structure is unstable. This is the basic idea in Ref. [1], which can be expressed by the following lemma:

> **Lemma B.4** *The member lengths are preserved by some affine motions if all member directions of the structure lie on the same conic at infinity.*

Proof From the affine motions of nodes i and j as defined in Eq. (B.34), the following equation holds if the length of member k, which is connected by nodes i and j, does not change by the affine motion:

$$\begin{aligned} |\bar{\mathbf{X}}_i - \bar{\mathbf{X}}_j|^2 - |\mathbf{X}_i - \mathbf{X}_j|^2 &= (\bar{\mathbf{X}}_i - \bar{\mathbf{X}}_j)^\top(\bar{\mathbf{X}}_i - \bar{\mathbf{X}}_j) - (\mathbf{X}_i - \mathbf{X}_j)^\top(\mathbf{X}_i - \mathbf{X}_j) \\ &= (\mathbf{X}_i - \mathbf{X}_j)\mathbf{T}^\top\mathbf{T}(\mathbf{X}_i - \mathbf{X}_j) - (\mathbf{X}_i - \mathbf{X}_j)^\top\mathbf{I}_d(\mathbf{X}_i - \mathbf{X}_j) \\ &= (\mathbf{X}_i - \mathbf{X}_j)(\mathbf{T}^\top\mathbf{T} - \mathbf{I}_d)(\mathbf{X}_i - \mathbf{X}_j) \\ &= 0, \end{aligned} \tag{B.37}$$

where $\mathbf{I}_d \in \mathbb{R}^{d\times d}$ is an identity matrix.

Denoting $\mathbf{N} = \mathbf{T}^\top\mathbf{T} - \mathbf{I}_d$ and comparing Eq. (B.37) with Eq. (B.36), the member directions $\mathbf{d}_k$ ($=\mathbf{X}_i - \mathbf{X}_j$) of the structure lie on the same conic at infinity defined by $\mathbf{N}$ ($=\mathbf{T}^\top\mathbf{T} - \mathbf{I}_d$) if all member lengths of the structure are preserved. □

The following lemma shows that Connelly's condition is equivalent to our stability condition for stability of tensegrity structures as presented in Lemma 4.6:

Lemma B.5 *Rank of the geometry matrix* $\mathbf{G}$ *is equal to* $(d^2 + d)/2$, *if and only if the member directions do not lie on the same conic at infinity.*

Proof Consider the three-dimensional case ($d = 3$). Since $\mathbf{N} \in \mathbb{R}^{d\times d}$, which defines the conic at infinity in Eq. (B.36), is a symmetric matrix, it can be written as a linear combination of $(d^2 + d)/2$ symmetric matrices

$$\mathbf{N} = \alpha_x \begin{pmatrix} 1 & 0 & 0 \\ 0 & 0 & 0 \\ 0 & 0 & 0 \end{pmatrix} + \alpha_y \begin{pmatrix} 0 & 0 & 0 \\ 0 & 1 & 0 \\ 0 & 0 & 0 \end{pmatrix} + \alpha_z \begin{pmatrix} 0 & 0 & 0 \\ 0 & 0 & 0 \\ 0 & 0 & 1 \end{pmatrix}$$
$$+ \alpha_{xy} \begin{pmatrix} 0 & 1 & 0 \\ 1 & 0 & 0 \\ 0 & 0 & 0 \end{pmatrix} + \alpha_{xz} \begin{pmatrix} 0 & 0 & 1 \\ 0 & 0 & 0 \\ 1 & 0 & 0 \end{pmatrix} + \alpha_{yz} \begin{pmatrix} 0 & 0 & 0 \\ 0 & 0 & 1 \\ 0 & 1 & 0 \end{pmatrix}. \tag{B.38}$$

Because $\mathbf{N}$ is a non-trivial matrix, the coefficients α_i cannot be zero at the same time.

The member direction $\mathbf{d}_k$ of member k connecting nodes i and j ($i < j$) is written as

$$\mathbf{d}_k = \mathbf{X}_i - \mathbf{X}_j = \begin{pmatrix} u_k \\ v_k \\ w_k \end{pmatrix}, \tag{B.39}$$

where u_k, v_k, and w_k are the kth entries of the coordinate difference vectors $\mathbf{u}$, $\mathbf{v}$, and $\mathbf{w}$, respectively.

Substituting Eqs. (B.38) and (B.39) into Eq. (B.36), we have

$$\begin{aligned} \mathbf{d}_k^\top \mathbf{N} \mathbf{d}_k &= \alpha_x u_k^2 + \alpha_y v_k^2 + \alpha_z w_k^2 + 2\alpha_{xy} u_k v_k + 2\alpha_{xz} v_k w_k + 2\alpha_{yz} v_k w_k \\ &= 0. \end{aligned} \tag{B.40}$$

If the member directions lie on the same conic at infinity, all member directions of the structure should satisfy Eq. (B.40), and the equations similar to Eq. (B.40) for all members k ($= 1, 2, \dots, m$) can be summarized in a matrix form as

$$\mathbf{B}\boldsymbol{\alpha} = \mathbf{0}, \tag{B.41}$$

where $\boldsymbol{\alpha} = (\alpha_x, \alpha_y, \alpha_z, 2\alpha_{xy}, 2\alpha_{xz}, 2\alpha_{yz})^\top$. It is easy to observe that

$$\mathbf{B} = \mathbf{G}, \tag{B.42}$$

where the geometry matrix $\mathbf{G}$ is defined in Eq. (4.110) for three-dimensional cases and in Eq. (4.111) for two-dimensional cases.

If the member directions do not lie on the same conic at infinity, then Eq. (B.41) has no non-trivial solution for the coefficient vector $\boldsymbol{\alpha}$. Hence, the rank of the matrix $\mathbf{B}$ or $\mathbf{G}$ is $(d^2 + d)/2$.

Conversely, if the rank of $\mathbf{G}$ is $(d^2 + d)/2$, then there exists no non-trivial solution $\boldsymbol{\alpha}$ for Eq. (B.41); i.e., there exists no matrix $\mathbf{N}$ satisfying Eq. (B.40) for all members; hence, the member directions do not lie on the same conic at infinity, which concludes the proof. □

The necessary condition for stability derived in Lemma 4.6 is considered to be more convenient to use than Connelly's descriptive condition, because only the rank of the well-established geometry matrix constructed from the nodal coordinates and connectivity of the structure needs to be investigated.

References

1. Connelly, R. (1982). Rigidity and energy. *Inventiones Mathematicae*, *66*(1), 11–33.
2. Gray, A. (1997). *Modern Differential Geometry of Curves and Surfaces with Mathematica* (2nd ed.). Boca Raton: CRC Press.

Appendix C
Tensegrity Tower

Abstract This chapter introduces connectivity and geometry of tensegrity towers, which are used as numerical examples in Chap. 5.

C.1 Introduction

Tensegrity tower is a special tensegrity structure, which is constructed by assembling simple tensegrity structures as unit cells along the vertical (z-)direction. The needle tower, designed by Kenneth Snelson and built in 1968 [1], may be one of the best-known tensegrity towers.

In this study, we adopt the prismatic tensegrity structures that have been extensively studied in Chaps. 3 and 6 as unit cells to construct tensegrity towers. Each unit cell in a tensegrity tower is regarded as one layer. Hence, a tensegrity tower consists of one or more layers, and there are at least three struts in each layer. Moreover, the nodes of each unit cell (layer) are located on two parallel planes: the top plan and the bottom plane of the layer.

Example C.1 An example tensegrity tower consisting of three layers as shown in Fig. C.1a.

Figure C.1a shows a three-layer tensegrity tower. It is constructed by assembling three unit cells (each is a one-layer tensegrity tower) as shown in Fig. C.1b in the z-direction. The unit cell is a prismatic structure with four struts. The struts are connected to the nodes that are located on two different parallel planes.

Note that part of the horizontal cables of the unit cells are replaced by saddle cables, and diagonal cables are added so as to make the unit cells work together. More details on connectivity of a tensegrity tower will be presented in the next section.

J.Y. Zhang and M. Ohsaki, *Tensegrity Structures*, Mathematics for Industry 6,
DOI 10.1007/978-4-431-54813-3

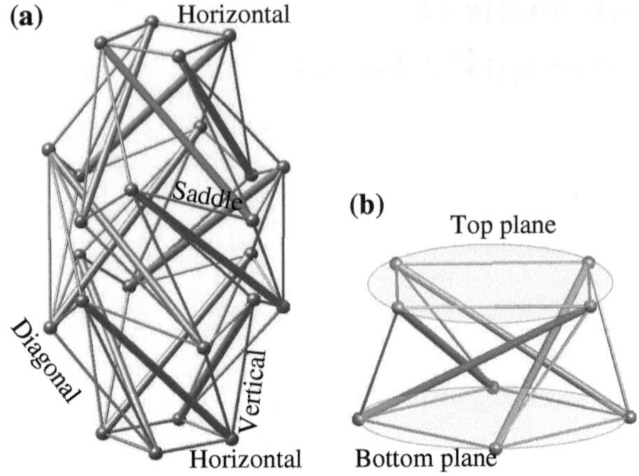

Fig. C.1 The three-layer tensegrity tower in (**a**), which is assembled from the unit cell in (**b**)

C.2 Nodes and Members

Suppose that a tensegrity tower has n^{L} layers and $\bar{n}$ struts in each layer. The nodes located on the same (top or bottom) plane of a layer have the same z-coordinate. Since there is no physical contact between any two struts, the total number n of nodes of a tensegrity tower is computed as

$$n = 2n^{\mathrm{L}}\bar{n}. \tag{C.1}$$

To connect different unit cells, some of the horizontal cables are replaced by saddle cables, and moreover, diagonal cables are added to let different unit cells work together. Accordingly, the cables of a tensegrity tower are classified into the following four types as shown in Fig. C.1a [2]:

Four types of cables of a tensegrity tower:

- *Horizontal cables* that connect the nodes on the same plane. They exist only on the bottom plane of the lowest layer as well as the top plane of the highest layer.
- *Vertical cables* that are connected by the nodes on the top and bottom planes of the same layer.
- *Saddle cables* that connect the nodes on different planes of the adjacent layers, e.g., the top plane of layer k and the bottom plane of layer $k+1$.
- *Diagonal cables* that connect the nodes on the same top (or bottom) planes of the adjacent layers, e.g., the top (or bottom) plane of layer k and the top (or bottom) plane of the layer $k+1$.

C.3 Elevation

Denote height of the kth layer by H_k ($k = 1, 2, \ldots, n^{\mathrm{L}}$), and the overlap between two adjacent layers k and $k-1$ by h_k ($k = 1, 2, \ldots, n^{\mathrm{L}}$). Note that we set $h_1 = 0$ for simplicity of formulation below. As discussed in Chap. 5, constraints on elevation of a tensegrity tower can be explicitly imposed in the process of form-finding. Hence, designers are free to design the elevation of the structure by assigning H_k and h_k. The total height H of the structure is computed by

$$H = \sum_{k=1}^{n^{\mathrm{L}}} (H_k - h_k), \quad \text{with} \quad h_1 = 0. \tag{C.2}$$

Suppose that we have assigned the heights H_k of the layers as well as overlaps h_k between them. The z-coordinate z_k^{t} of the nodes on the top plane of layer k ($k = 1, 2, \ldots, n^{\mathrm{L}}$) is determined as

$$z_k^{\mathrm{t}} = \sum_{i=1}^{k} (H_i - h_i), \tag{C.3}$$

and coordinate z_k^{b} of the nodes on the bottom plane of layer k is

$$z_k^{\mathrm{b}} = z_k^{\mathrm{t}} - H_k. \tag{C.4}$$

C.4 Connectivity

Connectivity of a tensegrity tower with any number of layers ($n^{\mathrm{L}} \geq 1$) and any number of struts ($\bar{n} \geq 3$) in each layer is given in this section.

The nodes on the bottom and top planes of layer k are respectively denoted by $\mathrm{P}_{k,j}^{\mathrm{b}}$ and $\mathrm{P}_{k,j}^{\mathrm{t}}$ and numbered as

$$\begin{aligned} \mathrm{P}_{k,j}^{\mathrm{b}} &= 2(k-1)\bar{n} + j, \\ \mathrm{P}_{k,j}^{\mathrm{t}} &= (2k-1)\bar{n} + j, \end{aligned} \quad (j = 1, \ldots, \bar{n}). \tag{C.5}$$

C.4.1 Struts

The jth strut $\mathrm{B}_{k,j}$ in layer k is connected by nodes $\mathrm{P}_{k,j}^{\mathrm{b}}$ and $\mathrm{P}_{k,j}^{\mathrm{t}}$ on different planes, and is denoted as

$$\mathrm{B}_{k,j} = [\mathrm{P}_{k,j}^{\mathrm{b}}, \mathrm{P}_{k,j}^{\mathrm{t}}], \tag{C.6}$$

where $[i, j]$ indicates that the pair of nodes i and j are connected to by a member.

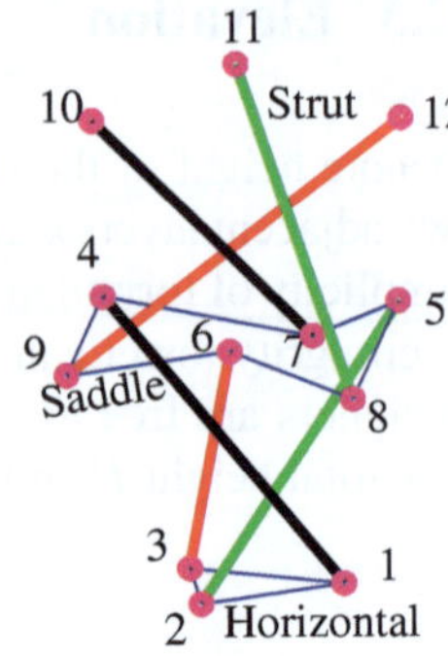

Fig. C.2 An example of connectivity of struts, horizontal cables, and saddle cables

Example C.2 Connectivity of struts of a tensegrity tower as shown in Fig. C.2 with three struts in each layer.

Figure C.2 shows part of a tensegrity tower, each layer of which has three struts; i.e., $\bar{n} = 3$. For clarity, the vertical and diagonal cables have been removed from the figure.

The nodes of this example structure are numbered as follows, following their definition in Eq. (C.5):

k	1			2		
j	1	2	3	1	2	3
$\mathrm{P}^{\mathrm{b}}_{k,j}$	1	2	3	7	8	9
$\mathrm{P}^{\mathrm{t}}_{k,j}$	4	5	6	10	11	12

(C.7)

The pairs of nodes connected by struts $\mathrm{B}_{1,j}$ $(j = 1, 2, 3)$ in layer 1 are

$$\mathrm{B}_{1,1} = [1,\ 4],\ \mathrm{B}_{1,2} = [2,\ 5],\ \mathrm{B}_{1,3} = [3,\ 6], \tag{C.8}$$

and those $\mathrm{B}_{2,j}$ $(j = 1, 2, 3)$ in layer 2 are

$$\mathrm{B}_{2,1} = [7,\ 10],\ \mathrm{B}_{2,2} = [8,\ 11],\ \mathrm{B}_{2,3} = [9,\ 12]. \tag{C.9}$$

C.4.2 Horizontal Cables and Saddle Sables

For the one-layer tensegrity towers ($n^{\mathrm{L}} = 1$), for example the one with four struts as shown in Fig. C.1b, it has been proved previously in Chap. 6 that they are super-stable if the horizontal cables are connected to the adjacent nodes, when they are of dihedral symmetry. To avoid any risk of resulting in unstable structures, the horizontal and saddle cables are connected to the adjacent nodes. Therefore, connectivity of the horizontal and saddle cables of a tensegrity tower is given as follows:

- *Horizontal cables* $\mathrm{H}_{1,j}$ and $\mathrm{H}_{n^{\mathrm{L}},j}$ are connected by adjacent nodes on the bottom and top planes of the lowest and highest layers, respectively, as

$$\begin{aligned} \mathrm{H}_{1,j} &= [\mathrm{P}^{\mathrm{b}}_{1,j},\ \mathrm{P}^{\mathrm{b}}_{1,j+1}], \\ \mathrm{H}_{n^{\mathrm{L}},j} &= [\mathrm{P}^{\mathrm{t}}_{n^{\mathrm{L}},j},\ \mathrm{P}^{\mathrm{t}}_{n^{\mathrm{L}},j+1}], \end{aligned} \quad (j = 1, 2, \ldots, \bar{n}). \tag{C.10}$$

- *Saddle cables* are connected by the nodes on the top plane of layer k and the bottom plane of layer $k + 1$ as

$$\begin{aligned} \mathrm{S}_{k,2j} &= [\mathrm{P}^{\mathrm{b}}_{k+1,j},\ \mathrm{P}^{\mathrm{t}}_{k,j+1}], \\ \mathrm{S}_{k,2j-1} &= [\mathrm{P}^{\mathrm{t}}_{k,j},\ \mathrm{P}^{\mathrm{b}}_{k+1,j}], \end{aligned} \quad (j = 1, 2, \ldots, \bar{n};\ k = 1, 2, \ldots, n^{\mathrm{L}} - 1). \tag{C.11}$$

Note that the following relations have been used in Eqs. (C.10) and (C.11).

$$\begin{aligned} \mathrm{P}^{\mathrm{b}}_{k,\bar{n}+1} &= \mathrm{P}^{\mathrm{b}}_{k,1}, \\ \mathrm{P}^{\mathrm{t}}_{k,\bar{n}+1} &= \mathrm{P}^{\mathrm{t}}_{k,1}. \end{aligned} \tag{C.12}$$

Example C.3 An example of connectivity of horizontal cables and saddle cables as shown in Fig. C.2.

For part of the tensegrity tower as shown in Fig. C.2, where the vertical cables and diagonal cables have been removed, the three horizontal cables located on the bottom plan of layer 1 are

$$\mathrm{H}_{1,1} = [1,\ 2],\quad \mathrm{H}_{1,2} = [2,\ 3],\quad \mathrm{H}_{1,3} = [1,\ 3]. \tag{C.13}$$

Moreover, the saddle cables connecting the nodes located on the top plane of layer 1 and bottom plane of layer 2 are

$$\begin{aligned} &\mathrm{S}_{1,1} = [4,\ 7],\ \mathrm{S}_{1,2} = [7,\ 5],\ \mathrm{S}_{1,3} = [5,\ 8], \\ &\mathrm{S}_{1,4} = [8,\ 6],\ \mathrm{S}_{1,5} = [6,\ 9],\ \mathrm{S}_{1,6} = [9,\ 4]. \end{aligned} \tag{C.14}$$

C.4.3 Vertical Cables and Diagonal Cables

Since connectivity of vertical and diagonal cables is not unique, we use the parameters c^{v} and c^{d} to define their connectivity as follows:

- *Vertical cable*: The connectivity of vertical cables in layer k (=1, 2, ..., n^{L}) is defined by using an integer c^{v} (=1, 2, ..., $\bar{n}$) as

$$\mathrm{V}_{k,j} = [\mathrm{P}^{\mathrm{t}}_{k,j},\ \mathrm{P}^{\mathrm{b}}_{k,j+c^{\mathrm{v}}}], \quad (j = 1, 2, \ldots, \bar{n}), \tag{C.15}$$

where $j + c^{\mathrm{v}} = j + c^{\mathrm{v}} - \bar{n}$ if $j + c^{\mathrm{v}} > \bar{n}$.

- *Diagonal cable*: The connectivity of diagonal cables connecting the nodes located on bottom or top planes of layers k and $k+1$ ($k = 1, 2, \ldots, n^{\mathrm{L}} - 1$) is defined by using an integer c^{d}(=0, 1, ..., $\bar{n} - 1$) as

$$\begin{aligned} \mathrm{D}^{\mathrm{b}}_{k,j} &= [\mathrm{P}^{\mathrm{b}}_{k,j},\ \mathrm{P}^{\mathrm{b}}_{k+1,j+c^{\mathrm{d}}}], \\ \mathrm{D}^{\mathrm{t}}_{k,j} &= [\mathrm{P}^{\mathrm{t}}_{k,j},\ \mathrm{P}^{\mathrm{t}}_{k+1,j+c^{\mathrm{d}}}], \end{aligned} \quad (j = 1, \ldots, \bar{n}), \tag{C.16}$$

where $j + c^{\mathrm{d}} = j + c^{\mathrm{d}} - \bar{n}$ is used if $j + c^{\mathrm{d}} > \bar{n}$.

Example C.4 An example of connectivity of vertical cables in layer k as shown in Fig. C.3.

Figure C.3 shows layer k of a tensegrity tower with five struts in each layer. The vertical cables connect the nodes on the bottom plane and those on the top plane of the same layer. For the parameter $c^{\mathrm{v}} = 1, 2$, we have the following different connectivity of vertical cables as follows:

$$\begin{aligned} c^{\mathrm{v}} = 1 : [\mathrm{P}^{\mathrm{t}}_{k,1},\ \mathrm{P}^{\mathrm{b}}_{k,2}]\ [\mathrm{P}^{\mathrm{t}}_{k,2},\ \mathrm{P}^{\mathrm{b}}_{k,3}]\ [\mathrm{P}^{\mathrm{t}}_{k,3},\ \mathrm{P}^{\mathrm{b}}_{k,4}]\ [\mathrm{P}^{\mathrm{t}}_{k,4},\ \mathrm{P}^{\mathrm{b}}_{k,5}]\ [\mathrm{P}^{\mathrm{t}}_{k,5},\ \mathrm{P}^{\mathrm{b}}_{k,1}], \\ c^{\mathrm{v}} = 2 : [\mathrm{P}^{\mathrm{t}}_{k,1},\ \mathrm{P}^{\mathrm{b}}_{k,3}]\ [\mathrm{P}^{\mathrm{t}}_{k,2},\ \mathrm{P}^{\mathrm{b}}_{k,4}]\ [\mathrm{P}^{\mathrm{t}}_{k,3},\ \mathrm{P}^{\mathrm{b}}_{k,5}]\ [\mathrm{P}^{\mathrm{t}}_{k,4},\ \mathrm{P}^{\mathrm{b}}_{k,1}]\ [\mathrm{P}^{\mathrm{t}}_{k,5},\ \mathrm{P}^{\mathrm{b}}_{k,2}]. \end{aligned} \tag{C.17}$$

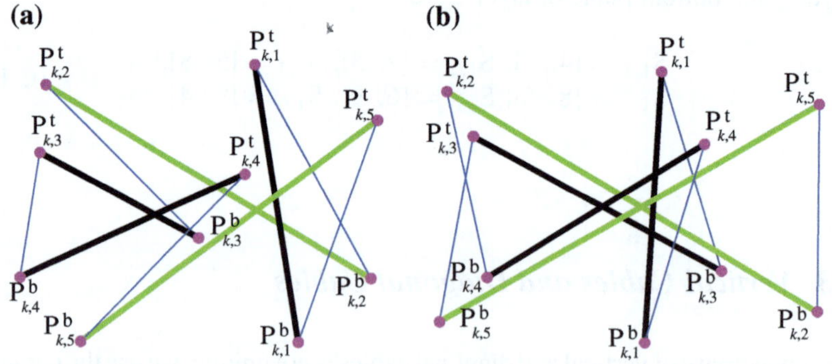

Fig. C.3 An example of connectivity of vertical cables in layer k. **a** $c^{v} = 1$, **b** $c^{v} = 2$

C.4.4 Number of Members

From the connectivity of members and nodes for a general n^{L}-layer tensegrity tower with $\bar{n}$ struts in each layer, the numbers of struts m^{b}, horizontal cables m^{h}, vertical cables m^{v}, saddle cables m^{s}, and diagonal cables m^{d} are respectively

$$\begin{aligned} m^{\mathrm{b}} &= n^{\mathrm{L}}\bar{n}, \\ m^{\mathrm{h}} &= 2\bar{n}, \\ m^{\mathrm{s}} &= 2(n^{\mathrm{L}} - 1)\bar{n}, \\ m^{\mathrm{v}} &= n^{\mathrm{L}}\bar{n}, \\ m^{\mathrm{d}} &= 2(n^{\mathrm{L}} - 1)\bar{n}, \end{aligned} \tag{C.18}$$

and the total number m of members of the tensegrity tower is

$$m = (6n^{\mathrm{L}} - 2)\bar{n}. \tag{C.19}$$

References

1. Snelson, K. *Tensegrity, Weaving and the Binary World*. Available online: http://kennethsnelson.net/tensegrity/.
2. Sultan, C., Corless, M., & Skelton, R. E. (2002). Symmetrical reconfiguration of tensegrity structures. *International Journal of Solids and Structures*, *39*(8), 2215–2234.

Appendix D
Group Representation Theory and Symmetry-adapted Matrix

Abstract This chapter introduces some important properties of group and its representation theory. Moreover, the analytical formulation for block-diagonalization of force density matrix for the structures with dihedral symmetry is presented. This formulation is extensively utilized for self-equilibrium analysis as well as presentation of super-stability conditions for the prismatic structures in Chap. 6, the star-shaped structures in Chap. 7, and the regular truncated tetrahedral structures in Chap. 8.

D.1 Group

In mathematics, a group is a set of elements together with an operation that combines any two of its elements to form a third element also in the set while satisfying four conditions called the group axioms, namely *closure*, *associativity*, *identity*, and *invertibility*:

1. Any two elements of the group must combine to give an element that is also a member of the group.
2. The associative law of combination must be satisfied.
3. The group must contain an element that commutes with all the other elements and also leaves them unchanged, which is call the *identity element*.
4. The inverse of every element in the group is also an element of the group.

Since we use group (representation) theory mainly to study the symmetric geometry of a structure, we restrict our description of group (representation) theory in geometry. There are basically two types of groups: point group and space group. *Point group* indicates that there is at least one point in the system which is not affected by any of the operations. If translational operations are allowed, the system can no longer be described by point symmetry. A symmetry group that contains translational elements is referred to as *space group*.

J.Y. Zhang and M. Ohsaki, *Tensegrity Structures*, Mathematics for Industry 6,
DOI 10.1007/978-4-431-54813-3

The dihedral group used in Chaps. 6 and 7, and the tetrahedral group in Chap. 8 are point groups.

D.2 Symmetry Operations

A symmetry operation is an element in the group, and it is an operation that transforms a structure to a position, which is physically indistinguishable from its initial position. The *order* of a group is the number of elements in the group. In a point group of order N, there are five different types of symmetry operations listed as follows:

1. E, *identity operation*:
   Nothing will be done to the structure, so the structure is unmoved. The corresponding symmetry operation is called the identity.
2. C_N, *rotation operation*:
   This is an operation that rotates the structure counter-clockwise or clockwise about an axis. If a rotation by $2\pi/N$ brings the structure into coincidence with itself, the structure is said to have an N-fold rotation axis.
3. σ, *reflection operation*:
   This is an operation of reflection about a plane.
   a. σ_{v}, *vertical plane of symmetry* that contains the principal axis;
   b. σ_{d}, *dihedral plane of symmetry* that contains the principal axis and bisects pairs of two-fold axes which are perpendicular to the principal axis;
   c. σ_{h}, *horizontal plane of symmetry* that is perpendicular to the principal axis.
4. S_N, *improper axis of rotation* (rotation-reflection operation):
   The operation can be represented by the product of reflection with respect to horizontal plane of symmetry σ_{h} and rotation C_N about the principal axis as
   $$\mathrm{S}_N = \sigma_{\mathrm{h}}\mathrm{C}_N. \tag{D.1}$$
5. i, *inversion operation*:
   If the origin of a Cartesian coordinate system is placed at the fixed point that does not move under any operation, then for every point (x, y, z) in the system there must be a symmetry related point at $(-x, -y, -z)$.

D.3 Character and Representation

A *group multiplication table* is a square array, the first row and the first column are the elements (symmetry operations), while the other entries in the array are the products of the corresponding operations.

Table D.1 Group multiplication table of group G of order N

Elements	G_1	G_2	G_3	...	G_N
G_1	G_1G_1	G_1G_2	G_1G_3	...	G_1G_N
G_2	G_2G_1	G_2G_2	G_2G_3	...	G_2G_N
G_3	G_3G_1	G_3G_2	G_3G_3	...	G_3G_N
$\vdots$	$\vdots$	$\vdots$	$\vdots$	$\vdots$	$\vdots$
G_N	G_NG_1	G_NG_2	G_NG_3	...	G_NG_N

Example D.1 Multiplication table of the group G of order N in Table D.1.

Suppose that the order of a group G is N, the elements of which are denoted by G_i $(i = 1, 2, \ldots, N)$. The product G_iG_j of two operations G_i and G_j defines the successive application of the two operations, where the one G_j on the *right* is carried out before the one G_i on the left.

There are in total N^2 different combinations (products) G_iG_j of any two operations G_i and G_j $(i, j = 1, 2, \ldots, N)$, which are summarized in Table D.1.

If R, P, and Q are the elements of a group and have the following relation

$$R = Q^{-1}PQ, \tag{D.2}$$

then we say that R is the *transform* of P by Q, or that P and R are *conjugate* to each other. The elements of a group which are conjugate to each other are said to be a *class*.

If a set of matrices form a group that obeys the group multiplication table for a given group, the matrices are said to form an *matrix representation* of that group. A representation which can be reduced to a sum of other representations is called a *reducible representation*. Otherwise, it is an *irreducible representation*.

We list the three properties of the irreducible representation of a group without proof as

1. If the irreducible representations of a group are one-dimensional, they must form a group by themselves;
2. The sum of the squares of the dimensions of the irreducible representations is equal to the order of the group.
3. The number of irreducible representations in a group is identical to the number of its classes.

A *character* is defined as the trace of an irreducible representation matrix, representing a given operation in a given group. The *character table* for a group lists the characters for the various operations associated with each irreducible representation.

Two groups are said to exhibit *isomorphism* if a one-to-one correspondence can be established between the elements of these two groups. In isomorphism, each element of one group is uniquely mirrored by an element of the other group. However, if two or more different elements of one group have the same image in the other group, then these two groups are said to exhibit *homomorphism*.

An extremely important property of the matrices, which multiply isomorphically to group operations, is that *their characters are invariant to a similarity transformation*.

D.4 Dihedral Group

Suppose that we have a regular N-gon on the xy-plane. Center of the N-gon is at the origin of coordinate system. Take the z-axis as the principal axis. When the N-gon is rotated about the z-axis through the angle $2i\pi/N$ ($i = 0, 1, \ldots, N-1$), it is transformed into itself. All of these N rotations, denoted by C_N^i ($i = 0, 1, \ldots, N-1$), form a *cyclic group* of order N. Note that the rotation corresponding to $i = 0$ is in fact the identity operation of the group.

If the two surfaces of the N-gon, the top and the bottom, are distinguished, it is usually called a *N-gonal dihedron*. Take any axis of an N-gonal dihedron joining a vertex with the opposite vertex if N is even, or with the mid-point of the opposite edge if N is odd. When it is rotated about this axis through the angle π, the N-gonal dihedron is carried into itself too. There are N of these symmetry operations, denoted by $\mathrm{C}_{2,i}$ ($i = 0, 1, \ldots, N-1$).

Therefore, the complete group carrying the N-gonal dihedron into itself consists of $2N$ symmetry operations as mentioned above. The group constructed by these symmetry operations is called the *dihedral group*, which is denoted by $\mathbf{D}_N$. Accordingly, order of the dihedral group $\mathbf{D}_N$ is $2N$.

The symmetry elements of any point group can be produced from its generators. Any of the four basic symmetry elements can be used as generators, either alone or in combination. At most, three of these are sufficient to describe the point symmetry of any system. Table D.2 lists the generators for some commonly used point groups including the dihedral group.

A dihedral group $\mathbf{D}_N$ consists of two one-dimensional irreducible representations, A_1 and A_2, if N is odd; and there are four one-dimensional irreducible

Table D.2 Generators for the various point groups

Group	$\mathbf{C}_N$	$\mathbf{S}_N$	$\mathbf{C}_{N\mathrm{v}}$	$\mathbf{C}_{N\mathrm{h}}$	$\mathbf{D}_N$	$\mathbf{D}_{N\mathrm{d}}$
Generators	C_N	S_N	$\mathrm{C}_N, \sigma_\mathrm{v}$	$\mathrm{C}_N, \sigma_\mathrm{h}$	$\mathrm{C}_N, \mathrm{C}_2$	$\mathrm{C}_N, \mathrm{C}_2, \sigma_\mathrm{d}$
Group	$\mathbf{T}$	$\mathbf{T}_\mathrm{d}$	$\mathbf{T}_\mathrm{h}$	$\mathbf{O}$	$\mathbf{O}_\mathrm{h}$	$\mathbf{D}_{N\mathrm{h}}$
Generators	$\mathrm{C}_2^z, \mathrm{C}_3^{xyz}$	$\mathrm{S}_4^z, \mathrm{C}_3^{xyz}$	$\mathrm{C}_2^z, \mathrm{C}_3^{xyz}, \mathrm{i}$	$\mathrm{C}_4^z, \mathrm{C}_3^{xyz}$	$\mathrm{C}_4^z, \mathrm{C}_3^{xyz}, \mathrm{i}$	$\mathrm{C}_N, \mathrm{C}_2, \sigma_\mathrm{h}$

representations, including A_1, A_2, B_1, and B_2, if N is even. Moreover, there are p two-dimensional irreducible representations, denoted by E_k ($k = 1, 2, \ldots, p$), where

$$p = \begin{cases} (N-1)/2, & \text{if } N \text{ is odd,} \\ (N-2)/2, & \text{if } N \text{ is even.} \end{cases} \tag{D.3}$$

All characters of the A_1 representation are 1, while the characters of the A_2 are 1 for the N-fold cyclic rotations C_N^i ($i = 0, 1, \ldots, N-1$) about z-axis, and -1 for the two-fold rotations $C_{2,i}$ ($i = 0, 1, \ldots, N-1$).

The characters of cyclic rotations of the B_1 and B_2 representations alternate between 1 and -1; and the characters for the two-fold rotations alternate between 1 and -1.

The character of the two-dimensional representation E_k corresponding to the cyclic rotation C_N^i is $\cos(2ik\pi/N)$; and the character corresponding to the two-fold rotation $C_{2,i}$ is zero. The two generators $\mathbf{R}_0^{E_k}$ and $\mathbf{R}_N^{E_k}$ of the two-dimensional irreducible representation matrices can be written as follows:

$$\begin{aligned} \mathbf{R}_0^{E_k} &= \begin{pmatrix} \cos(2k\pi/N) & -\sin(2k\pi/N) \\ \sin(2k\pi/N) & \cos(2k\pi/N) \end{pmatrix}, \\ \mathbf{R}_N^{E_k} &= \begin{pmatrix} 1 & 0 \\ 0 & -1 \end{pmatrix}. \end{aligned} \tag{D.4}$$

All the two-dimensional irreducible representation matrices of the dihedral group can be generated by using the above two generators as

$$\mathbf{R}_{i+Nj}^{E_k} = \left(\mathbf{R}_0^{E_k}\right)^i \left(\mathbf{R}_N^{E_k}\right)^j, \quad (i = 0, 1, \ldots, N-1;\ j = 0, 1). \tag{D.5}$$

In summary, the irreducible matrix representations of a dihedral group $\mathbf{D}_N$ are listed in Table D.3.

D.5 Symmetry-Adapted Force Density Matrix

This section presents the direct strategy for the symmetry-adapted force density matrix, of which the independent blocks (sub-matrices) with much smaller sizes lie in its leading diagonal. The prismatic tensegrity structures that are of dihedral symmetry are taken as example structures for the presentation of the symmetry-adapted force density matrix. Note that, in a symmetric prismatic structure, its nodes belong to a regular obit; i.e., they have one-to-one correspondence to the symmetry operations of the dihedral group.

Table D.3 Irreducible representation matrices $\mathbf{R}_i^{\mu}$ of dihedral group $\mathbf{D}_N$

$\mathbf{D}_N$	C_N^i	$C_{2,i}$		
A_1	1	1		
A_2	1	-1	z, R_z	
(B_1)	$(-1)^i$	$(-1)^i$		N even
(B_2)	$(-1)^i$	$(-1)^{(i+1)}$		N even
E_1	$\begin{pmatrix} C_i & -S_i \\ S_i & C_i \end{pmatrix}$	$\begin{pmatrix} C_i & S_i \\ S_i & -C_i \end{pmatrix}$	(x, y) (R_x, R_y)	
E_k	$\begin{pmatrix} C_{ik} & -S_{ik} \\ S_{ik} & C_{ik} \end{pmatrix}$	$\begin{pmatrix} C_{ik} & S_{ik} \\ S_{ik} & -C_{ik} \end{pmatrix}$		$k = 2, 3, \ldots, p$
	$\mathbf{R}_i^{\mu}$	$\mathbf{R}_{N+i}^{\mu}$		$i = 0, 1, \ldots, N-1$

The first column denotes the irreducible representations μ of the group, the first row denotes its symmetry operations with i running from 0 to $N-1$. C_{ik} and S_{ik} respectively denote $\cos(2ik\pi/N)$ and $\sin(2ik\pi/N)$. x, y, z and R_x, R_y, R_z respectively stand for symmetry operations of the corresponding coordinates and rotations about the corresponding axes

D.5.1 Force Density Matrix

A prismatic tensegrity structure with dihedral symmetry, denoted by $\mathbf{D}_N$, consists of $2N$ nodes, $2N$ horizontal cables, N vertical cables, and N struts. The definition of its connectivity can be found in Chap. 3.

We assume that cables carry tension and struts carry compression. Nodes of a prismatic structure lie on two parallel planes; horizontal cables connect the nodes on the same plane, and vertical cables and struts connect those on different planes. The nodes and horizontal cables have one-to-one correspondence to the symmetry operations of the dihedral group, while struts and vertical cables have one-to-two correspondence.

Every node of a prismatic tensegrity structure is connected by three different types of members: two horizontal cables, one vertical cable, and one strut; and each type of members has the same prestress and length. The nodes on the top plane of the structure are numbered from 0 to $N-1$, and those on the bottom are N to $2N-1$. We use the notation $\mathbf{D}_N^{h,v}$ to describe the connectivity of a prismatic tensegrity with $\mathbf{D}_N$ symmetry: h and v respectively describe the connectivity of the horizontal and vertical cables, while that of struts is fixed.

Let q_h, q_v and q_s denote the force densities (prestress to length ratios) of horizontal cables, vertical cables, and struts, respectively. From the numbering and connectivity of nodes, force density matrix $\mathbf{E} \in \mathbb{R}^{2N\times 2N}$ of the structure can be written as

$$\mathbf{E} = \begin{pmatrix} \mathbf{E}_1 & \mathbf{E}_2 \\ \mathbf{E}_2^{\top} & \mathbf{E}_1 \end{pmatrix}. \tag{D.6}$$

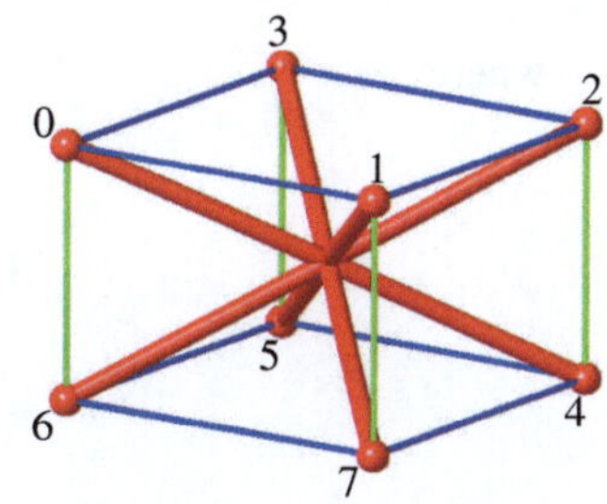

Fig. D.1 Prismatic tensegrity structure $\mathbf{D}_4^{1,2}$ with dihedral symmetry $\mathbf{D}_4$. The struts do not mechanically contact with each other, although it is physically inevitable as shown in the figure

Denote $q = 2q_{\rm h} + q_{\rm s} + q_{\rm v}$, and let $\hat{\mathbf{I}}_a \in \mathbb{R}^{N\times N}$ be a matrix with only one non-zero entry $\hat{I}_{(i,i+a)}$ in each row and column defined as

$$\hat{I}_{(i,i+a)} = 1, \quad (i = 1, 2, \dots, N), \tag{D.7}$$

where a (=0, 1, ..., $N-1$) is a constant positive integer, and we set $i + a := i + a - N$ if $i + a > N$. The sub-matrices $\mathbf{E}_1$ and $\mathbf{E}_2$ ($\in\mathbb{R}^{N\times N}$) in $\mathbf{E}$ are written as

$$\begin{aligned} \mathbf{E}_1 &= q\hat{\mathbf{I}}_0 - q_{\rm h}\hat{\mathbf{I}}_h - q_{\rm h}\hat{\mathbf{I}}_{N-h}, \\ \mathbf{E}_2 &= -q_{\rm s}\hat{\mathbf{I}}_0 - q_{\rm v}\hat{\mathbf{I}}_v, \end{aligned} \tag{D.8}$$

where $\hat{\mathbf{I}}_0$ is actually an N-by-N identity matrix $\mathbf{I}_N$.

Example D.2 Force density matrix of the prismatic tensegrity structure $\mathbf{D}_4^{1,2}$ as shown in Fig. D.1.

The structure $\mathbf{D}_4^{1,2}$ as shown in Fig. D.1 is of $\mathbf{D}_4$ symmetry, and its connectivity is defined by $h = 1$ for horizontal cables and $v = 2$ for vertical cables. From the definition of the force density matrix in Eq. (D.8), we have

$$\begin{aligned} \mathbf{E}_1 &= q\hat{\mathbf{I}}_0 - q_{\rm h}\hat{\mathbf{I}}_1 - q_{\rm h}\hat{\mathbf{I}}_3 \\ &= \begin{pmatrix} q & -q_{\rm h} & 0 & 0 - q_{\rm h} \\ -q_{\rm h} & q & -q_{\rm h} & 0 \\ 0 & -q_{\rm h} & q & -q_{\rm h} \\ -q_{\rm h} & 0 & -q_{\rm h} & q \end{pmatrix}, \\ \mathbf{E}_2 &= -q_{\rm s}\hat{\mathbf{I}}_0 - q_{\rm v}\hat{\mathbf{I}}_2 \\ &= \begin{pmatrix} -q_{\rm s} & 0 & -q_{\rm v} & 0 \\ 0 & -q_{\rm s} & 0 & -q_{\rm v} \\ -q_{\rm v} & 0 & -q_{\rm s} & 0 \\ 0 & -q_{\rm v} & 0 & -q_{\rm s} \end{pmatrix}, \end{aligned} \tag{D.9}$$

where

$$\hat{\mathbf{I}}_h = \hat{\mathbf{I}}_1 = \begin{pmatrix} 0 & 1 & 0 & 0 \\ 0 & 0 & 1 & 0 \\ 0 & 0 & 0 & 1 \\ 1 & 0 & 0 & 0 \end{pmatrix},$$

$$\hat{\mathbf{I}}_{N-h} = \hat{\mathbf{I}}_3 = \begin{pmatrix} 0 & 0 & 0 & 1 \\ 1 & 0 & 0 & 0 \\ 0 & 1 & 0 & 0 \\ 0 & 0 & 1 & 0 \end{pmatrix},$$

$$\hat{\mathbf{I}}_v = \hat{\mathbf{I}}_2 = \begin{pmatrix} 0 & 0 & 1 & 0 \\ 0 & 0 & 0 & 1 \\ 1 & 0 & 0 & 0 \\ 0 & 1 & 0 & 0 \end{pmatrix}. \tag{D.10}$$

D.5.2 Symmetry-Adapted Form

In conventional methods, the symmetry-adapted version $\tilde{\mathbf{E}}$ of the force density matrix $\mathbf{E}$ is usually obtained using the unitary transformation matrix $\mathbf{T} \in \mathbb{R}^{2N \times 2N}$:

$$\tilde{\mathbf{E}} = \mathbf{T}\mathbf{E}\mathbf{T}^{\top}, \tag{D.11}$$

where $\mathbf{T}\mathbf{T}^{\top}$ is an identity matrix:

$$\mathbf{T}\mathbf{T}^{\top} = \mathbf{I}_{2N}. \tag{D.12}$$

Although the transformation matrix is not needed to derive the blocks $\tilde{\mathbf{E}}^{\mu}$ in our direct strategy as presented later in Eq. (D.20), it is necessary for the proof of its formulation. Hence, we introduce the details of $\mathbf{T}$ for obtaining $\tilde{\mathbf{E}}$ as in Eq. (D.11) before presentation of its direct formulation in Lemma D.1.

Because the nodes of a symmetric prismatic tensegrity structure have one-to-one correspondence to the symmetry operations; i.e., any node can be transformed to another by only one symmetry operation of that group, the transformation matrix $\mathbf{T}$ can be easily obtained from the irreducible matrix representations.

For the one-dimensional representation μ, which is $\mathbf{A}_1$, $\mathbf{A}_2$, $\mathbf{B}_1$, or $\mathbf{B}_2$, the row $\mathbf{T}^{\mu} \in \mathbb{R}^{1 \times 2N}$ of $\mathbf{T}$ corresponding to μ is

$$\mathbf{T}^{\mu} = \frac{1}{\sqrt{2N}} \left(\mathbf{R}_0^{\mu}, \ \mathbf{R}_1^{\mu}, \ldots, \mathbf{R}_j^{\mu}, \ldots, \mathbf{R}_{2N-1}^{\mu} \right), \tag{D.13}$$

where $\mathbf{R}_j^{\mu}$ is the character of the one-dimensional representation μ, and $\mathbf{T}^{\mu}$ is normalized as $\mathbf{T}^{\mu}(\mathbf{T}^{\mu})^{\top} = 1$ with the coefficient $1/\sqrt{2N}$.

Example D.3 Transformation matrix $\mathbf{T}^{\mathrm{A}_2}$ for the one-dimensional representation A_2 of the structure with $\mathbf{D}_3$ symmetry.

According to Eq. (D.13) and representation matrices (characters in Table D.3), the transformation matrix $\mathbf{T}^{\mathrm{A}_2}$ corresponding to the one-dimensional representation A_2 of a structure with $\mathbf{D}_3$ symmetry is

$$\mathbf{T}^{\mathrm{A}_2} = \frac{1}{\sqrt{6}}(1,\ 1,\ 1,\ -1,\ -1,\ -1). \tag{D.14}$$

For a two-dimensional representation E_k, there are four rows in $\mathbf{T}^{\mathrm{E}_k} \in \mathbb{R}^{4\times 2N}$. The irreducible representation matrix $\mathbf{R}_j^{\mathrm{E}_k}$ of the jth symmetry operation corresponding to representation E_k is

$$\mathbf{R}_j^{\mathrm{E}_k} = \begin{pmatrix} R_{j(1,1)}^{\mathrm{E}_k} & R_{j(1,2)}^{\mathrm{E}_k} \\ R_{j(2,1)}^{\mathrm{E}_k} & R_{j(2,2)}^{\mathrm{E}_k} \end{pmatrix}. \tag{D.15}$$

The four entries of ${}_j\mathbf{R}^{\mathrm{E}_k}$ are located in the jth column of $\mathbf{T}^{\mathrm{E}_k}$ as follows to construct the transformation matrix $\mathbf{T}^{\mathrm{E}_k}$

$$\mathbf{T}^{\mathrm{E}_k} = \frac{1}{\sqrt{N}} \begin{pmatrix} R_{0(1,1)}^{\mathrm{E}_k} & \cdots & R_{j(1,1)}^{\mathrm{E}_k} & \cdots & R_{2N-1(1,1)}^{\mathrm{E}_k} \\ R_{0(1,2)}^{\mathrm{E}_k} & \cdots & R_{j(1,2)}^{\mathrm{E}_k} & \cdots & R_{2N-1(1,2)}^{\mathrm{E}_k} \\ R_{0(2,1)}^{\mathrm{E}_k} & \cdots & R_{j(2,1)}^{\mathrm{E}_k} & \cdots & R_{2N-1(2,1)}^{\mathrm{E}_k} \\ R_{0(2,2)}^{\mathrm{E}_k} & \cdots & R_{j(2,2)}^{\mathrm{E}_k} & \cdots & R_{2N-1(2,2)}^{\mathrm{E}_k} \end{pmatrix} \tag{D.16}$$

which is written as

$$\mathbf{T}^{\mathrm{E}_k} = \frac{1}{\sqrt{N}} \begin{pmatrix} \mathbf{C}_{jk}, & \mathbf{C}_{jk} \\ -\mathbf{S}_{jk}, & \mathbf{S}_{jk} \\ \mathbf{S}_{jk}, & \mathbf{S}_{jk} \\ \mathbf{C}_{jk}, & -\mathbf{C}_{jk} \end{pmatrix}, \tag{D.17}$$

where $\mathbf{C}_{jk}$ and $\mathbf{S}_{jk}$ ($\in\mathbb{R}^N$) are row vectors defined as follows:

$$\begin{aligned} \mathbf{C}_{jk} &= \left(C_0,\ C_k, \ldots, C_{jk}, \ldots, C_{(N-1)k}\right), \\ \mathbf{S}_{jk} &= \left(S_0,\ S_k, \ldots, S_{jk}, \ldots, S_{(N-1)k}\right), \quad (j = 0, \ldots, N-1). \end{aligned} \tag{D.18}$$

Example D.4 Transformation matrix $\mathbf{T}^{\mathrm{E}_1}$ for the two-dimensional representation $\mathbf{E}_1$ of a structure with $\mathbf{D}_3$ symmetry.

According to Eq. (D.17), the transformation matrix $\mathbf{T}^{\mathrm{E}_1}$ for the structure with $\mathbf{D}_3$ symmetry is

$$\mathbf{T}^{\mathrm{E}_1} = \frac{1}{\sqrt{3}} \begin{pmatrix} C_0 & C_1 & C_2 & C_0 & C_1 & C_2 \\ -S_0 & -S_1 & -S_2 & S_0 & S_1 & S_2 \\ S_0 & S_1 & S_2 & S_0 & S_1 & S_2 \\ C_0 & C_1 & C_2 & -C_0 & -C_1 & -C_2 \end{pmatrix}$$
$$= \frac{1}{2\sqrt{3}} \begin{pmatrix} 2 & -1 & -1 & 2 & -1 & -1 \\ 0 & -\sqrt{3} & \sqrt{3} & 0 & \sqrt{3} & -\sqrt{3} \\ 0 & \sqrt{3} & -\sqrt{3} & 0 & \sqrt{3} & -\sqrt{3} \\ 2 & -1 & -1 & -2 & 1 & 1 \end{pmatrix}. \quad \text{(D.19)}$$

Combining $\mathbf{T}^{\mu}$ for all representations to obtain $\mathbf{T}$, it is easy to verify from the great orthogonality theorem [1] that $\mathbf{T}$ is a unitary transformation matrix satisfying Eq. (D.12). Substituting $\mathbf{T}$ into Eq. (D.11), the force density matrix can be block-diagonalized with the structure as in Eq. (6.18). Super-stability investigation and self-equilibrium analysis are then significantly simplified by dealing with these blocks, dimensions of which are only one or two no matter how complex the structure is.

Size of the transformation matrix $\mathbf{T}$ increases in proportion to the number of its nodes. Therefore, it is difficult to derive analytical symmetry-adapted force density matrix in this numerical way for complex structure that has a large number of nodes. Moreover, the symmetry-adapted formulation by Eq. (D.11) can only deal with each specific structure, but not all structures with similar symmetry properties.

To have a more systematic solution, Lemma D.1 below presents a direct way for deriving the symmetry-adapted force density matrix of the structures with dihedral symmetry. It should be notable that the direct formulation is applicable to the cases that the nodes of the structure have one-to-one correspondence to the symmetry operations of the corresponding group.

In Lemma D.1, only a reference node, e.g., node 0, is needed to present the blocks $\tilde{\mathbf{E}}^{\mu}$ of $\tilde{\mathbf{E}}$ corresponding to the representation μ. Consider the structure $\mathbf{D}_N^{h,v}$ in general. Irreducible representation matrices corresponding to the nodes connected to the reference node 0 are denoted by $\mathbf{R}_0^{\mu}$, $\mathbf{R}_h^{\mu}$, $\mathbf{R}_{N-h}^{\mu}$, $\mathbf{R}_N^{\mu}$, and $\mathbf{R}_{N+v}^{\mu}$, which are listed in Table D.4. In the following lemma, we show that blocks $\tilde{\mathbf{E}}^{\mu}$ of each representation μ can be directly written as sum of products of the force densities and their corresponding irreducible representation matrices.

Table D.4 Selected irreducible representation matrices corresponding to the nodes connecting to the reference node 0

		Horizontal cable		Strut	Vertical
μ	$\mathbf{R}_0^\mu$	$\mathbf{R}_h^\mu$	$\mathbf{R}_{N-h}^\mu$	$\mathbf{R}_N^\mu$	$\mathbf{R}_{N+v}^\mu$
A_1	1	1	1	1	1
A_2	1	1	1	-1	-1
B_1	1	$(-1)^h$	$(-1)^{N-h}$	1	$(-1)^v$
B_2	1	$(-1)^h$	$(-1)^{N-h}$	-1	$(-1)^{v+1}$
E_k	$\mathbf{I}_2$	$\begin{pmatrix} C_{hk} & -S_{hk} \\ S_{hk} & C_{hk} \end{pmatrix}$	$\begin{pmatrix} C_{hk} & S_{hk} \\ -S_{hk} & C_{hk} \end{pmatrix}$	$\begin{pmatrix} 1 & 0 \\ 0 & -1 \end{pmatrix}$	$\begin{pmatrix} C_{vk} & S_{vk} \\ S_{vk} & -C_{vk} \end{pmatrix}$

Lemma D.1 *The block $\tilde{\mathbf{E}}^\mu$ corresponding to representation μ of the symmetry-adapted force density matrix $\tilde{\mathbf{E}}$ can be written in a general form as*

$$\tilde{\mathbf{E}}^\mu = q\mathbf{R}_0^\mu - q_\mathrm{h}\mathbf{R}_h^\mu - q_\mathrm{h}\mathbf{R}_{N-h}^\mu - q_\mathrm{s}\mathbf{R}_N^\mu - q_\mathrm{v}\mathbf{R}_{N+v}^\mu, \tag{D.20}$$

where $q = 2q_\mathrm{h} + q_\mathrm{s} + q_\mathrm{v}$.

Proof Using components $\mathbf{T}^\mu$ of $\mathbf{T}$ corresponding to representation μ, the block $\tilde{\mathbf{E}}^\mu$ corresponding to representation μ can be computed as

$$\tilde{\mathbf{E}}^\mu = \mathbf{T}^\mu \mathbf{E} (\mathbf{T}^\mu)^\top. \tag{D.21}$$

1. *One-dimensional blocks*
For the one-dimensional representations, $\mathbf{T}^\mu \in \mathbb{R}^{1\times 2N}$ is a row vector denoted as $\mathbf{T}^\mu = [\boldsymbol{\alpha}_1, \boldsymbol{\alpha}_2]$. From Eq. (D.6), Eq. (D.21) becomes

$$\tilde{\mathbf{E}}^\mu = \boldsymbol{\alpha}_1\mathbf{E}_1\boldsymbol{\alpha}_1^\top + \boldsymbol{\alpha}_2\mathbf{E}_1\boldsymbol{\alpha}_2^\top + 2\boldsymbol{\alpha}_1\mathbf{E}_2\boldsymbol{\alpha}_2^\top. \tag{D.22}$$

Consider representation A_1 for example. All irreducible representation matrices (equal to their characters) are equal to 1, hence, all the entries in $\mathbf{T}^{A_1}$ (also in $\boldsymbol{\alpha}_1$ and $\boldsymbol{\alpha}_2$) are $1/\sqrt{2N}$. Therefore, we have

$$\begin{aligned} \boldsymbol{\alpha}_1\hat{\mathbf{I}}_a\boldsymbol{\alpha}_1^\top + \boldsymbol{\alpha}_2\hat{\mathbf{I}}_a\boldsymbol{\alpha}_2^\top &= \frac{2N}{2N} = 1 \\ &= \mathbf{R}_a^{A_1}, \quad (a = 0, h, N-h). \end{aligned} \tag{D.23}$$

In a similar manner, we can also have

$$\begin{aligned} 2\boldsymbol{\alpha}_1\hat{\mathbf{I}}_b\boldsymbol{\alpha}_2^\top &= 2\frac{N}{2N} = 1 \\ &= \mathbf{R}^{\mathrm{A}_1}_{N+b}, \qquad (b = 0, v). \end{aligned} \tag{D.24}$$

From Eqs. (D.8) and (D.22)–(D.24), we have

$$\begin{aligned} \tilde{\mathbf{E}}^{\mathrm{A}_1} &= \boldsymbol{\alpha}_1\mathbf{E}_1\boldsymbol{\alpha}_1^\top + \boldsymbol{\alpha}_2\mathbf{E}_1\boldsymbol{\alpha}_2^\top + 2\boldsymbol{\alpha}_1\mathbf{E}_2\boldsymbol{\alpha}_2^\top \\ &= \boldsymbol{\alpha}_1(q\hat{\mathbf{I}}_0 - q_\mathrm{h}\hat{\mathbf{I}}_h - q_\mathrm{h}\hat{\mathbf{I}}_{N-h})\boldsymbol{\alpha}_1^\top + \boldsymbol{\alpha}_2(q\hat{\mathbf{I}}_0 - q_\mathrm{h}\hat{\mathbf{I}}_h - q_\mathrm{h}\hat{\mathbf{I}}_{N-h})\boldsymbol{\alpha}_2^\top \\ &\quad + 2\boldsymbol{\alpha}_1(-q_\mathrm{s}\hat{\mathbf{I}}_0 - q_\mathrm{v}\hat{\mathbf{I}}_v)\boldsymbol{\alpha}_2^\top \\ &= q(\boldsymbol{\alpha}_1\hat{\mathbf{I}}_0\boldsymbol{\alpha}_1^\top + \boldsymbol{\alpha}_2\hat{\mathbf{I}}_0\boldsymbol{\alpha}_2^\top) - q_\mathrm{h}(\boldsymbol{\alpha}_1\hat{\mathbf{I}}_h\boldsymbol{\alpha}_1^\top + \boldsymbol{\alpha}_2\hat{\mathbf{I}}_h\boldsymbol{\alpha}_2^\top) \\ &\quad - q_\mathrm{h}(\boldsymbol{\alpha}_1\hat{\mathbf{I}}_{N-h}\boldsymbol{\alpha}_1^\top + \boldsymbol{\alpha}_2\hat{\mathbf{I}}_{N-h}\boldsymbol{\alpha}_2^\top) - q_\mathrm{s}(2\boldsymbol{\alpha}_1\hat{\mathbf{I}}_0\boldsymbol{\alpha}_2^\top) - q_\mathrm{v}(2\boldsymbol{\alpha}_1\hat{\mathbf{I}}_v\boldsymbol{\alpha}_2^\top) \\ &= q\mathbf{R}^{\mathrm{A}_1}_0 - q_\mathrm{h}\mathbf{R}^{\mathrm{A}_1}_h - q_\mathrm{h}\mathbf{R}^{\mathrm{A}_1}_{N-h} - q_\mathrm{s}\mathbf{R}^{\mathrm{A}_1}_N - q_\mathrm{v}\mathbf{R}^{\mathrm{A}_1}_{N+v}, \\ &\qquad (a = 0, h, N-h;\ b = 0, v). \end{aligned} \tag{D.25}$$

Hence, Eq. (D.20) holds for the one-dimensional representation A_1.

For other one-dimensional representations A_2, B_1, and B_2, the proof is summarized as follows:

$$\begin{aligned} \mathrm{A}_2 : \boldsymbol{\alpha}_1\hat{\mathbf{I}}_a\boldsymbol{\alpha}_1^\top + \boldsymbol{\alpha}_2\hat{\mathbf{I}}_a\boldsymbol{\alpha}_2^\top &= \frac{1}{2N}\sum_{i=0}^{N-1}[1 + (-1)(-1)] = 1 = \mathbf{R}^{\mathrm{A}_2}_a, \\ 2\boldsymbol{\alpha}_1\hat{\mathbf{I}}_b\boldsymbol{\alpha}_2^\top &= \frac{2}{2N}\sum_{i=0}^{N-1}[1 \times (-1)] = -1 = \mathbf{R}^{\mathrm{A}_2}_{N+b}, \\ \mathrm{B}_1 : \boldsymbol{\alpha}_1\hat{\mathbf{I}}_a\boldsymbol{\alpha}_1^\top + \boldsymbol{\alpha}_2\hat{\mathbf{I}}_a\boldsymbol{\alpha}_2^\top &= \frac{2}{2N}\sum_{i=0}^{N-1}(-1)^{i+(i+a)} = (-1)^a = \mathbf{R}^{\mathrm{B}_1}_a, \\ 2\boldsymbol{\alpha}_1\hat{\mathbf{I}}_b\boldsymbol{\alpha}_2^\top &= \frac{2}{2N}\sum_{i=0}^{N-1}(-1)^{i+(i+b)} = (-1)^b = \mathbf{R}^{\mathrm{B}_1}_{N+b}, \\ \mathrm{B}_2 : \boldsymbol{\alpha}_1\hat{\mathbf{I}}_a\boldsymbol{\alpha}_1^\top + \boldsymbol{\alpha}_2\hat{\mathbf{I}}_a\boldsymbol{\alpha}_2^\top &= \frac{2}{2N}\sum_{i=0}^{N-1}[(-1)^{i+(i+a)} + (-1)^{(i+1)+(i+1+a)}] \\ &= (-1)^a = \mathbf{R}^{\mathrm{B}_2}_a, \\ 2\boldsymbol{\alpha}_1\hat{\mathbf{I}}_b\boldsymbol{\alpha}_2^\top &= \frac{2}{2N}\sum_{i=0}^{N-1}(-1)^{i+(i+1+b)} = (-1)^{b+1} = \mathbf{R}^{\mathrm{B}_2}_{N+b}. \end{aligned} \tag{D.26}$$

Therefore, the lemma is true for the blocks $\tilde{\mathbf{E}}^\mu$ corresponding the one-dimensional representations of the dihedral group $\mathbf{D}_N$.

2. *Two-dimensional blocks*
   Let $\mathbf{T}^{\mathrm{E}_k}_r$ and $\mathbf{T}^{\mathrm{E}_k}_s$ $(r, s = 1, 2, 3, 4)$ respectively denote the rth and sth rows of $\mathbf{T}^{\mathrm{E}_k} \in \mathbb{R}^{4\times4}$. Denoting $\mathbf{T}^{\mathrm{E}_k}_r = [\boldsymbol{\alpha}_1, \boldsymbol{\alpha}_2]$ and $\mathbf{T}^{\mathrm{E}_k}_s = [\boldsymbol{\beta}_1, \boldsymbol{\beta}_2]$, the (r, s)th entry $\tilde{E}^{\mathrm{E}_k}_{(r,s)}$ of $\tilde{\mathbf{E}}^{\mathrm{E}_k}$ can be computed as follows from Eq. (D.6):

$$
\begin{aligned}
\tilde{E}^{\mathrm{E}_k}_{(r,s)} &= \mathbf{T}_r^{\mathrm{E}_k}\mathbf{E}(\mathbf{T}_s^{\mathrm{E}_k})^\top \\
&= (\boldsymbol{\alpha}_1\mathbf{E}_1\boldsymbol{\beta}_1^\top + \boldsymbol{\alpha}_2\mathbf{E}_1\boldsymbol{\beta}_2^\top) + (\boldsymbol{\alpha}_1\mathbf{E}_2\boldsymbol{\beta}_2^\top + \boldsymbol{\beta}_1\mathbf{E}_2\boldsymbol{\alpha}_2^\top) \\
&= (q\sigma^0_{(r,s)} - q_\mathrm{h}\sigma^h_{(r,s)} - q_\mathrm{h}\sigma^{N-h}_{(r,s)}) + (-q_\mathrm{s}\tau^0_{(r,s)} - q_\mathrm{v}\tau^v_{(r,s)}),
\end{aligned} \tag{D.27}
$$

where $\boldsymbol{\alpha}_2\mathbf{E}_2^\top\boldsymbol{\beta}_1^\top = \boldsymbol{\beta}_1\mathbf{E}_2\boldsymbol{\alpha}_2^\top$ has been applied. Moreover, in Eq. (D.8), we have

$$
\begin{aligned}
\sigma^a_{(r,s)} &= \boldsymbol{\alpha}_1\hat{\mathbf{I}}_a\boldsymbol{\beta}_1^\top + \boldsymbol{\alpha}_2\hat{\mathbf{I}}_a\boldsymbol{\beta}_2^\top, \quad (a = 0, h, N-h), \\
\tau^b_{(r,s)} &= \boldsymbol{\alpha}_1\hat{\mathbf{I}}_b\boldsymbol{\beta}_2^\top + \boldsymbol{\beta}_1\hat{\mathbf{I}}_b\boldsymbol{\alpha}_2^\top, \quad (b = 0, v).
\end{aligned} \tag{D.28}
$$

Consider the case of $(r, s) = (1, 1)$ for example. From Eq. (D.28), we have the following equations for $a(=0, h, N-h)$ and $b(=0, v)$ since $\boldsymbol{\alpha}_1 = \boldsymbol{\alpha}_2 = \boldsymbol{\beta}_1 = \boldsymbol{\beta}_2 = \big(C_0, C_k, \ldots, C_{jk}, \ldots, C_{(N-1)k}\big)$:

$$
\begin{aligned}
\sigma^a_{(1,1)} &= 2\boldsymbol{\alpha}_1\hat{\mathbf{I}}_a\boldsymbol{\alpha}_1^\top = 2\sum_{i=0}^{N-1}\frac{C_{ik}}{\sqrt{N}}\frac{C_{(i+a)k}}{\sqrt{N}} = \frac{2}{N}\sum_{i=0}^{N-1} C_{ik}(C_{ik}C_{ak} - S_{ik}S_{ak}) \\
&= \frac{C_{ak}}{N}\sum_{i=0}^{N-1}(1 + C_{2ik}) - \frac{S_{ak}}{N}\sum_{i=0}^{N-1} S_{2ik} \\
&= C_{ak},
\end{aligned} \tag{D.29}
$$

$$
\begin{aligned}
\tau^b_{(1,1)} &= 2\boldsymbol{\alpha}_1\hat{\mathbf{I}}_b\boldsymbol{\alpha}_1^\top = 2\frac{1}{N}\sum_{i=0}^{N-1} C_{ik}C_{(i+b)k} \\
&= \frac{2}{N}\sum_{i=0}^{N-1} C_{ik}(C_{ik}C_{bk} - S_{ik}S_{bk}) \\
&= C_{bk},
\end{aligned} \tag{D.30}
$$

where $\sum_{i=0}^{N-1} S_{2ik} = \sum_{i=0}^{N-1} C_{2ik} = 0$ has been applied.

In a similar way, we have the following results for the entries $\sigma^a_{(r,s)}$ of $\boldsymbol{\sigma}^a$:

$$
\boldsymbol{\sigma}^a = \begin{pmatrix} C_{ak} & 0 & S_{ak} & 0 \\ 0 & C_{ak} & 0 & -S_{ak} \\ S_{ak} & 0 & C_{ak} & 0 \\ 0 & -S_{ak} & 0 & C_{ak} \end{pmatrix}, \quad (a = 0, h, N-h), \tag{D.31}
$$

and for the entries $\tau^b_{(r,s)}$ of $\boldsymbol{\tau}^b$, we have

$$
\boldsymbol{\tau}^b = \begin{pmatrix} C_{bk} & S_{bk} & 0 & 0 \\ S_{bk} & -C_{bk} & 0 & 0 \\ 0 & 0 & C_{bk} & S_{bk} \\ 0 & 0 & S_{bk} & -C_{bk} \end{pmatrix}, \quad (b = 0, v). \tag{D.32}
$$

Therefore, we have

$$\begin{pmatrix} \tilde{\mathbf{E}}^{\mathrm{E}_k} & \mathbf{O}_2 \\ \mathbf{O}_2 & \tilde{\mathbf{E}}^{\mathrm{E}_k} \end{pmatrix} = q \begin{pmatrix} 1 & 0 & 0 & 0 \\ 0 & 1 & 0 & 0 \\ 0 & 0 & 1 & 0 \\ 0 & 0 & 0 & 1 \end{pmatrix}$$

$$- q_{\mathrm{h}} \begin{pmatrix} C_{hk} & 0 & S_{hk} & 0 \\ 0 & C_{hk} & 0 & -S_{hk} \\ S_{hk} & 0 & C_{hk} & 0 \\ 0 & -S_{hk} & 0 & C_{hk} \end{pmatrix}$$

$$- q_{\mathrm{h}} \begin{pmatrix} C_{(N-h)k} & 0 & S_{(N-h)k} & 0 \\ 0 & C_{(N-h)k} & 0 & -S_{(N-h)k} \\ S_{(N-h)k} & 0 & C_{(N-h)k} & 0 \\ 0 & -S_{(N-h)k} & 0 & C_{(N-h)k} \end{pmatrix}$$

$$- q_{\mathrm{s}} \begin{pmatrix} 1 & 0 & 0 & 0 \\ 0 & -1 & 0 & 0 \\ 0 & 0 & -1 & 0 \\ 0 & 0 & 0 & -1 \end{pmatrix}$$

$$- q_{\mathrm{v}} \begin{pmatrix} C_{vk} & S_{vk} & 0 & 0 \\ S_{vk} & -C_{vk} & 0 & 0 \\ 0 & 0 & C_{vk} & S_{vk} \\ 0 & 0 & S_{vk} & -C_{vk} \end{pmatrix}, \tag{D.33}$$

where $\mathbf{O}_2 \in \mathbb{R}^{2\times 2}$ is a zero matrix. Finally, we can prove that the following equation holds

$$\tilde{\mathbf{E}}^{\mathrm{E}_k} = q\mathbf{R}_0^{\mathrm{E}_k} - q_{\mathrm{h}}\mathbf{R}_h^{\mathrm{E}_k} - q_{\mathrm{h}}\mathbf{R}_{N-h}^{\mathrm{E}_k} - q_{\mathrm{s}}\mathbf{R}_N^{\mathrm{E}_k} - q_{\mathrm{v}}\mathbf{R}_{N+v}^{\mathrm{E}_k}, \tag{D.34}$$

since $C_h + C_{N-h} = 2C_h$, $S_h + S_{N-h} = 0$ and

$$\begin{aligned} \mathbf{R}_0^{\mathrm{E}_k} &= \begin{pmatrix} 1 & 0 \\ 0 & 1 \end{pmatrix}, \\ \mathbf{R}_h^{\mathrm{E}_k} + \mathbf{R}_{N-h}^{\mathrm{E}_k} &= 2\begin{pmatrix} C_{hk} & 0 \\ 0 & C_{hk} \end{pmatrix}, \\ \mathbf{R}_N^{\mathrm{E}_k} &= \begin{pmatrix} 1 & 0 \\ 0 & -1 \end{pmatrix}, \\ \mathbf{R}_{N+v}^{\mathrm{E}_k} &= \begin{pmatrix} C_{vk} & S_{vk} \\ S_{vk} & -C_{vk} \end{pmatrix}. \end{aligned} \tag{D.35}$$

Therefore, the lemma also holds for the two-dimensional representations.

In summary, the lemma is proved. □

D.6 Remarks

For the structures with dihedral symmetry, the nodes of which have one-to-one correspondence to the dihedral group, we have presented a direct strategy for the analytical derivation of their symmetry-adapted force density matrices. The symmetry-adapted form can significantly simplify stability investigation of the structures, because sizes of the independent blocks become much smaller than those of the original matrices; and more importantly, it provides us the possibility to have further insight into the stability of the whole class of structures with similar symmetry properties.

The independent blocks of the symmetry-adapted force density matrix of the structures with dihedral symmetry are only 1-by-1 or 2-by-2 matrices, such that positive semi-definiteness of them can be easily verified. Using the analytical form of symmetry-adapted force density matrix, Chap. 6 discusses the condition of super-stability for symmetric prismatic structures, showing that they are super-stable when their horizontal cables are connected to adjacent nodes.

Furthermore, Chap. 7 presents the super-stability condition for star-shaped structures: the structures are super-stable if they have odd number of struts, and moreover, the struts are as close to each other as possible.

The formulations presented in this chapter are for the structures with dihedral symmetry, however, the same approach is applicable to the structures with other point group symmetry. See, for example, the structures with tetrahedral symmetry in Chap. 8.

Reference

1. Kettle, S. F. A. (2007). *Symmetry and structure: Readable group theory for chemists*. New York: Wiley.

Index

J.Y. Zhang and M. Ohsaki, *Tensegrity Structures*, Mathematics for Industry 6,
DOI 10.1007/978-4-431-54813-3

MIX
Papier aus verantwortungsvollen Quellen
Paper from responsible sources
FSC® C105338

If you have any concerns about our products,
you can contact us on
ProductSafety@springernature.com

In case Publisher is established outside the EU,
the EU authorized representative is:
Springer Nature Customer Service Center GmbH
Europaplatz 3, 69115 Heidelberg, Germany

Printed by Libri Plureos GmbH
in Hamburg, Germany